Study Notes in System Dynamics

MIT Press/Wright-Allen Series in System Dynamics

Jay W. Forrester, editor

Industrial Dynamics, Jay W. Forrester, 1961

Growth of a New Product: Effects of Capacity-Acquisition Policies, Ole C. Nord, 1963

Resource Acquisition in Corporate Growth, David W. Packer, 1964

Principles of Systems*, Jay W. Forrester, 1968

Urban Dynamics, Jay W. Forrester, 1969

Dynamics of Commodity Production Cycles,* Dennis L. Meadows, 1970

World Dynamics,* Jay W. Forrester, 1971

The Life Cycles of Economic Development,* Nathan B. Forrester, 1973

Toward Global Equilibrium: Collected Papers,* Dennis L. Meadows and Donella H. Meadows, eds., 1977

Study Notes in System Dynamics,* Michael R. Goodman, 1974

Readings in Urban Dynamics (vol. 1)*, Nathaniel J. Mass, ed., 1974

Dynamics of Growth in a Finite World,* Dennis L. Meadows, William W. Behrens III, Donella H. Meadows, Roger F. Naill, Jorgen Randers, and Erich K. O. Zahn, 1974

Collected Papers of Jay W. Forrester,* Jay W. Forrester, 1975

Economic Cycles: An Analysis of Underlying causes,* Nathaniel J. Mass, 1975

Readings in Urban Dynamics (vol. 2)*, Walter W. Schroeder III, Robert E. Sweeney, and Louis Edward Alfeld, eds., 1975

Introduction to Urban Dynamics,* Louis Alfeld and Alan K. Graham, 1976

DYNAMO User's Manual (5th ed.), Alexander L. Pugh III, 1976

Managerial Applications of System Dynamics, Edward B. Roberts, ed., 1978

Elements of the System Dynamics Method, Jorgen Randers, ed., 1980

Ecosystem Succession: A General Hypothesis and a Test Model of a Grassland, Luis T. Gutierrez and William R. Fey, 1980

Corporate Planning and Policy Design: A System Dynamics Approach, James M. Lyneis, 1980

*Originally published by Wright-Allen Press and now distributed by the MIT Press

Study Notes in System Dynamics

Michael R. Goodman

The MIT Press
Cambridge, Massachusetts,
and London, England

Second printing by The MIT Press, 1980

Library of Congress catalog card number 80-82677
ISBN 0-262-57051-3

Foreword

The principles and the mechanics of system dynamics were first worked out in the 1940's and 1950's. Since then the technique has been used to understand such diverse problems as technical obsolescence, urban decay, drug addiction, commodity price fluctuation, environmental deterioration, and population growth. The literature of system dynamics applications now comprises more than 12 books and many dozens of papers published with translations into as many as 26 languages. System dynamics groups are active in Japan, the United States, and several East and West European countries.

Over the past two decades there has also been significant expansion in the teaching programs and materials devoted to system dynamics. Degree programs emphasizing this methodology are now available in several universities, and several introductory textbooks have been published. In addition, the system dynamics faculties at M.I.T. and Dartmouth have developed a wealth of course materials to describe, illustrate, and provide practice in the application of simple dynamic principles to real and interesting problems. *Study Notes in System Dynamics* presents in one place and in standard format many of these materials.

No one book can provide a comprehensive introduction to the total set of procedures and rules required to understand the behavior of complex systems. Indeed, theoretical understanding is still growing rapidly. However, this text does significantly extend the teaching materials available in published form. Thus it helps move the field of system dynamics toward greater maturity and toward more effective, interesting teaching and learning.

Dennis L. Meadows

Dartmouth, New Hampshire
August 5, 1974

Preface

This volume contains the material used in the special one semester course in system dynamics given at the Massachusetts Institute of Technology for the Fellows at the Center for Advanced Engineering Study (C.A.E.S.) during the Spring term, 1973.[1] The introductory course attempted to familiarize the class with basic principles of system dynamics and provide a broad exposure to its applications.

The first half of the course centered on methodology, with emphasis on the relationship between feedback loop structure and behavior. Students began to identify and represent relevant feedback loops of various systems through construction of causal-loop diagrams, flow diagrams, and DYNAMO equations. Computer simulation of model behavior completed this part of the course.

The second half of the course dealt with the applications of system dynamics to diverse systems. Applications included global, urban, industrial, economic, ecological, and biological phenomena. Each student also pursued a modeling project selected according to interest and modeling skill.

A mixed student audience took the course. Most had experience in management but little or no experience in mathematical modeling. The formal educational background of the participants generally included

---

[1]The material was also used in a two-quarter introductory course taught at Northeastern University in the Graduate School of Industrial Engineering during the Fall, 1972.

limited graduate work in various fields such as management and engineering.

Study Notes in System Dynamics collects the material presented in the first half of the C.A.E.S. course. Part I (Simple Structures) draws heavily on my masters thesis, "Elementary System Dynamics Structures," written under the supervision of Professor J.W. Forrester (Massachusetts Institute of Technology, 1972). Part II (Simple Structures Exercises) contains exercises developed while I served as teaching assistant for the "Principles of Systems" course taught by Professor Dennis L. Meadows in Spring, 1972 at M.I.T. Professor Meadows also contributed to this section. Part III (Analysis and Conceptualization Exercises) contains advanced material gathered from several system dynamics courses taught at the Massachusetts Institute of Technology in recent years. Professor Dennis L. Meadows, Edwin K. Jarmain, Naren K. Patni, and Kjell Kalgraff contributed to this section.

Study Notes supplements the introductory system dynamics material contained in Jay W. Forrester's Principles of Systems[2]. These notes should facilitate self-study or teaching when read in conjunction with Principles of Systems or, to a lesser extent, Professor Forrester's Industrial Dynamics.[3] Published applications such as Urban Dynamics,[4] World Dynamics,[5] and Toward Global Equilibrium provide appropriate avenues to more advanced work.

---

[2]Jay W. Forrester, Principles of Systems (Cambridge: Wright-Allen Press, 1968).

[3]Jay W. Forrester, Industrial Dynamics (Cambridge: The M.I.T. Press, 1968).

[4]Jay W. Forrester, Urban Dynamics (Cambridge: The M.I.T. Press, 1969).

[5]Jay W. Forrester, World Dynamics (Cambridge: Wright-Allen Press, 1970).

[6]Dennis L. Meadows and Donella H. Meadows, eds., Toward Global Equilibrium (Cambridge: Wright-Allen Press, 1972).

Study Notes overlaps Principles of Systems but takes a different orientation. It focuses primarily on the relationship between model structure and model behavior through analysis of generic feedback structures common to most complex systems. These feedback structures, referred to as simple or elementary structures, include positive feedback, negative feedback, and combined positive and negative feedback processes. Principles of Systems introduces elementary structures, but does not offer the same emphasis or detail. In addition, Principles of Systems offers few applications of these structures.

Elementary structures serve as building blocks for both understanding and constructing elaborate models. A study of these structures should help the beginning student recognize that simple structures often adequately represent complex systems. The logical beginning point for conceptualization of a system dominated by a known mode of behavior is identification of basic feedback processes capable of producing that real world behavior. A basic structure also often explains a variety of phenomena exhibiting a common behavior mode. This transferability of simple structures from one phenomena (and field) to another can greatly facilitate communication across academic disciplines and has important educational value. The case examples illustrate this structure transferability.

HOW TO USE STUDY NOTES:

Although designed for the beginning student, Study Notes assumes some exposure to system dynamics and familiarity with the mechanics of flow diagramming and DYNAMO equation-writing. Principles of Systems or Industrial Dynamics furnishes a suitable background. The exercises indicate the most appropriate readings in Principles of Systems and alternative readings in Industrial Dynamics. The material in this volume requires little mathematical background. A knowledge of basic calculus would be helpful but is not necessary. However, the book includes material for readers desiring more extensive analytical treatment.

The notes have been divided into three parts. Part I deals exclusively with behavior and applications of generic feedback structures. Part II provides exercises to test the understanding of simple structures.

These model structure and behavior exercises also serve as an underlying theme for the Part III exercises. Part III contains model analysis and formulation exercises involving complex structures.

In Part I, at the beginning of each chapter is a section covering:

1. Recommended background readings and exercises.
2. Chapter purpose.
3. Recommended follow-up or practice exercises.

These introductions help guide the reader through the chapters and serve as the basis for designing a course in system dynamics.

Each exercise in Part II begins with a statement of purpose and a rough estimate of completion time. The exercises follow a variety of formats depending upon their content. Except for Exercise 7, they require only simple algebraic computations. A solution, accompanied by an explanation where necessary, follows each exercise. Some exercises ask the reader to check his results by simulating behavior on a computer. Access to a computer, however, is not a prerequisite for these exercises. Careful scrutiny of solution simulations will prove adequate.

Part III exercises are both more lengthy and more difficult than those in Part II. The exercises fall into two areas: analysis (Exercises 9-11), and conceptualization (Exercises 12-15).

Exercises 9 through 11 investigate exponential delays and multilevel feedback networks. The exercises help identify familiar generic structures embedded in these networks. Exercises 9 and 11 depart from the self-contained nature of the _Study Notes_ when used in conjunction with _Principles of Systems_. Exercise 9 requires background knowledge on time-delay representations in system dynamics; this material is not sufficiently covered in _Principles of Systems_. The exercise recommends appropriate readings in _Industrial Dynamics_ and also provides supplementary notes to complete the necessary background. Exercise 11 draws upon Jay W. Forrester's "Market Growth as Influenced by Capital Investment,"[7] not included in this volume.

---

[7] Jay W. Forrester, "Market Growth as Influenced by Capital Investment" _Industrial Management Review_ 9 (1968), pp. 83-105. Also reprinted in Jay W. Forrester, _Collected Papers of Jay W. Forrester_ (Cambridge: Wright-Allen Press, forthcoming 1974).

Exercises 12 through 15 comprise the most challenging material in this volume. These exercises require creative applications of accumulated knowledge in all phases of model construction. Given a verbal description of a problem or a behavioral phenomenon, the reader constructs, simulates, and tests his model. This phase of study often proves most revealing as well as most frustrating. Careful study of the sample model solutions should yield substantial value to the reader lacking access to a computer. Increasing difficulty of model formulation governs the order of the Part III modeling exercises. The extent of explanation provided in the solution also distinguishes the exercises. For example, the first exercise solution contains a detailed explanation and analysis of model behavior while the last contains only the model and computer simulation.

ACKNOWLEDGEMENTS:

I could not hope to include an exhaustive list of all contributors to or supporters of the book. However, Professor Dennis L. Meadows contributed substantially to Part II. William K. Horton furnished substantial input to Chapter 2 while Richard O. Foster and Louis E. Alfeld made significant contributions to Chapter 5. Robert E. Sweeney patiently provided extensive editorial work to the entire effort. Diane K. Leonard painstakingly developed the more than 300 figures. Nathaniel J. Mass and Gilbert W. Low offered useful technical criticism. Jane H. Foster handled the copy editing. Mary E. Ritchie typed the final manuscript. Special thanks go to those who provided moral and financial support. Professor Jay W. Forrester at the Sloan School of Management and Dr. Paul Brown at the Center for Advanced Engineering Study stand out in this respect. A special note of gratitude must go to Naren K. Patni of Wright-Allen Press whose encouragement helped make Study Notes a reality.

Contents

Part One **Simple Structures**

Chapter 1 Causal-Loop Diagramming 3

Chapter 2 Positive Feedback Structure 13

Chapter 3 Negative Feedback Structure 35

Chapter 4 S-Shaped Growth Structure 67

Chapter 5 Review of Simple Structures: Industrial Land-Use Model 93

Part Two **Exercises in Simple Structures**

Exercise 1 Causal-Loop Diagramming 137
Solution 141

Exercise 2 Graphical Integration 149
Solution 155

Exercise 3 Flow Diagramming 161
Solution 165

Exercise 4 Positive Feedback 167
Solution 171

Exercise 5 Negative Feedback: Application to Population Decay 175
Solution 179

Exercise 6 Negative Feedback: Application to Inventory Control 183
Solution 189

Exercise 7 First-Order Linear Systems 193
(with Dennis L. Meadows)
Solution 203

Exercise 8 Simple Structures 209
Solution 211

Part Three Exercises in Analysis and Conceptualization

Exercise 9 Delays: Exercise and Supplementary Notes 219
by Dennis L. Meadows
Solution 249

Exercise 10 Commodity Production Cycle Model 257
by Dennis L. Meadows
Solution 267

Exercise 11 Analysis of Market Growth Model 281
by Narendra K. Patni
Solution 299

Exercise 12 Residential Community Model 309
by Michael R. Goodman
Solution 313

Exercise 13 Future Electronics Model 349
by Edwin N. Jarmain
Solution 353

Exercise 14 Yellow-Fever Model 365
by Kjell Kalgraf
Solution 369

Exercise 15 Kaibab Plateau Model 377
by Michael R. Goodman
Solution 381

Part One
Simple Structures

Chapter 1
Causal-Loop Diagramming

Preparation: *Before beginning Chapter 1, the reader should have had some exposure to system dynamics modeling. Chapters 1 and 4 in Jay W. Forrester's Principles of Systems or Chapters 5 and 6 in Industrial Dynamics offer a general background. Chapter 7 in Principles of Systems or Chapter 8 in Industrial Dynamics provides familiarity with flow diagramming.*

Purpose: *This chapter introduces causal-loop diagramming through the development of a simple, multi-loop model. Positive and negative feedback loops appear in terms of their causal representations. Chapters 2 and 3 develop flow diagrams and equations in order to investigate the behavior of the positive and negative loops in detail. The background provided by this chapter should facilitate understanding of more complex model representations used in this book.*

Practice: *Exercise 1.*

1.1 INTRODUCTION 5

1.2 CAUSAL-LOOP EXAMPLE 5

1.3 REPRESENTING CAUSAL-LOOP RELATIONSHIPS 6

1.4 CAUSAL LOOPS 9

1.5 PITFALLS 11

1.6 SUMMARY 11

1.1 INTRODUCTION

System dynamics focuses on the structure and behavior of systems composed of interacting feedback loops. DYNAMO flow diagrams and causal-loop diagrams offer a convenient way to represent loop structures before development of system equations. Flow diagrams consist of rates, levels, and auxiliary elements organized into a consistent network. Causal-loop diagrams identify the principal feedback loops without distinguishing between the nature of the interconnected variables. Causal-loop diagrams play two important roles in system dynamics studies. First, during model development, they serve as preliminary sketches of causal hypotheses. Second, causal-loop diagrams can simplify illustration of a model. In both capacities, causal-loop diagrams allow the analyst to quickly communicate the structural assumptions underlying his model.

Since development of causal-loop diagrams does not require knowledge of flow diagramming, the beginning student of system dynamics can readily adopt this technique. Causal-loop diagramming encourages the modeler to conceptualize real world systems in terms of feedback loops. While serving as convenient communication devices, however, causal-loop diagrams have several shortcomings. These are discussed in section 1.5.

1.2 CAUSAL-LOOP EXAMPLE

In Figure 1-1, a causal-loop diagram describes feedback relationships between migration M and job availability JA. The diagram in-

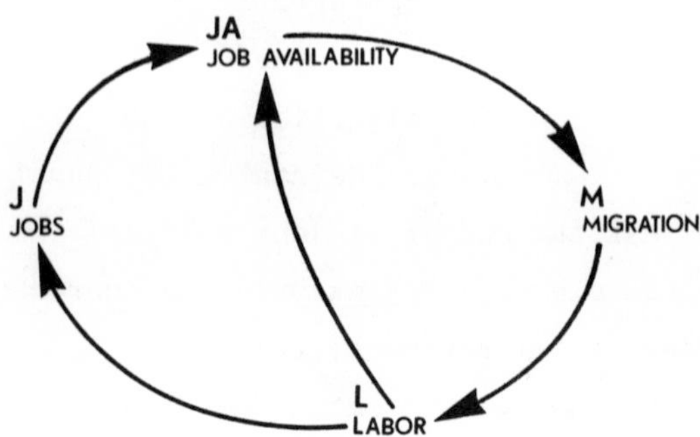

FIGURE 1-1 A causal-loop diagram

corporates simple causal hypotheses relating two feedback loops underlying urban behavior. These hypotheses include:

1. Job availability JA attracts migrants M to the city.

2. New arrivals to the city expand the labor population L.

3. Population absorbs available jobs, decreasing the amount of job availability JA.

4. In the long run, as labor also creates demand for additional goods, urban services and facilities, a further increase in the total number of jobs J in the area comes about.

5. More jobs J increase job availability JA.

For simplicity, Figure 1-1 ignores the type of employment involved, demographic characteristics of migrants, information and perception delays, and such other possible determinants of migration as housing, location, and taxes. Our step-by-step development of this diagram will illustrate the mechanics of loop diagramming.

To diagram the loop structure of the system and identify the type of polarity of each loop, we must first establish the pairwise relationships of relevant variables. Section 1.3 presents a test to ascertain the polarity of the causal pairs. Section 1.4 fits together the causal pairs into closed loops and develops the test of loop polarity.

1.3 REPRESENTING CAUSAL-LOOP RELATIONSHIPS

Definitions of respective variables are as follows:

Variables	Definition
Jobs (J)	Total number of vacant and filled jobs in the urban area
Job Availability (JA)	Number of vacant jobs
Migration (M)	Net migration of labor into urban area
Labor (L)	Total resident labor population in area

We assume that the number of jobs available modulates the flow of people into an urban area. For example, an increase in job availability causes an increase in migration to the area. A decrease in available employment has the opposite effect. The causal representation for this assumption appears below:

FIGURE 1-2 Positive job availability-migration link

The arrow indicates the direction of influence; the sign (plus or minus), the type of influence. An increase in job availability JA should produce an increase in migration M. Therefore, the relationship has a plus sign signifying the "positive" character of the link. More generally, if, all other things being equal, a change in one variable generates a change in the same direction in the second variable relative to its prior value, then the relationship between the two variables is positive. To apply this definition or test, we must consider only adjacent pairs of variables. We will also apply this definition with slight modification to determine the polarity of closed (feedback) loops.

The next example of a positive relationship involves migration M and labor L. We represent the relationship in the same manner as the JA-M relationship. Figure 1-3 shows the M-L relationship. An increase in migration rate M brings in additional laborers who increase the resident labor population.

A negative relation denoted by a minus sign occurs when a change in one variable produces a change in the opposite direction in the second variable. Figure 1-4 illustrates a negative relationship.

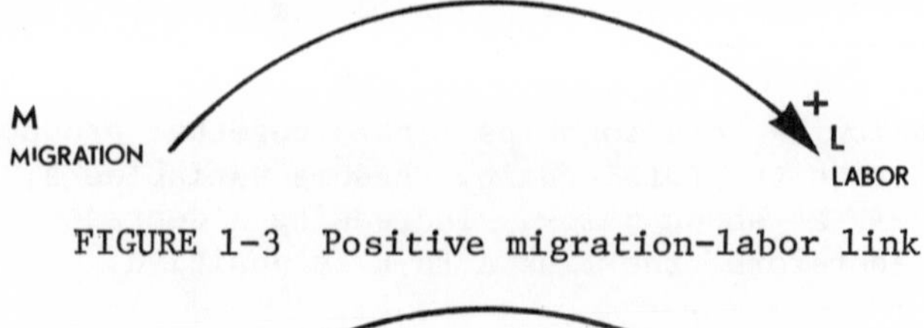

FIGURE 1-3 Positive migration-labor link

FIGURE 1-4 Negative labor-job availability link

Figure 1-4, like Figure 1-2, embodies a causal assumption. Figure 1-4 assumes that an increase in the resident labor population will eventually decrease the number of available jobs in the urban area. The new laborers in the city will occupy available jobs and thereby reduce job availability JA. If the labor population should decline we assume that more jobs would become available. Increasing or decreasing labor produces an opposite change in JA.[1]

Let us turn our attention to an assumption in the preceding examples. To determine the polarity of a pairwise relationship we keep all other factors constant. For example, examining the link between labor L and job availability JA, we ignore the fact that jobs also change. A change in the total number of jobs in the area, of course, can change job availability JA. Figure 1-5 illustrates this simple positive relationship. An increase in the total number of jobs J, assuming all else (including the labor force) remains constant, can increase job availability JA. Similarly, when JA changes in Figure 1-2, we assume that M changes regardless of other determinants of migration or the number of available potential in-migrants. The assumption that all other impinging variables remain constant during determination of causal polarity is central to causal analysis.

FIGURE 1-5 Positive jobs-job availability link

[1]Two negative pairwise relationships linked together produce a positive relationship across the total chain. Assume variables A, B, and C are linked negatively as shown below. Increasing A decreases B which increases C. Therefore, the link A to C is positive.

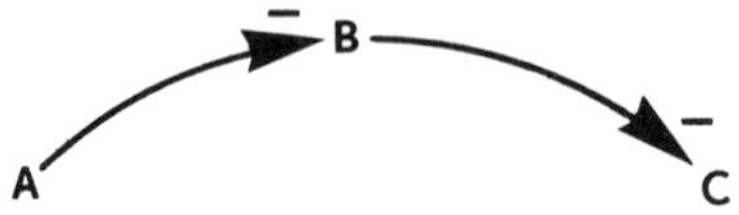

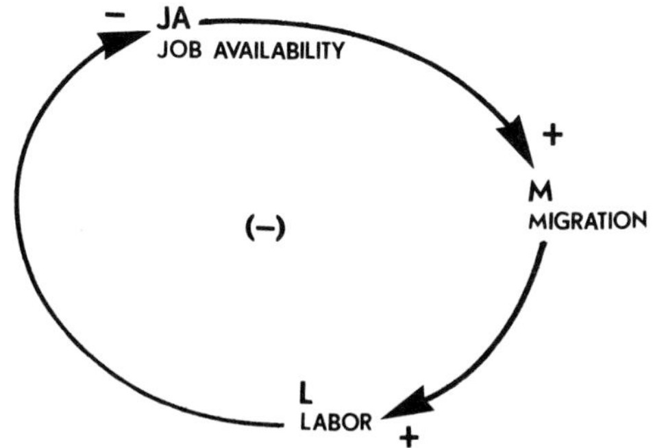

FIGURE 1-6 Single causal loop

1.4 CAUSAL LOOPS

Figure 1-6 combines the pairwise relationships among JA, M, and L. The feedback loop portrays the response of migration and labor to job availability for a given number of jobs. The determinants of the job level lie outside the loop boundary for the present.

To determine the polarity of the entire loop, we trace the consequences of an arbitrary change in one loop variable. Assume, for example, a sudden rise in job availability JA. The rise in JA attracts a migrant stream into the city which in turn increases the urban labor population; JA increases L. But an increase in the labor force <u>decreases</u> job availability as laborers occupy newly available jobs. The externally-caused increase in JA has triggered a set of internal reactions and adjustments in the system. These changes create pressure in opposition to the change in JA. The loop attempts to maintain JA at a fixed value or <u>goal</u> despite external influences to the contrary. <u>When a feedback loop response to a variable change opposes the original perturbation, the loop is negative or goal-seeking.</u>[2] <u>When a loop response reinforces the original perturbation, the loop is positive.</u>[3] As a shortcut method of determining loop polarity, add up the number of negative signs around the path: a) if even, the loop is positive;

---

[2]Chapter 3 discusses the behavioral implications of a negative loop.

[3]Chapter 2 covers positive feedback behavior.

b) if odd, the loop is negative. Figure 1-6 shows a negative feedback loop, denoted by the negative sign in the center of the loop.

When more than one loop comprises a system, we first determine the sign of each closed path in the preceding manner, holding constant all other variables (and hence loops) outside the closed path. Eventually, each closed path receives a loop polarity. For the sake of illustration, Figure 1-7 adds a positive link between labor L and jobs J. The link assumes that increasing the labor force will eventually increase employment since demand for urban services, housing, construction, and entertainment facilities will grow. The model now contains two closed loops: the familiar negative loop composed of JA, M, and L; and a new positive (outer) loop involving all four variables. The new link has not affected the polarity of the loop containing JA, M, and L. We can ascertain the polarity of the new loop by tracing the effect of a transient increase in job availability JA around the loop and ignoring all other links outside this path. The increase in job availability induces migrants into the urban area and thereby expands the labor force. The expanded labor population eventually increases the employment (job) base of the area which increases the job availability. The transient increase in JA is reinforced by the internal causal mechanism of the system; so the loop is positive. Note the plus sign in the diagram.

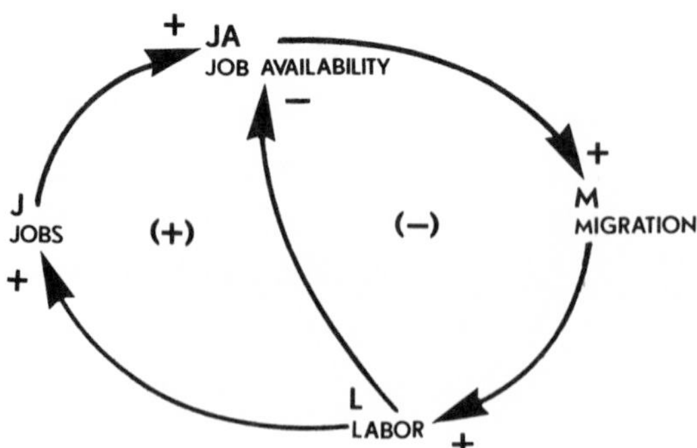

FIGURE 1-7 Two-loop diagram

1.5 PITFALLS

Causal-loop diagrams can be most useful during the early stages of model conceptualization as they help identify and organize principal components and feedback loops of the system under study. However, DYNAMO flow diagrams are indispensible in later stages of model construction and equation writing. Causal-loop diagrams lack the precision and detail of the rate, level, and auxiliary elements found in flow diagrams. Hence, flow diagrams bring out conceptual errors that do not readily show up in causal-loop diagrams. For example, a causal-loop diagram may inadvertently contain closed loops without levels. Such an error would be easily detected in a flow diagram.

Flow diagrams yield considerably more information than causal-loop diagrams about system structure and behavior. Because they depict the rate-level structure of a system, flow diagrams can often indicate possible types or modes of system behavior. For example, the order of a system (number of system levels) limits the possible behavior modes of that system. Therefore, knowledge of the order of the system is essential in predicting behavior. Such information is explicitly contained in flow diagrams but not in causal-loop diagrams.

Causal-loop diagrams often also obscure information necessary for understanding the behavior of an isolated feedback loop structure. For simplicity causal-loop diagrams frequently omit model structure such as delays and averaging processes. However, these processes embody negative feedback loops that could have considerable effect on system behavior. The flow diagram would call attention to these additional loops.

1.6 SUMMARY

The causal-loop diagramming process begins with identification of the relationship between individual pairs of variables. When a change in one variable produces a change in the same direction in a second variable the relationship is defined as positive. When the change in the second variable runs in the opposite direction, the relationship is defined as negative. The variables are linked together to form the feedback loops of the system. The polarity of a loop is determined by assuming all else remains constant and tracing the results of an arbitrary change around the loop: 1) Reinforcement of the change indicates a positive feedback loop; 2) Opposition to the change indicates a negative feedback loop.

Causal-loop diagramming simplifies the transformation of verbal description into feedback structure. Such diagramming also readily reveals the loop structure of complex models to people unfamiliar with flow diagrams or DYNAMO notation. Although useful as communication tools, causal-loop diagrams cannot substitute for detailed flow diagrams which must first be constructed before simulation analysis can proceed further.

Chapter 2
Positive Feedback Structure

Preparation: Before studying the feedback structures in the following chapters, the reader should be familiar with the following mechanics of system dynamics modeling:

1. *Causal-loop diagramming;*
2. *Graphical integration;*
3. *Flow diagramming;*
4. *DYNAMO equation writing.*

Exercises 1 through 4 review the above aspects of system dynamics.

Purpose: Chapter 1 explored the use of causal-loop diagrams in representing the feedback-loop structure of systems. It also introduced positive and negative feedback loops without considering their behavior. Chapter 2 surveys in detail behavioral characteristics of the positive loop. Chapter 3 will examine the characteristics of negative feedback loops. Chapter 4 will investigate simple structures combining both positive and negative feedback loops.

Chapter 2 offers examples of familiar positive feedback processes. It focuses on the simplest form of positive feedback, the single-level structure. Through numerical simulation, we illustrate how a single-level structure can produce exponential growth. The basic concepts of the time constant and doubling time of a dynamic system are also developed. These concepts are applied extensively throughout this volume. Two simplified examples of positive feedback structures, drug addiction and population growth, complete the chapter.

Practice: Section 5.1 of Chapter 5 reinforces the material in this chapter. The student can easily undertake this section before completing the remaining chapters.

2.1 INTRODUCTION 15

2.2 GENERAL STRUCTURE 18

2.3 EXPONENTIAL GROWTH EQUATION 21

2.4 TIME HORIZON OF EXPONENTIAL GROWTH 24

2.5 SUPEREXPONENTIAL GROWTH 26

2.6 SUMMARY 26

2.7 EXAMPLE 1--DRUG ADDICTION 28

2.8 EXAMPLE 2--POPULATION GROWTH 31

2.1 INTRODUCTION

In a positive feedback process, a variable continually feeds back upon itself to reinforce its own growth or collapse. Several familiar phrases characterize the phenomenon of positive feedback. For example, the "band wagon effect" creates an image of individuals adopting a cause or candidate--"jumping on the band wagon"--in response to the number of others already "on the band wagon." As a political movement grows, its popularity and ability to attract support also grow.

The "snowball effect" relates the growth of certain ideas, or modes of action, to the growth of a snowball as it rolls down a mountainside. As a rolling snowball picks up snow, its mass and circumference increase, which causes the snowball to grow even faster.

Both "vicious circles" and "virtuous circles" are synonyms for positive feedback. In a vicious circle a worsening of one element in a causal chain brings about further degradation of the element. Conversely, in a virtuous circle, positive changes in a system element trigger further improvement. The viciousness or virtuousness of a positive feedback system depends on whether the elements of the loop mutually deteriorate or mutually improve.

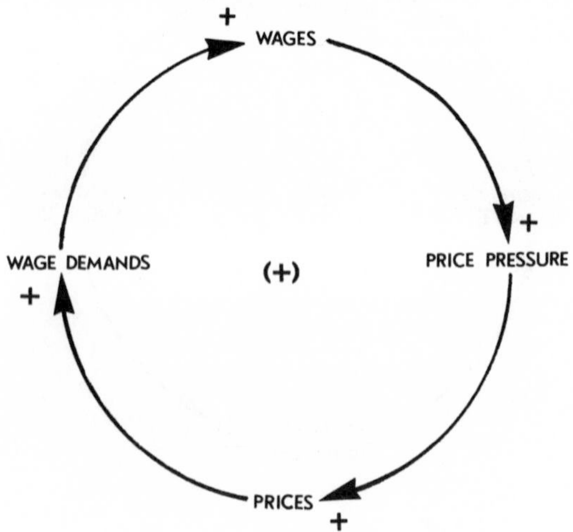

FIGURE 2-1 Causal-loop diagram of wage-price spiral

The inflationary wage-price spiral provides a classic example of a vicious circle. Workers worried by high prices demand higher wages. Manufacturers raise prices to cover the increased cost of labor. Workers perceive that prices have risen again and demand higher wages to cover these increases. This further demand closes the loop. Figure 2-1 shows the causal-loop diagram of the spiral. Each variable in the loop in Figure 2-1, exhibits the same qualitative behavior. Within the system, an increase in one variable leads to an increase in the other three. This latter increase leads in turn to an increase in the first variable and so forth. All four variables are mutually enhancing and tend to increase without limit. The plus (+) sign in the center of the loop symbolizes the positive feedback nature of the loop.

The arms race serves as another example of positive feedback. As one nation increases its armaments, it also increases the threat felt by an enemy. The enemy may respond to the threat by increasing its own armaments. As a result, the first nation feels more threatened and once again manufactures more armaments. Figure 2-2 illustrates the vicious circle.

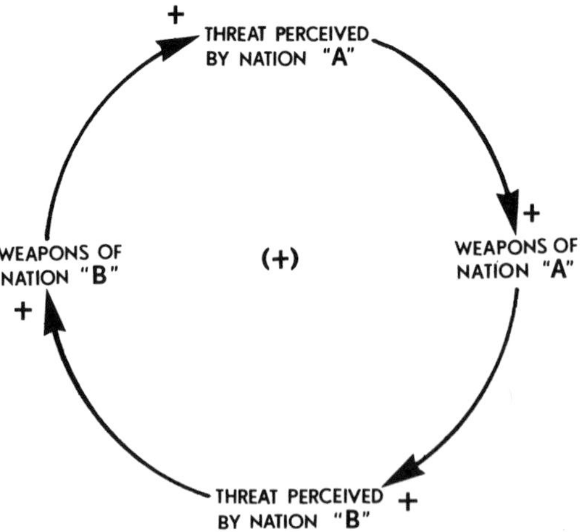

FIGURE 2-2 Causal-loop diagram of arms proliferation

The exponential growth curve in Figure 2-3 characterizes most positive feedback systems.[1] World population, food production, industrialization, pollution and consumption of nonrenewable natural resources all exhibit exponential growth.[2] Such growth, effects of which seem minimal at first, skyrockets within a short time. The fable of the jeweller's assistant provides another revealing illustration of the explosive nature of exponential growth.

The assistant offered to work for the jeweller for one year on condition that his salary start at one lire (0.2¢) and double each week. The jeweller, elated at the thought of cheap labor, quickly agreed. By the fifth week the assistant's pay was 16 lire; the tenth, 512; and the fifteenth, 16,384. By the twenty-fifth week his salary was 16,777,220 lire. Two weeks later the jeweller had to sell his shop just to pay his assistant.

---

[1]Another less common form of behavior characteristic of positive feedback is accelerated decay or exponential collapse pictured below. This chapter, however, will not explicitly deal with the collapse mode.

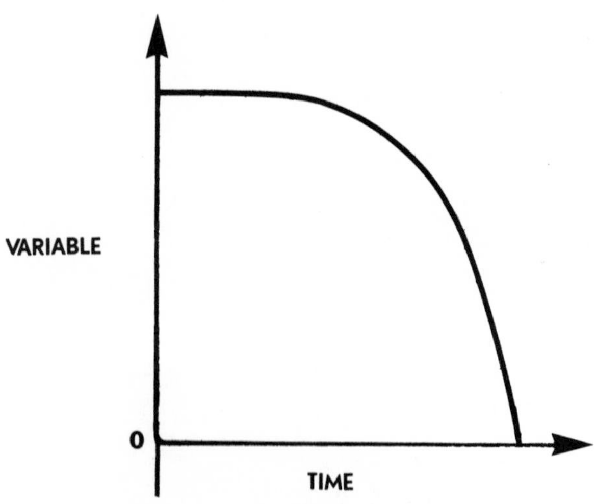

Exponential collapse mode

[2]D.L. Meadows, et al., The Limits to Growth (New York: Universe Books, 1972), p. 25.

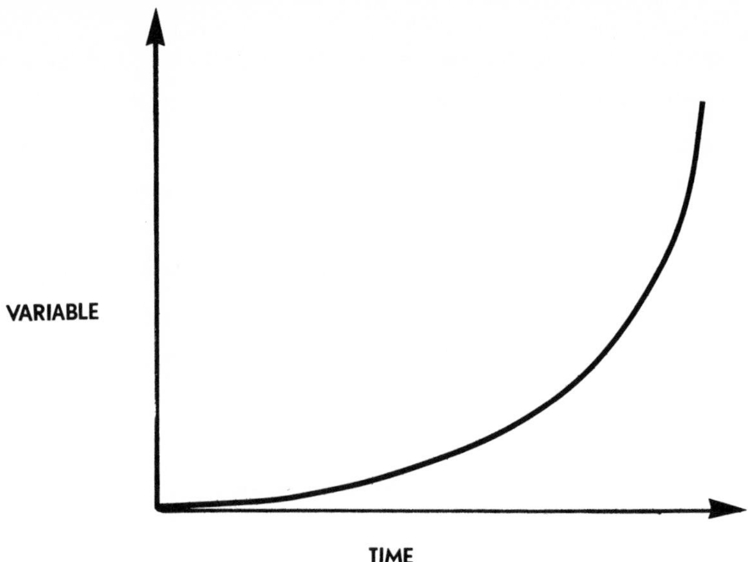

FIGURE 2-3 Exponential growth curve

This chapter explores the basic positive feedback loop. It uses a single level and single rate structure to explain why exponential growth occurs, and to describe the mechanisms governing the growth rate. World population growth and the rise of drug addiction exemplify the importance of the simple positive loop in interesting occurrences of exponential behavior.

2.2 GENERAL STRUCTURE

Figure 2-4 shows a very simple example of positive feedback. A one-way flow of material accumulates in the level LEV. In turn, information about the quantity in the level at any time controls the flow into LEV via the rate RT. RT is proportionately related to LEV by a constant CONST. The equations for the system with an arbitrary CONST value of 0.2 and an initial LEV value of 1 are:

```
LEV.K=LEV.J+(DT)(RT.JK)                          1, L
LEV=1                                            1.1, N
    LEV     - LEVEL (UNITS)
    RT      - RATE (UNITS/YEAR)

RT.KL=CONST*LEV.K                                2, R
CONST=.2                                         2.1, C
DT=1                                             2.2, C
    RT      - RATE (UNITS/YEAR)
    CONST   - CONSTANT (FRACTION/YEAR)
    LEV     - LEVEL (UNITS)
```

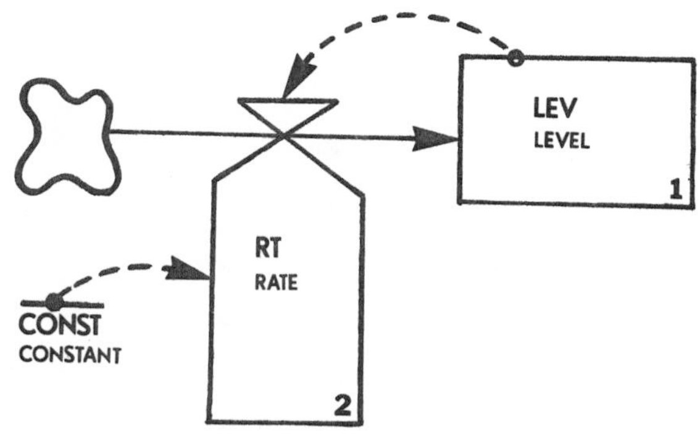

FIGURE 2-4 Flow diagram of positive feedback

To discover why exponential growth results from the structure in Figure 2-4, we "numerically" simulate the system behavior. The simple computation procedure, using rate and level equations, follows:

1. Compute the first RT value with the aid of the initial value of the level LEV.

2. Multiply the RT value by DT, the time interval over which we compute the rate of flow.

3. Add the product of RT and DT to the initial LEV value to produce a new LEV value. Replace the initial LEV value and with the new LEV value repeat the process for the desired number of DT intervals in the time span of the simulation.

Figure 2-5 illustrates the numerical simulation procedure. Figure 2-5 might, for example, represent the growth of a level of savings accruing interest at 20 percent per year (CONST) compounded every year (DT = 1) for 15 years. The level accumulates all past values of the rate and thus acts as a "memory." The rate continually increases because RT is proportional to LEV; each new increment to the level is larger than the previous increment.[3] If LEV is plotted as a function of time, the expected exponential curve seen in Figure 2-6 results. RT, therefore, also grows exponentially.

---

[3]Exponential growth contrasts with linear growth which increases by a fixed increment each time period. The structure in Figure 2-4 can never produce linear growth.

FIGURE 2-5

NUMERICAL SIMULATION

Time	Level	Rate Value	New Increment
0	1.00	0.20	0.20
1	1.20	0.24	0.24
2	1.44	0.29	0.29
3	1.73	0.35	0.35
4	2.08	0.42	0.42
5	2.50	0.50	0.50
6	3.00	0.60	0.60
7	3.60	0.72	0.72
8	4.32	0.86	0.86
9	5.18	1.04	1.04
10	6.22	1.24	1.24
11	7.46	1.49	1.49
12	8.95	1.79	1.79
13	10.74	2.15	2.15
14	12.89	2.58	2.58
15	15.47	3.09	3.09

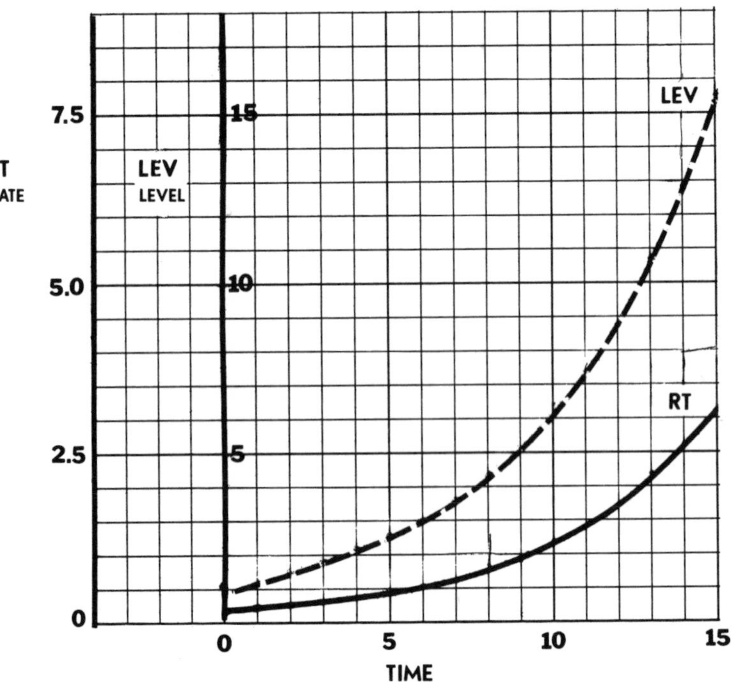

FIGURE 2-6 Exponential growth of LEV and RT

2.3 EXPONENTIAL GROWTH EQUATION

We can represent our system analytically by the equation (2.1):[4]

$$LEV(t) = LEV(0)e^{(CONST*t)} \tag{2.1}$$

where

LEV(t) = level value at time t

LEV(0) = initial level value

CONST = constant of proportionality

t = time

e = base of natural logarithm

Equation (2.1) allows one to find the LEV value at any point in time with a single computation. Equation (2.1) can also be used to derive some important measures of exponential growth: the time constant T and the doubling time T_d.

---

[4]Equation (2.1) can be derived as follows:

$$LEV.K = LEV.J+(DT)(RT.JK)$$

or,

$$\frac{LEV.K-LEV.J}{DT} = RT.JK.$$

By dropping the DYNAMO notation and taking the limit as DT approaches zero we get:

$$\frac{dLEV(t)}{dt} \doteq RT(t)$$

but,

$$RT(t) = CONST*LEV(t).$$

So by substitution,

$$\frac{dLEV(t)}{dt} = CONST*LEV(t).$$

By separating variables and integrating both sides,

$$\int_{LEV(0)}^{LEV(t)} \frac{dLEV(\tau)}{LEV(\tau)} = \int_0^t CONST*d\tau \qquad \text{(where } \tau \text{ is a dummy variable)}$$

we get,

$$\ln \frac{LEV(t)}{LEV(0)} = CONST*t$$

or,

$$LEV(t) = LEV(0)e^{CONST*t}.$$

The time constant T is defined as the reciprocal of CONST or T = 1/CONST. The time constant T has the dimension of units of time. If an interval of time equal to T passes, then from the equation (2.1) the value of LEV at time T becomes:

$$\begin{aligned} LEV(T) &= LEV(0)e^{(\frac{1}{T}*T)} \\ &= LEV(0)e^{1} \\ &= 2.72*LEV(0) \end{aligned}$$

After the passage of one time constant, LEV(T) is 2.72 times larger than its initial value LEV(0). Likewise, in the next T interval, LEV(2T) will be 2.72*2.72 or a factor of 7.40 greater than LEV(0). feedback system. A larger T (or smaller CONST) produces a flatter LEV growth curve.

The doubling time T_d is related directly to the time constant. The doubling time T_d is the time interval required for an exponentially growing variable to double in value. Using the equation (2.1), we can relate T_d to the time constant T:

$$\begin{aligned} 2*LEV(0) &= LEV(0)e^{CONST*T_d} \\ 2 &= e^{CONST*T_d} \end{aligned}$$

or,

$$\ln(2) = CONST*T_d.$$

In terms of the time constant T,

$$0.69 = (1/T)T_d$$

and,

$$T_d = 0.69*T$$

The doubling time is approximately 70 percent of the time constant. Every period of time equal to T_d, LEV doubles in value. Figure 2-7 shows the relationship between the time constant T and the doubling time T_d.

Equation (2.1) indicates that in a single rate-level system, CONST can have any positive value and still produce exponential growth. Equivalently, as long as the slope between the rate RT and the level LEV remains positive, as sketched in Figure 2-8, the system exhibits a positive feedback process characterized by exponential

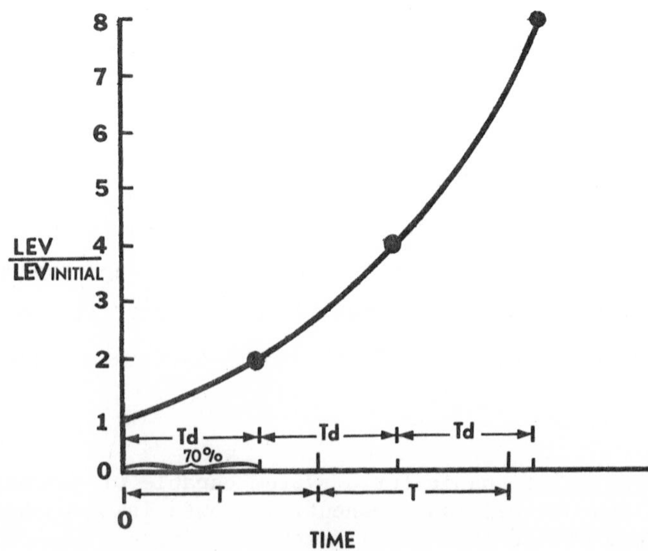

FIGURE 2-7 Time constant T and doubling time T_d relationship

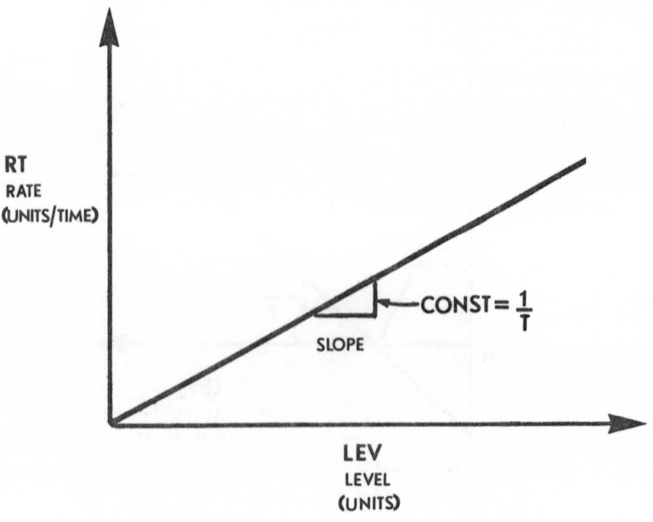

FIGURE 2-8 Rate-level graph

growth. Any level value initially greater than zero produces a positive rate value which in the next DT period produces a still larger level value. This tendency continues indefinitely. Both the level and rate migrate away from the origin on the rate-level graph of Figure 2-8 regardless of the steepness of the slope.[5] The slope of the curve equals the inverse of the time constant.

2.4 TIME HORIZON OF EXPONENTIAL GROWTH

The time period or horizon over which exponential growth occurs may seem to alter the character of growth even though the underlying system remains the same. The curves in Figure 2-9 illustrate this point. Curve (a) shows level growth over a time span equal to one-

---

[5]The rate-level relationship for a system capable of producing exponential collapse as well as exponential growth is shown below. Note that the slope of the line still remains positive. When the system operates in the region to the left of the unstable point, continually larger negative rate values produce the collapse mode.

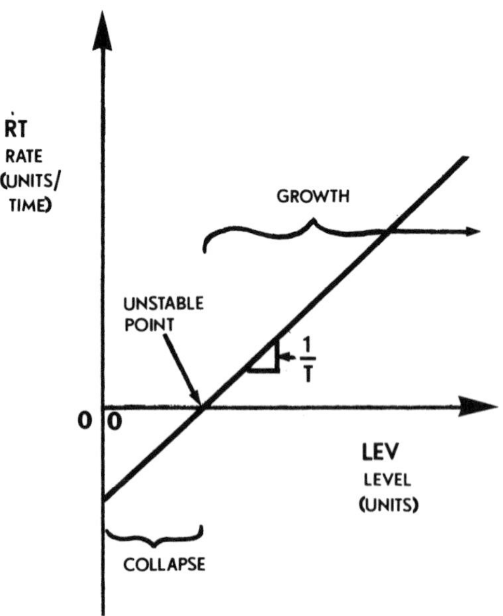

Rate-level graph for collapse or growth mode

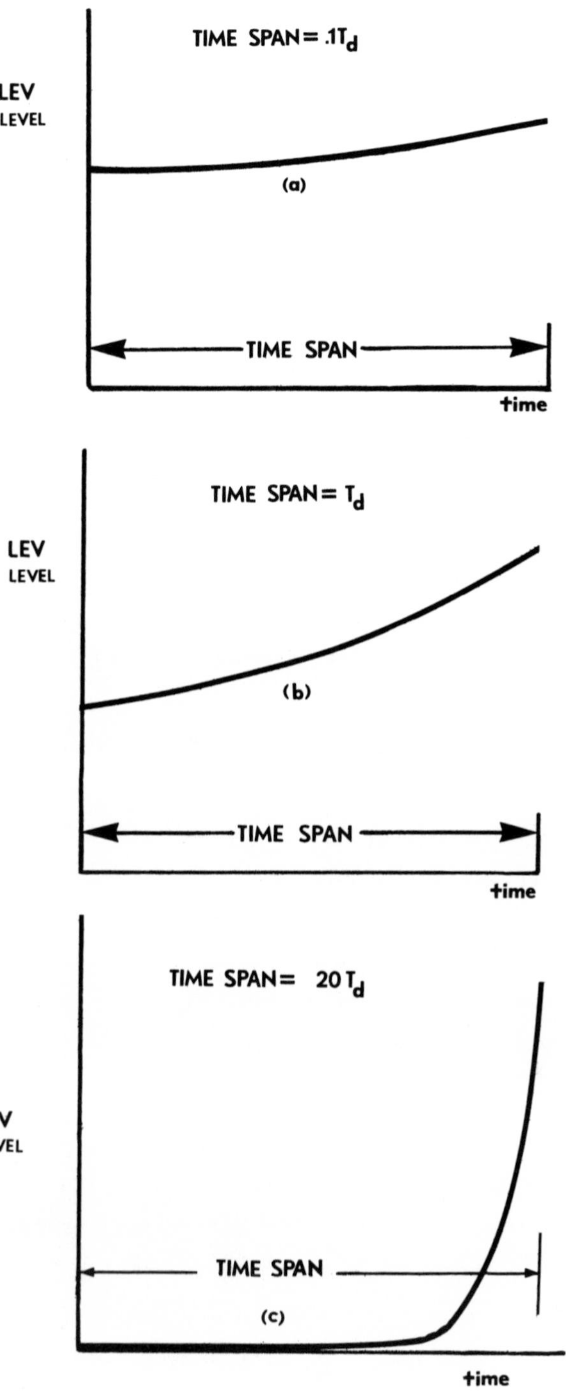

FIGURE 2-9 Time horizon of exponential growth

tenth the doubling time. Growth appears minimal. The short observation period is deceptive, however. Over a time span equal to a doubling time, the same level grows as in curve (b). Over a time span equal to twenty times the doubling time, the growth pattern in curve (c) obviously takes on quite a different character. Because of the scaling, the level shows little absolute growth during the first fifteen doubling times. The apparently inactive variable, however, suddenly changes character, and increases significantly.

2.5 SUPEREXPONENTIAL GROWTH

Many positive feedback systems are characterized by doubling times that decrease, rather than remain fixed, as the values of system levels increase. World population growth illustrates a real occurrence of this phenomena as will be seen in section 2.8. Instead of the linear (fixed slope) rate-level relationship sketched in Figure 2-8, a nonlinear (changing slope) relationship characterizes the system shown in Figure 2-10. The rate in Figure 2-10 increases more than in fixed proportion to the level because the slope, the inverse of the time constant, is increasing. A computer simulation of a positive feedback system containing such a nonlinear rate-level relationship appears in Figure 2-11. The exponential growth curve of the linear rate-level structure is included with the simulation in Figure 2-11 for comparative purposes. The nonlinear system has more pronounced growth than the linear system. Such a system is said to demonstrate "super" or "supra" exponential growth.

2.6 SUMMARY

Positive feedback characterizes a process in which change in one element of a closed loop causes further change in the same direction in an unending process. The simplest model of a positive feedback system contains a single level which accumulates a rate directly proportional to the level. The most common mode of behavior of both the level and rate is exponential growth although accelerated collapse may also occur in some systems.

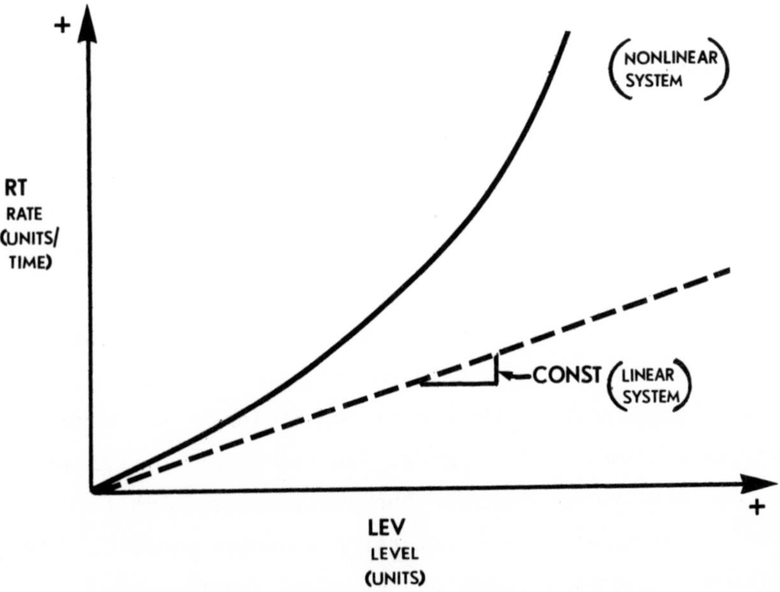

FIGURE 2-10 Nonlinear rate-level graph

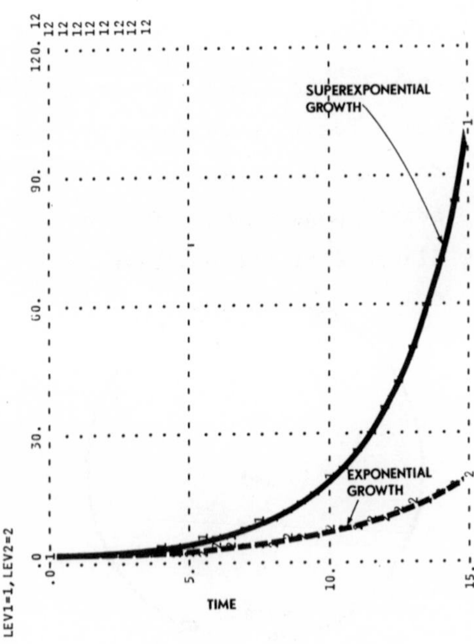

FIGURE 2-11 Superexponential growth

Exponential growth is distinguished by a constant doubling time, the length of time required for a level to double in magnitude. Superexponential growth is marked by a decreasing doubling time. Viewed over a time span of a few doubling times, the absolute growth of an exponentially increasing level appears insignificant. However, within a few more doublings, the level grows to enormous proportions. This potential for explosive growth is present in every system experiencing exponential growth.

Of course, no variable can grow forever in a limited environment. Within a real-world system, exponential growth occurs only as long as the growth forces within the system can dominate retarding forces. However, as system variables grow, "negative" or controlling feedback forces must eventually overtake the positive feedback forces as determinants of system behavior. Chapter 4 explores in detail the process of shifting dominance between positive and negative feedback loops. The remainder of this chapter investigates examples of positive feedback.

2.7 EXAMPLE 1--DRUG ADDICTION

The explosive growth in the number of heroin addicts in a community might result from a positive feedback process. For example, we can picture drug addiction as a closed loop with addicts drawing non-addicts into the addict population. Figure 2-12 contains the causal-loop and flow diagrams of the positive loop. For simplicity,

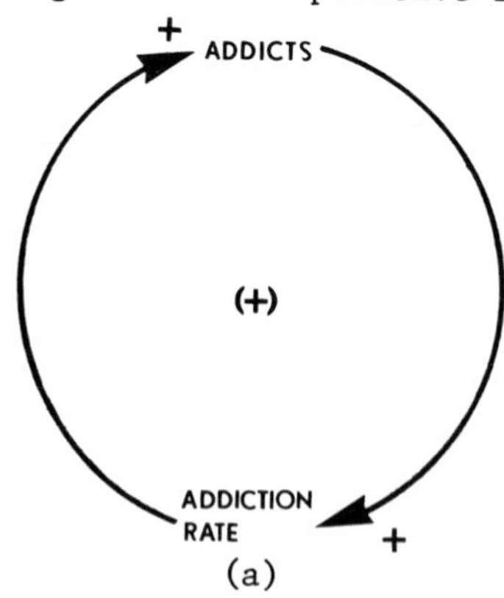

FIGURE 2-12 Causal-loop and flow diagrams of drug addiction model

we assume that the addiction rate depends entirely on the level of addicts and is not restrained by the number of potential addicts. We also assume that each addict brings one non-addict into the addict pool every three years (the addict population increases by a fractional addition rate FAR of 0.33 per year). We ignore outflow of addicts from the level through arrests, drop-outs, and rehabilitation. The model, although obviously oversimplified, illustrates some interesting behavior.

The system equations are:

```
ADCTS.K=ADCTS.J+(DT)(AR.JK)                                  1, L
ADCTS=10                                                     1.1, N
    ADCTS  - ADDICTS (PEOPLE)
    AR     - ADDICTION RATE (PEOPLE/YEAR)
AR.KL=ADCTS.K*FAR                                            2, R
FAR=.33                                                      2.1, C
    AR     - ADDICTION RATE (PEOPLE/YEAR)
    ADCTS  - ADDICTS (PEOPLE)
    FAR    - FRACTIONAL ADDICTION RATE (FRACTION/YEAR)
```

Starting in year zero with ten addicts, the addict population grows with a doubling time of 2.1 years (Figure 2-13). Over the twenty doubling times of simulation, the number of addicts doubles in small increments at first, but skyrockets in the last few years. The same model structure governs the entire simulation, yet the behavior seems to undergo a transformation from no drug problem into a drug crisis during the last five years.

The rapidity with which exponential growth thrusts a level from awareness to crisis proportions has important policy implications.

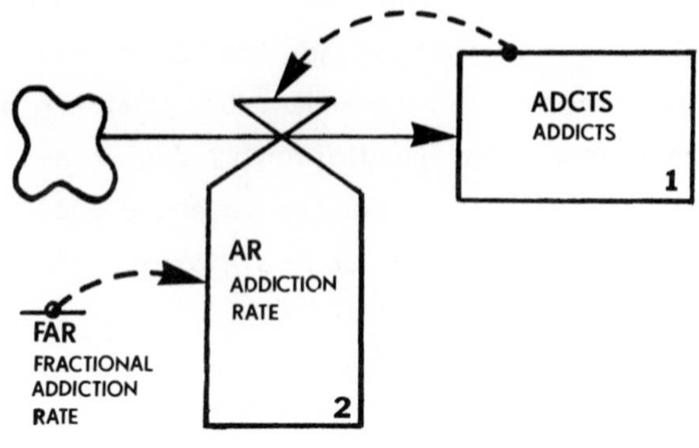

FIGURE 2-12 (continued)

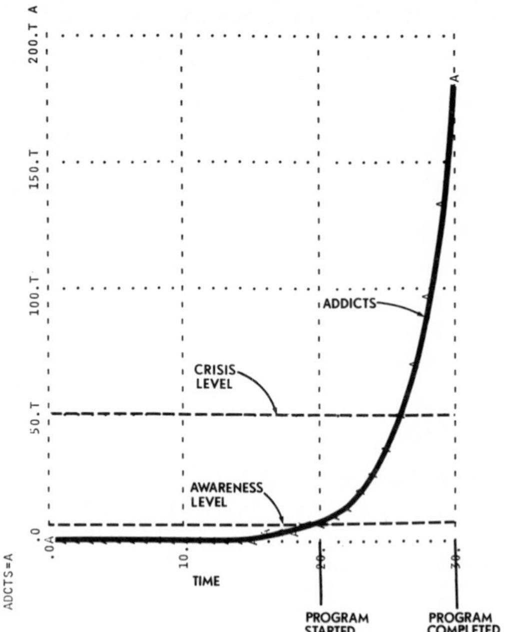

FIGURE 2-13 Addicts growth trend

For example, suppose that the awareness level of drug addiction--the point at which community resources are mobilized into a treatment program--is at the level shown in Figure 2-13. Assume also a minimum of ten years to build an effective drug program for handling an anticipated crisis roughly eight times as severe as conditions prevailing at time of awareness. The crisis stage also appears in Figure 2-13. What addiction problem will the program actually face at the end of ten years?

Since the addict population doubles roughly every two years, after only three doublings or six years beyond the awareness level, the crisis breaks out. In another two doublings, the additional time needed for a fully operational treatment program, the addict population exceeds the crisis stage population by four times. The treatment program must handle four times its intended load. Waiting until an exponentially growing problem reaches alarming dimensions may mean that late solutions cannot avert disaster. The program in this example would need to start well in advance of the awareness stage for it to be effective.

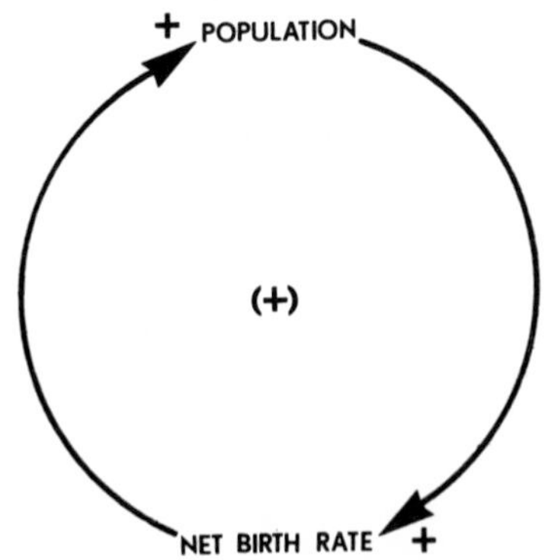

FIGURE 2-14 Population-birth loop

2.8 EXAMPLE 2--POPULATION GROWTH

The first-order positive feedback structure can explain much, but not all, of the basic character of human population growth. Figure 2-14 contains a simple causal representation of the relationship between the population POP and net birth rate NBR (births per year less deaths per year). As population POP increases, net birth rate NBR increases.

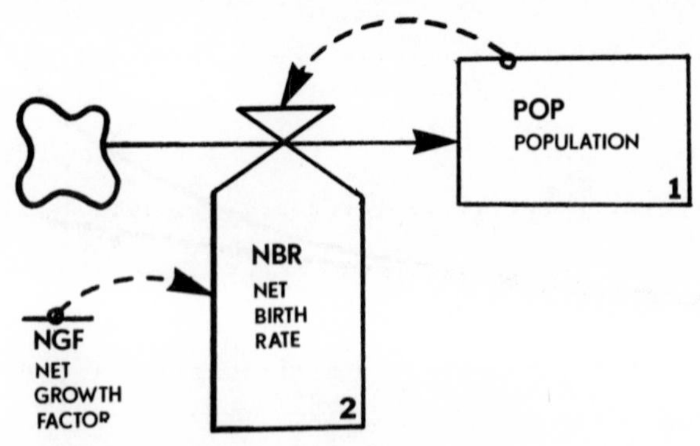


```
POP.K=POP.J+(DT)(NBR.JK)                                   1, L
POP=.5E9                                                   1.1, N
     POP     - POPULATION (PEOPLE)
     NBR     - NET BIRTH RATE (PEOPLE/YEAR)

NBR.JK=NGF*POP.K                                           2, R
NGF=.003                                                   2.1, C
     NBR     - NET BIRTH RATE (PEOPLE/YEAR)
     NGF     - NET GROWTH FACTOR (FRACTION/YEAR)
     POP     - POPULATION (PEOPLE)
```

FIGURE 2-15 Population-birth loop

We assume that births always exceed deaths each year and that a fixed percentage of the population reproduces new population each year. An increase in net birth rate increases the population which in turn increases the net birth rate in a positive feedback manner.

Figure 2-15 contains a flow diagram and equations for the causal loop. The net growth factor NGF, the percentage growth rate, is set at 0.3 percent per year. The doubling time (0.7*1/NGF) is approximately 233 years. The model, simulated over 300 years in Figure 2-16, begins with 0.5 billion persons. Figure 2-16 also plots the actual historical growth trend of the world population from 1650 to 1950.[6] Comparison of the historical curve to the model-generated curve clearly distinguishes exponential growth from historical superexponential growth. While the historical trend for the first 150 years

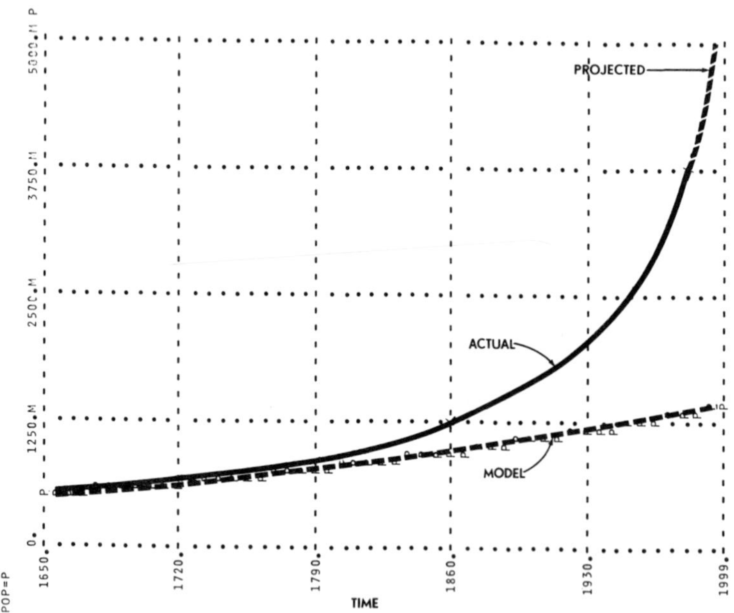

FIGURE 2-16 Actual world population compared with model population growth

[6] Meadows, et al., <u>The Limits to Growth</u>, p. 33.

apparently had a 233-year doubling time, this doubling time did not remain constant. For example, population doubling time between 1940 and 1970 was 33 years. To more faithfully replicate world population growth, the simple population model would require additional structure accounting for historical decline in deaths per year. Improvements in medical technology, sanitation, and nutrition, for example, could lower death rates.

Chapter 3
Negative Feedback Structure

Preparation: The reader should attempt Exercise 5 before reading Chapter 3. Exercise 6 should also be attempted before examining the inventory example in section 3.10.

Purpose: Chapter 2 introduced positive feedback structure capable of producing exponential growth (or collapse) over time. The explosive behavior of a system dominated by positive feedback cannot persist indefinitely. Eventually, goal-seeking processes retard growth (or cause collapse). This chapter investigates negative feedback structures capable of generating such retarding pressures.

The chapter first introduces the general decision process associated with negative feedback. A single-level (first-order) model which embodies the goal-seeking characteristics of negative feedback follows. Both analytical and graphical (rate-level) analysis outline model behavior. The analysis includes the role of initial conditions in transient response, the effect of the time constant, and the response of feedback structure to exogenous inputs.

The second half contains three different examples of first-order negative structures. The first example investigates the simple inventory control system introduced in Exercise 6. The second example explores the thermal adjustment of a liquid equilibrating to the temperature of its surrounding environment. The final example, which examines a one-level pollution system, illustrates the effects on system behavior of nonlinear time constants. These three examples demonstrate the prevalence of basic negative feedback processes within environmental and industrial systems.

Chapter 4 combines negative feedback with positive feedback processes to form the basic S-shaped growth structure typical of many social and physical life cycles.

Practice: Exercise 7.

3.1 INTRODUCTION 37
3.2 CAUSAL-LOOP DIAGRAM OF SIMPLE NEGATIVE FEEDBACK 38
3.3 FLOW DIAGRAM 40
3.4 SYSTEM BEHAVIOR 41
3.5 SLOPE AND TIME CONSTANT 44
3.6 INITIAL CONDITIONS 45
3.7 ZERO-VALUE GOAL STRUCTURE 45
3.8 SYSTEM COMPENSATION 49
3.9 SUMMARY 51
3.10 EXAMPLE 1--INVENTORY CONTROL SYSTEM 51
3.11 EXAMPLE 2--LIQUID COOLING 54
3.12 EXAMPLE 3--POLLUTION ABSORPTION 57
Basic Linear Pollution Model 57
Response of Basic Model to a Constant POLGR 57
Nonlinear Pollution Model 60
Response of Nonlinear Model to a Constant POLGR 61

3.1 INTRODUCTION

Negative feedback is characterized by goal-directed or goal-oriented behavior. Such terms as self-governing, self-regulating, self-equilibrating, homeostatic, or adaptive, all implying the presence of a goal, define negative feedback systems.

Thermostatically controlled heating is a common self-governing system. A causal-loop diagram of the system appears in Figure 3-1. The thermostat system attempts to maintain a preselected desired room temperature. The decision-making unit, the thermostat, senses a disparity between desired and actual room temperature and activates the heating unit. The addition of heat eventually raises room temperature to the desired level. Then the thermostat automatically shuts off the heater.

The system description in Figure 3-1 applies equally well to a thermostatically controlled oven, the electric eye of a camera, the automatic pilot of an airplane, or the speed governor of an engine. These systems, developed in the field of control engineering, pursue and attempt to maintain specific objectives. The concept of control itself entails goal orientation.

Mechanically controlled systems have equivalents in the biological world also. The human body contains numerous self-regulating physiological processes to maintain a relatively constant internal environment. This self-regulation, called homeostasis, is necessary for survival. The temperature regulation system, for example, maintains normal body temperature through continual alteration of metabolic activities and blood flow rates. Digestion, blood-sugar regulation, and waste removal provide additional examples of homeostatic processes.

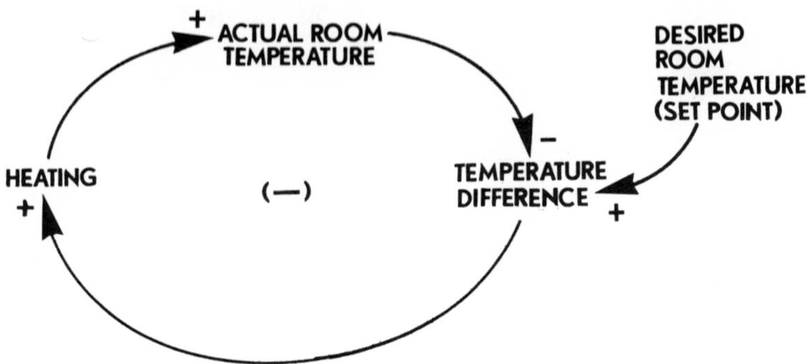

FIGURE 3-1 Thermostat heating system

Goal-directed action is fundamental to human social behavior. Consider the socialization of a child at home. Parents transmit their values, attitudes, and expectations to the child through a process of negative feedback. When a discrepancy arises between desired and perceived behavior of the child, the parents initiate corrective action in the form of punishment. The child, in turn, learns how to appease his parents and/or obtain desired rewards. "The behavior with which a parent controls his child, either aversively or through positive reinforcement, is shaped and maintained by the child's responses."[1]

How is negative feedback represented in terms of rates and levels? What characteristic behavior modes are associated with simple negative feedback? What role do parameters play in a negative feedback system? This chapter investigates these questions and presents three case studies to illustrate negative feedback.

3.2 CAUSAL-LOOP DIAGRAM OF SIMPLE NEGATIVE FEEDBACK

The four basic elements of negative feedback systems appear in Figure 3-2: the desired state (goal), the discrepancy, the action (rate), and the system state (level). The thermostat system in Figure 3-1 is a specific example of the feedback loop in Figure 3-2.

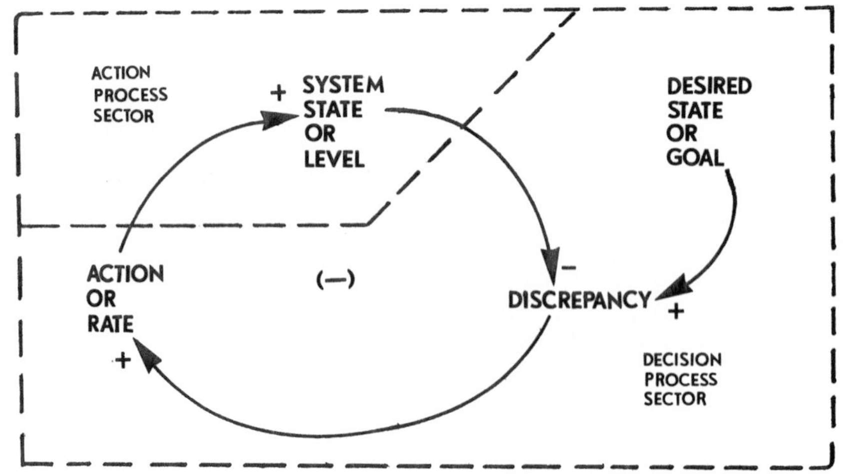

FIGURE 3-2 Causal-loop diagram--negative feedback

---

[1] B.F. Skinner, Beyond Freedom and Dignity (New York: A.A. Knopf, 1971), p. 169.

The difference between the negative feedback loop in Figure 3-2 and a positive feedback loop resides entirely in the decision-process sector: the information sensing, comparison, and decision-making components intervening between the system state and action.[2] The decision-process sector specifies how the level controls the rate:

> ...a decision process is one that controls any system action. It can be a clear explicit human decision. It can be a subconscious decision. It can be the governing processes in biological development. It can be the natural consequences of the physical structure of the system.[3]

The simplest negative feedback system contains a goal and discrepancy element in the decision-process sector. The goal serves as a reference or guideline on which the system bases action. The goal is determined externally or exogenously. That is, a direct causal link from the system to the goal is lacking. The goal acts as an input to the system. The "set-temperature" affects the thermostat system in this manner.

The discrepancy between the goal and the state of the system determines the magnitude and direction of the corrective action taken. The thermostat performs this function in the heating system. A sensor, such as a thermocouple, monitors room temperature and compares it to the thermostat setting. A temperature discrepancy causes the thermostat to activate the heating unit.

The decision process sector in man is the brain. For example, through his visual perception a driver senses the lateral position (state) of his vehicle relative to a curb. Corrective action occurs when the auto moves to an undesirable position. The resulting mechanical alteration of the steering wheel modifies the position of the automobile.

Figure 3-2 illustrates how the decision-process sector of negative feedback completes the circularity of the system and acts in concert with the action process to produce goal-seeking behavior. An arbitrary increase (or decrease) in the level produces a discrep-

---

[2] Systems without a decision sector are "open-loop" systems. Action taken does not depend on the state.

[3] Forrester, Principles of Systems, p. 36.

ancy between the goal and level of the system. To minimize the discrepancy, the system initiates action to decrease (increase) the level. This system response from the closed loop tends to counter the outside change in the system level. The same scenario occurs in any negative feedback system.

3.3 FLOW DIAGRAM

Figure 3-3 shows the generalized flow diagram of negative feedback. The rate RT, modulating the flow into or out of the level LEV, provides the action component of the system. The discrepancy DISC, an auxiliary variable, and the fraction of the discrepancy acted upon per time unit FPT, a constant, control the rate RT. DISC is simply the difference between the level LEV and goal GL, a constant. The equations below represent a negative feedback structure.

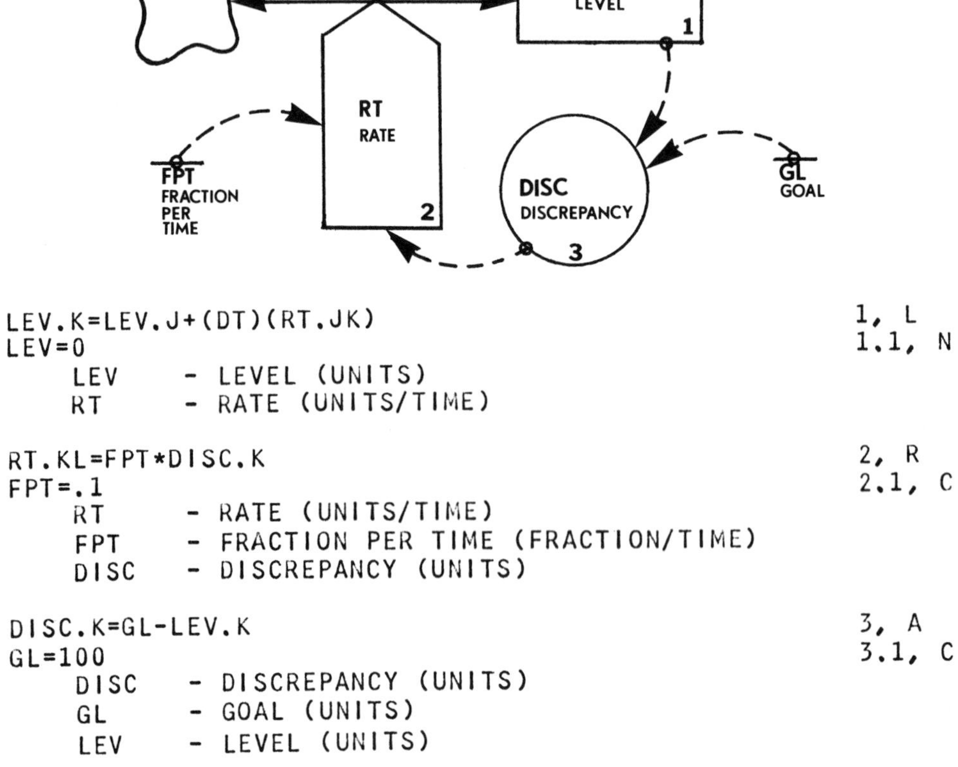


```
LEV.K=LEV.J+(DT)(RT.JK)                                   1, L
LEV=0                                                     1.1, N
     LEV     - LEVEL (UNITS)
     RT      - RATE (UNITS/TIME)

RT.KL=FPT*DISC.K                                          2, R
FPT=.1                                                    2.1, C
     RT      - RATE (UNITS/TIME)
     FPT     - FRACTION PER TIME (FRACTION/TIME)
     DISC    - DISCREPANCY (UNITS)

DISC.K=GL-LEV.K                                           3, A
GL=100                                                    3.1, C
     DISC    - DISCREPANCY (UNITS)
     GL      - GOAL (UNITS)
     LEV     - LEVEL (UNITS)
```

FIGURE 3-3 Flow diagram--First-order negative feedback

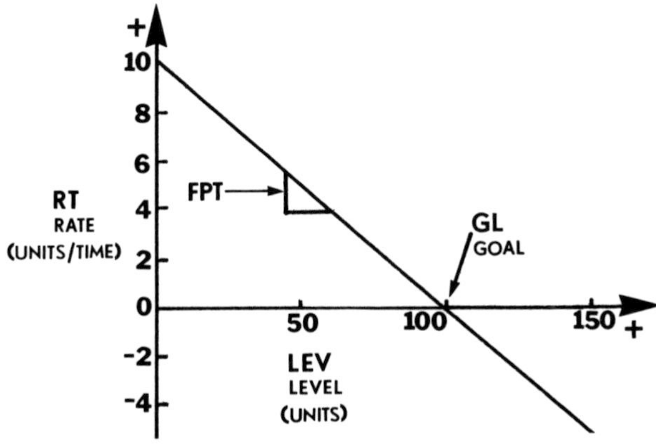

FIGURE 3-4 Rate-level graph

By graphing rate RT versus level LEV, we gain deeper insight into the behavior of the system. Substitute the DISC auxiliary equation into the rate equation:

R RT.KL = FPT(GL-LEV.K)

Figure 3-4 graphs RT as a function of LEV. The static relationship in Figure 3-4 does not explicitly depend on time. The slope and horizontal axis intercept depend solely on the values of the constants, FPT and GL. The influence of these constants on system behavior will be examined in sections 3.5 and 3.7.

3.4 SYSTEM BEHAVIOR

The behavior of the system with respect to time can be graphically "simulated" by using Figure 3-4 and constructing a level versus time plot. The simulation procedure takes the following steps:[4]

1. Use the rate-level graph (Figure 3-4) to determine the initial RT value from the initial LEV value.

2. Add the product of the RT value and the time interval DT to the level value to produce a new value of LEV. Plot the new LEV on a level LEV versus time graph one DT to the right of the last calculation.

3. Given the new level value, compute a new rate value for each DT until the time span under consideration is covered.

---

[4] Graphical simulation follows the same steps as numerical simulation in Chapter 2 but is easier to visualize.

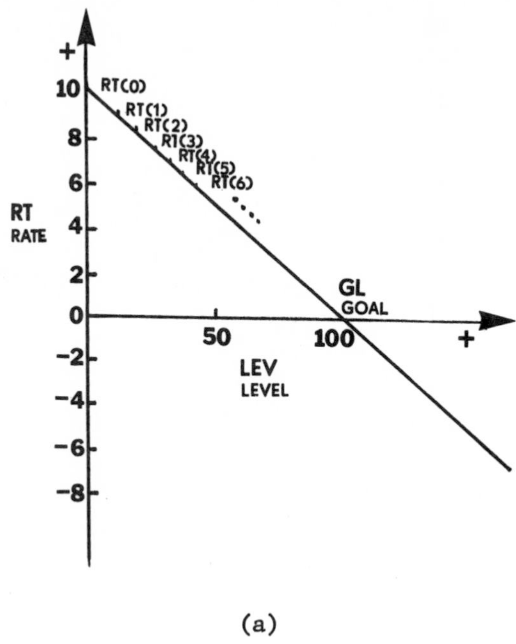

(a)

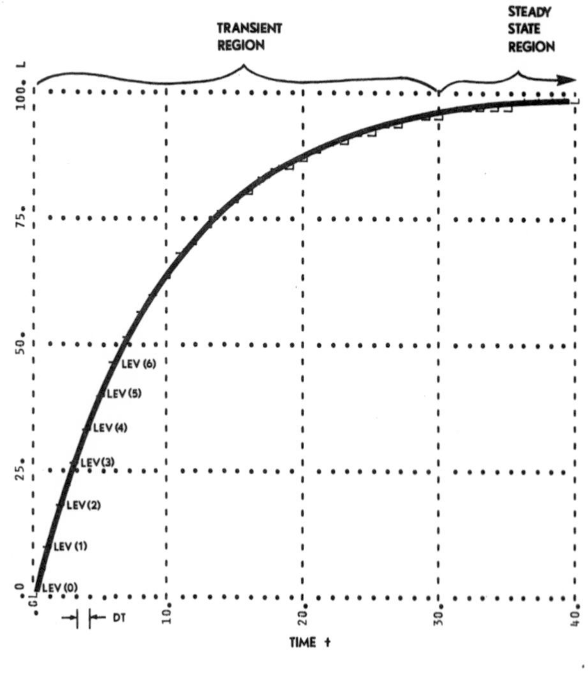

(b)

FIGURE 3-5 Simulation of negative feedback

Figure 3-5 illustrates the simulation procedure and resulting behavior of LEV over time. The parenthetical numbers, such as LEV(0) or RT(6), following LEV and RT correspond to the time interval at which LEV and RT are computed. At time equals zero, the level LEV(0) yields a rate RT(0) shown in Figure 3-5 (a). The product of DT and RT(0) added to LEV(0) produces LEV(1) at time equal to one DT unit (Figure 3-5 (b)). LEV(1) in turn determines RT(1). When multiplied by DT and added to LEV(1), RT(1) produces LEV(2) at time equal to two DT units. Each new increment to LEV becomes smaller and smaller as LEV increases. The increments to LEV eventually terminate as LEV asymptotically approaches goal GL and RT approaches zero. Connect the points generated in Step 2, and the characteristic curve of a simple negative feedback system appears in Figure 3-5 (b).

The behavior in Figure 3-5 (b) contains two separable regions: the transient and the steady state. In the transient region the level value differs from the goal value. The behavior of the level is characteristically goal-seeking and transitory. The steady state region, on the other hand, is characterized by goal attainment and persistent behavior. The rate in steady state approaches zero.[5]

We can derive an analytical expression for LEV as a function of time and the constants FPT and GL:[6]

$$LEV(t) = GL+[LEV(0)-GL]e^{-FPT*t} \qquad (3.1)$$

where

LEV(t) = level value at time t

GL = goal

LEV(0) = level initial value

e = base of natural logarithm

FPT = slope of rate-level graph (Figure 3-4)

t = time

---

[5]Theoretically LEV never reaches GL precisely and RT never reaches zero. However, we assume that LEV closely approximates GL and RT becomes zero in the steady state region. The separation between transient and steady state regions, however, is somewhat arbitrary.

[6]The differential equation of the system is:

(Footnote continued to next page.)

3.5 SLOPE AND TIME CONSTANT

The inverse of the slope of the rate-level graph is the time constant T of the system as shown in Chapter 2 (section 2.3). The time constant of the negative feedback system indicates the length of time required for LEV to attain 63% of the discrepancy between the goal GL and the initial level value LEV(0). Equation (3.1) shows how we obtain this figure:

$$\begin{aligned} LEV(t = T = 1/FPT) &= GL+[LEV(0)-GL]e^{-1} \\ &= GL+[LEV(0)-GL]/2.73 \\ &= LEV(0)+0.632[GL-LEV(0)] \end{aligned}$$

In another time interval of length 1/FPT, 63% of the discrepancy between GL and LEV(t = 1/FPT) is made up. In the third 1/FPT interval, LEV attains 95% of its goal GL. Figure 3-6 shows that, in a period of time roughly equal to three time constants, the simple negative feedback system attains its goal.

---

6 (cont.)

$$\frac{dLEV(t)}{dt} = FPT[GL-LEV(t)]$$

Separation of variables leads to

$$\frac{dLEV(t)}{GL-LEV(t)} = FPT*dt$$

Integrating both sides from LEV(0) to LEV(t) and from time 0 to time t:

$$\int_{LEV(0)}^{LEV(t)} \frac{dLEV(\tau)}{GL-LEV(\tau)} = \int_0^t FPT*d\tau \qquad \text{(where } \tau \text{ is a dummy variable)}$$

$$-\ln[GL-LEV(\tau)]\Big|_0^t = FPT*\tau\Big|_0^t$$

or,

$$\ln[GL-LEV(t)/GL-LEV(0)] = -FPT*t.$$

Taking the antilog of both sides:

$$\frac{GL-LEV(t)}{GL-LEV(0)} = e^{-FPT*t}.$$

Rearranging terms:

$$GL-LEV(t) = [GL-LEV(0)]e^{-FPT*t}$$

$$LEV(t) = GL-[GL-LEV(0)]e^{-FPT*t}$$

or,

$$LEV(t) = GL+[LEV(0)-GL]e^{-FPT*t}.$$

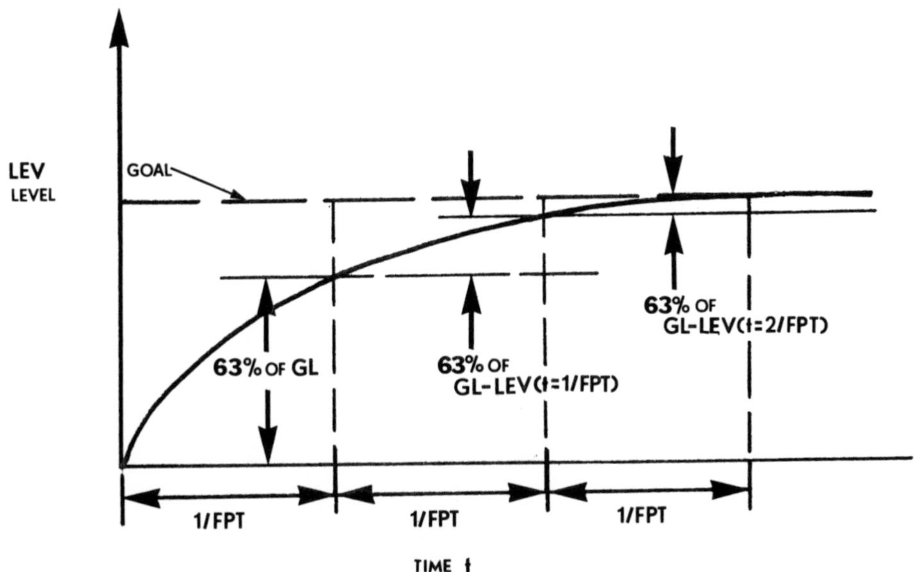

FIGURE 3-6 Transient response of negative feedback

3.6 INITIAL CONDITIONS

In the simple negative feedback system, two general responses of the level to an initial discrepancy can take place: asymptotic growth toward a goal or exponential decay toward a goal. The initial value of the level relative to the goal determines which behavior occurs. Figure 3-4 shows that an initial level value less than the goal produces flow into the level. The inward flow slows as the level asymptotically approaches the goal over time. An initial value greater than the goal specifies a flow out of the level. The rate decreases as the level decays toward the goal and equilibrium. Of course, no response occurs when the initial level value equals the goal.

3.7 ZERO-VALUE GOAL STRUCTURE

Referring to Figure 3-4, the point of intercept between the rate-level line and the horizontal axis determines the goal or final equilibrium value of the system. Any goal value is possible. Many systems may have goal values at zero. Physical processes such as radioactive decay pursue zero-value goals. The depreciation of capital and the decline of a population with more deaths than births offer additional examples. Figure 3-7 illustrates a causal loop, a flow diagram, and equations for a structure with a goal of zero.

Two basic elements of the generalized loop in Figure 3-2 disappear: the exogenous goal and the discrepancy. Figure 3-8 contains the rate-level graph for the zero-value goal system. In Figure 3-8, the rate RT always remains negative and, thus, flow is always out of the level LEV since the goal value is zero. The transient behavior of the system occurs only with a non-zero initial level value. Figure 3-9 depicts such a transient behavior mode.

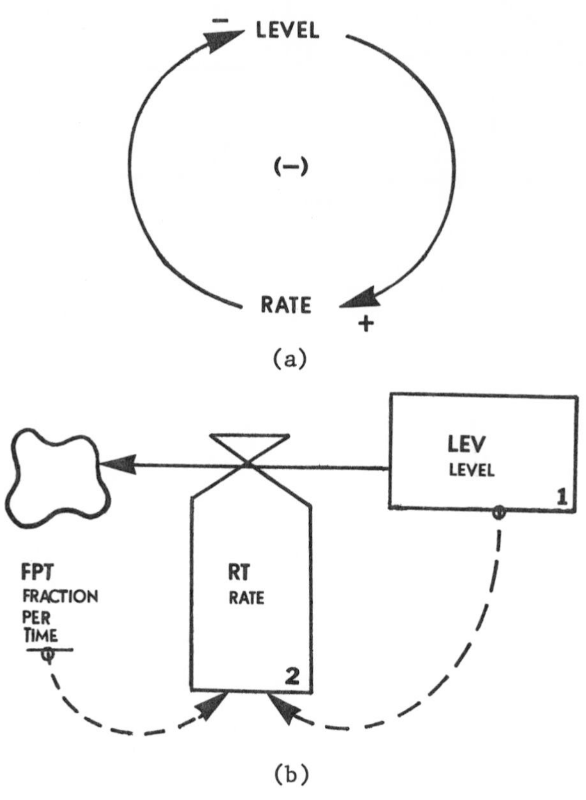


```
LEV.K=LEV.J+(DT)(RT.JK)                                 1, L
LEV=100                                                 1.1, N
     LEV     - LEVEL (UNITS)
     RT      - RATE (UNITS/TIME)

RT.KL=-FPT*LEV.K                                        2, R
FPT=.1                                                  2.1, C
     RT      - RATE (UNITS/TIME)
     FPT     - FRACTION PER TIME (FRACTION/TIME)
     LEV     - LEVEL (UNITS)
```

FIGURE 3-7 Negative feedback with zero-value goal

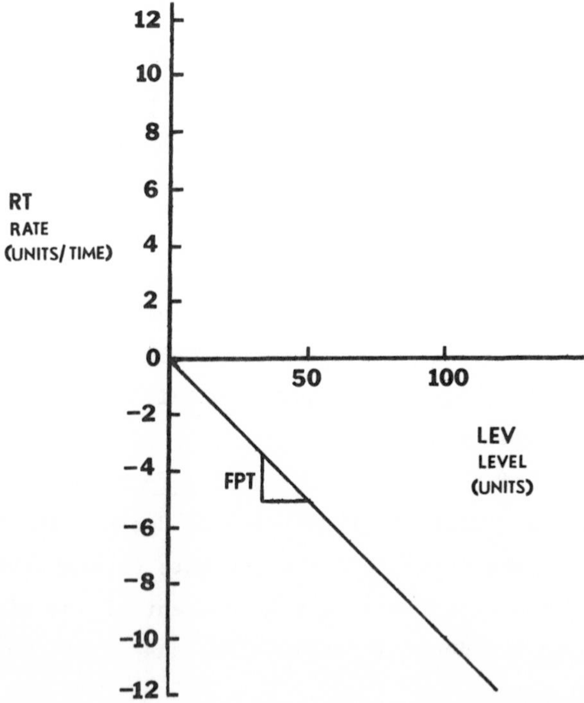

FIGURE 3-8 Rate-level graph when goal = 0

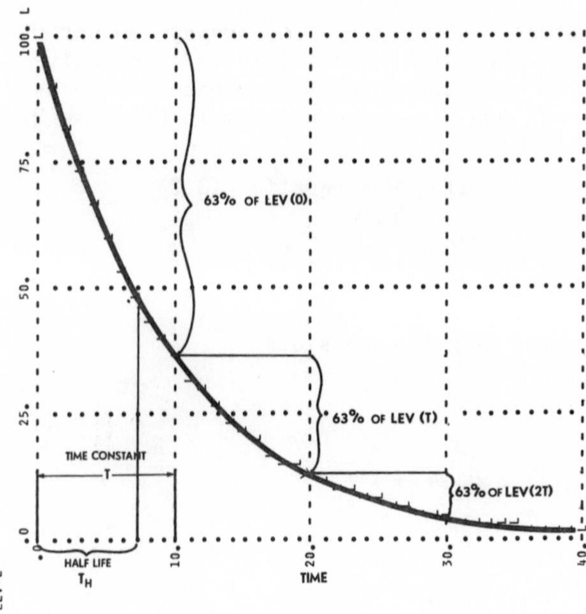

FIGURE 3-9 Level response when goal = 0

We can derive an analytical expression for the behavior from equation (3.1) by letting GL equal zero:

$$LEV(t) = LEV(0)e^{(-1/T)(t)} \qquad (3.2)$$

The preceding equation reproduces exactly the exponential equation of positive feedback except for the negative sign in the exponent. When t = T,

$$LEV(t = T) = 0.368*LEV(0).$$

In a time interval equal to T, 63% of the initial contents of the level disappear. In the next T interval, 63% of the contents at t = T are removed. Figure 3-9 shows how this on-going decline occurs over the given time interval.[7]

Besides the time constant, the <u>half-life</u> noted in Figure 3-9 provides a useful indicator of the rate of decay. The half-life is time interval during which half the contents of the level are removed. The half-life equals approximately 70 percent of the time constant.[8] The half-life in exponential decay is the analog of the doubling time in exponential growth.

---

[7]The time constant T also equals the average amount of time a given unit remains in the level of a single-level negative loop. This property allows the structure in Figure 3-7 to be used to represent a delay. Delays are discussed briefly in Exercise 9.

[8]To derive the half-life, use equation (3.2):

$$(1/2)LEV(0) = LEV(0)e^{-T_H/T}$$

where T_H = half-life

Taking the natural log of both sides:

$$\ln(1)-\ln(2) = -T_H/T$$

And,

$$T_H/T = \ln(2) = 0.69$$

Or,

$$T_H = 0.69*T.$$

3.8 SYSTEM COMPENSATION

A negative feedback system often attempts to maintain a goal in the presence of a constant inflow or outflow rate over which the feedback system exerts no control. What effect does the constant exogenous rate have on system behavior? Figure 3-10 presents a negative feedback structure with such an additional rate. The equations for the system reproduce those in Figure 3-3 except for the addition of RT2.

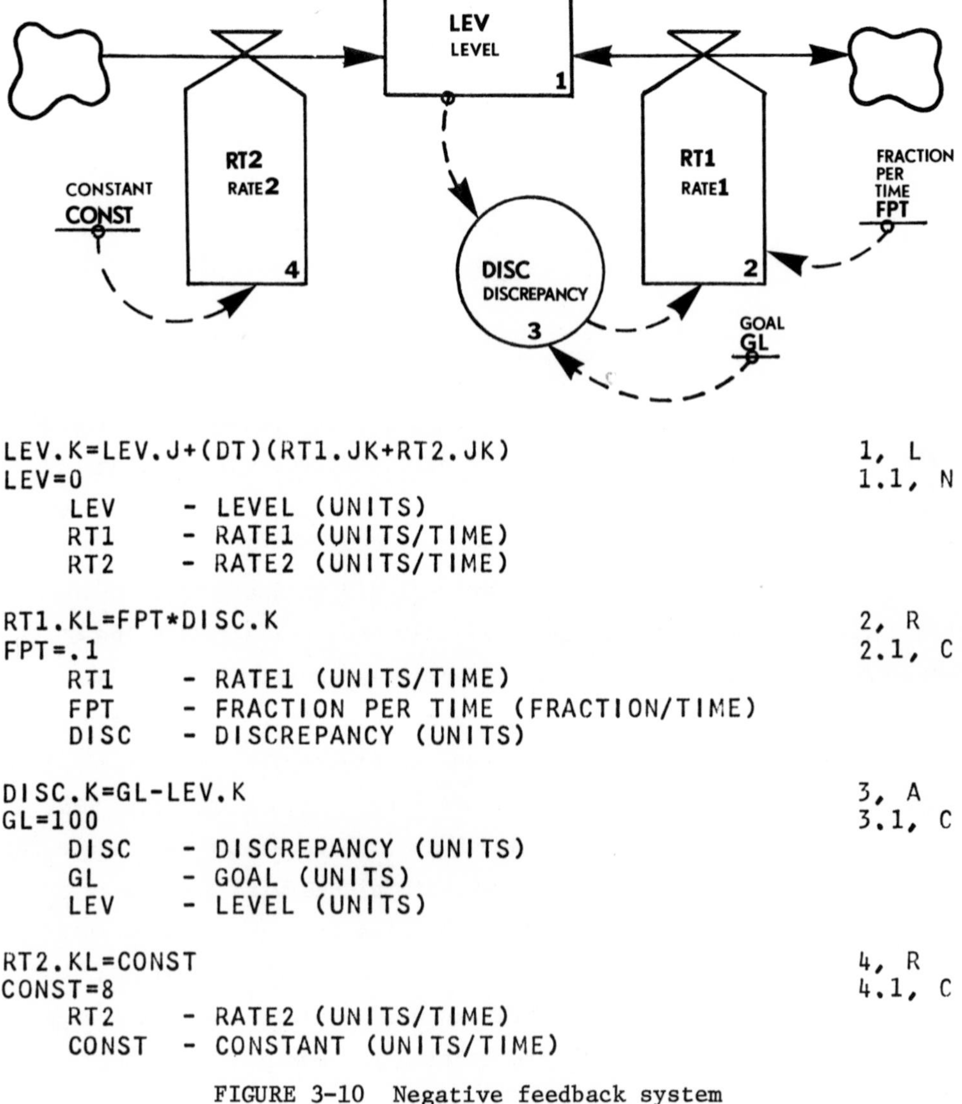


```
LEV.K=LEV.J+(DT)(RT1.JK+RT2.JK)                      1, L
LEV=0                                                1.1, N
    LEV     - LEVEL (UNITS)
    RT1     - RATE1 (UNITS/TIME)
    RT2     - RATE2 (UNITS/TIME)

RT1.KL=FPT*DISC.K                                    2, R
FPT=.1                                               2.1, C
    RT1     - RATE1 (UNITS/TIME)
    FPT     - FRACTION PER TIME (FRACTION/TIME)
    DISC    - DISCREPANCY (UNITS)

DISC.K=GL-LEV.K                                      3, A
GL=100                                               3.1, C
    DISC    - DISCREPANCY (UNITS)
    GL      - GOAL (UNITS)
    LEV     - LEVEL (UNITS)

RT2.KL=CONST                                         4, R
CONST=8                                              4.1, C
    RT2     - RATE2 (UNITS/TIME)
    CONST   - CONSTANT (UNITS/TIME)
```

FIGURE 3-10 Negative feedback system with constant inflow rate

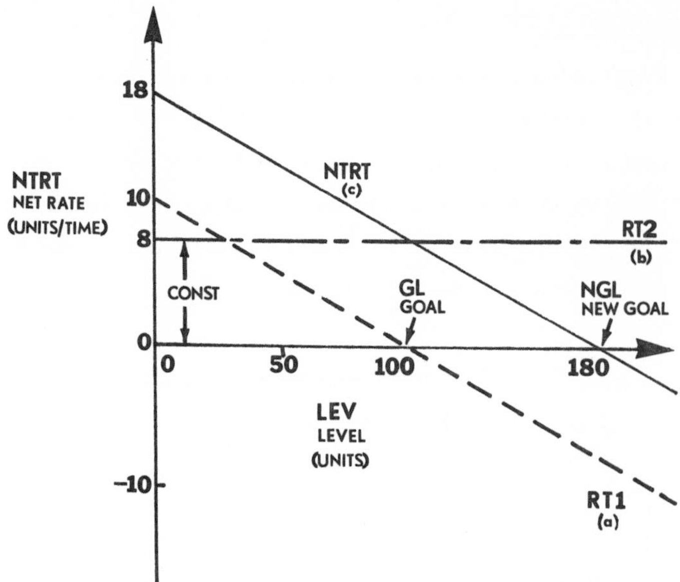

FIGURE 3-11 System-rate-level-graph -- constant input

Figure 3-11 contains the rate-level graph of the system.

Curve (a), RT1, in Figure 3-11 depicts the rate-level graph for the system without RT2. Curve (b), RT2, depicts the constant inflow rate alone. Curve (c), the net rate NTRT, depicts the addition of RT1 and RT2 or,

NTRT.KL = FPT(GL-LEV.K)+CONST.

The rate-level curves in Figure 3-11 aid in analysis of system behavior. Assume the level LEV initially equals its goal value GL when the inflow rate RT2 begins. RT2 causes an initial increase in LEV. The new LEV value produces a negative outflow rate RT1 since LEV now exceeds GL. However, as shown in Figure 3-11, RT1 is smaller than RT2. This rate difference causes a net gain in the level LEV. LEV continues to increase over time but at a decreasing rate since the difference between RT2 and RT1, NTRT, diminishes with increasing LEV. Eventually, the outflow rate RT1 compensates for the inflow rate RT2 and equilibrium is established. In the process, the system has accumulated a net inflow rate NTRT and has acquired a new larger goal value NGL. We can compute NGL by setting the sum of RT1 and RT2 equal to zero or,

RT2+RT1 = 0,

CONST = -FPT(GL-LEV),

and

LEV = NGL = GL+CONST/FPT.

In terms of the time constant T:

NGL = GL+(T)(CONST).

The new equilibrium value of the system exceeds the desired goal GL by the product of the time constant T and input rate CONST. A constant outflow rate yields a NGL value less than goal GL by the same product. A negative feedback system compensates for the additional inflow or outflow rate by attaining an equilibrium level different from the desired one.

3.9 SUMMARY

Negative feedback tends to keep a given system at equilibrium. In the most simple system, a single rate based on discrepancies between the desired goal and the level drives the system to its goal value. Inflow or outflow rates readjust until the level reaches the desired value. The time constant determines the speed at which the system reacts to changes in the level. We will examine three simple examples of negative feedback systems: an inventory control system, a cooling liquid system, and a pollution dissipation system.

3.10 EXAMPLE 1--INVENTORY CONTROL SYSTEM

A dealer likes to maintain a desired level of inventory. When stock falls below the desired level, the dealer places orders to the factory to replenish the supply. Orders stop when stock reaches the desired level. With too much inventory, the dealer sends the excess back to the factory. The flow diagram of Figure 3-12 represents this inventory control system.

Sales, which deplete the inventory INV, depend on market conditions outside the system boundary. Assume that the dealer has no influence on demand for his goods.

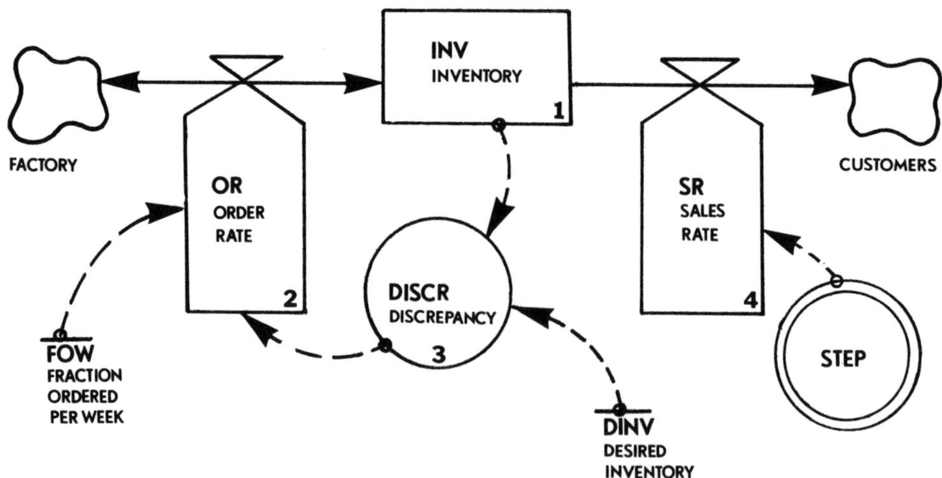

FIGURE 3-12 Simple inventory control system

The order rate OR replenishes the inventory stock and depends on the dealer's ordering policy. The flow diagram represents one simple ordering scheme. The order rate OR depends on the discrepancy DISCR between inventory INV and desired inventory DINV and on the fraction ordered per week FOW. The fraction ordered per week FOW measures how quickly the dealer responds to a discrepancy between INV and DINV. The time constant of the system, the inverse of FOW, indicates the number of weeks required to make up a sudden stock shortage through orders. A two week time constant (1/FOW = 2) produces a six week delay before orders (approximately) equal sales and equilibrium is re-established.

The equations for the model in Figure 3-12 appear below:

```
INV.K=INV.J+(DT)(OR.JK-SR.JK)                            1, L
INV=DINV                                                 1.1, N
    INV     - INVENTORY (UNITS)
    OR      - ORDER RATE (UNITS/WEEK)
    SR      - SALES RATE (UNITS/WEEK)
    DINV    - DESIRED INVENTORY (UNITS)
OR.KL=FOW*DISCR.K                                        2, R
FOW=.5                                                   2.1, C
    OR      - ORDER RATE (UNITS/WEEK)
    FOW     - FRACTION ORDERED PER WEEK (FRACTION/WEEK)
    DISCR   - DISCREPANCY (UNITS)
DISCR.K=DINV-INV.K                                       3, A
DINV=200                                                 3.1, C
    DISCR   - DISCREPANCY (UNITS)
    DINV    - DESIRED INVENTORY (UNITS)
    INV     - INVENTORY (UNITS)
SR.KL=STEP(20,4)                                         4, R
    SR      - SALES RATE (UNITS/WEEK)
```

The model assumes no material delays (delays in shipping and handling goods) nor information delays (delays in recognizing and acting on a discrepancy). An order placed is immediately filled. The dealer knows the inventory status at all times.

Since the inventory INV equals the desired inventory DINV of 200 units, the system is initially at equilibrium. In the fourth week, we simulate a sudden rise in sales from zero to 20 items/week to test the ordering policy's ability to maintain desired inventory.

Figure 3-13 shows that the simple discrepancy ordering policy proves inadequate for maintaining a desired level of inventory. The actual inventory INV falls 20 percent below the desired inventory DINV. Figure 3-13 demonstrates that, in the presence of a constant inflow rate, the final equilibrium value never meets the desired value.[9] The final value differs from the desired value by a quantity equal to the product of the time constant and the constant flow rate, as we saw in section 3.8.

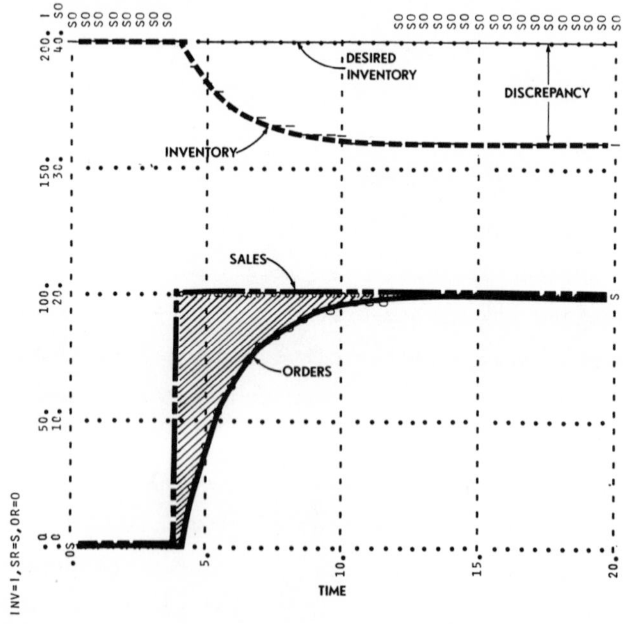

FIGURE 3-13 System response to step change in sales

[9] A control scheme that is proportional to the error between the desired state and actual state of a physical system proves inadequate in the presence of a constant input since an error or discrepancy always exists. This property of proportional-control systems is well known in the engineering control literature.

Figure 3-13 provides evidence of this failure. The difference between the sales rate SR and the order rate OR for the six weeks following the rise in sales indicates a new flow of goods out of inventory. The net flow falls to zero once the order rate OR compensates for the sales rate SR at approximately week 10. The cross-hatched area in Figure 3-13 represents the total accumulated inventory loss, an amount equal to the product of the time constant (1/FOW) and the input (SR).

Unless the time constant of the system becomes zero (an unlikely situation), the dealer's ordering policy never maintains a desired level of inventory under constant sales conditions. With information and material delays, the disparity between final inventory and desired inventory would further increase.

The general first-order negative feedback structure represents a common inventory maintenance policy. Despite simplifying assumptions about delays in the system, the structure yields considerable insight into the compensatory nature of negative feedback. The order rate responds to a sudden rise in sales but cannot maintain the desired inventory level.

3.11 EXAMPLE 2--LIQUID COOLING

A common first-order feedback process is the cooling of hot liquid. The process belongs to a general class of thermal phenomena in which an object at one temperature enters an environment with a different temperature. All the elements of a goal-seeking system are present. Hot coffee cooling to room temperature offers a familiar specific example.

Heat flows between the environment (room) and the coffee cup. The temperature reflects the amount of stored heat. A well-known physical law describes the direction and amount of heat transfer: the heat transfer rate HTR is proportional to the difference DISC between the ambient (room) temperature RTP and the coffee temperature CTP. The constant of proportionality C2 depends upon such physical properties of the coffee cup as its volume and insulation material. A model of the system appears in Figure 3-14.

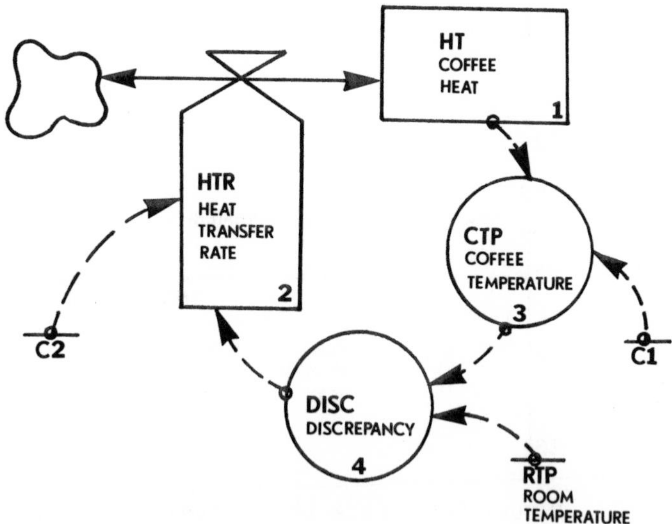

FIGURE 3-14 Flow diagram: coffee cooling

Dynamo equations for the system are:

```
HT.K=HT.J+(DT)(HTR.JK)                                     1, L
HT=TI/C1                                                   1.1, N
TI=200                                                     1.2, C
C1=1                                                       1.3, C
    HT     - COFFEE HEAT (BTU)
    HTR    - HEAT TRANSFER RATE (BTU/MIN)
    TI     - TEMPERATURE,INITIAL (°F)
    C1     - HEAT-TO-TEMPERATURE CONVERSION CONSTANT (°
               F/BTU)
HTR.KL=C2*DISC.K                                           2, R
C2=.1                                                      2.1, C
    HTR    - HEAT TRANSFER RATE (BTU/MIN)
    C2     - HEAT TRANSFER CONSTANT (BTU/°F/MIN)
    DISC   - DISCREPANCY (°F)
CTP.K=C1*HT.K                                              3, A
    CTP    - COFFEE TEMPERATURE (°F)
    C1     - HEAT-TO-TEMPERATURE CONVERSION CONSTANT (°
               F/BTU)
    HT     - COFFEE HEAT (BTU)
DISC.K=RTP-CTP.K                                           4, A
RTP=78                                                     4.1, C
    DISC   - DISCREPANCY (°F)
    RTP    - ROOM TEMPERATURE (°F)
    CTP    - COFFEE TEMPERATURE (°F)
```

Since the initial temperature of the coffee exceeds the room temperature, heat flows outward as the rate-level graph in Figure 3-15 indicates. The outflow rate persists until equilibrium with the room

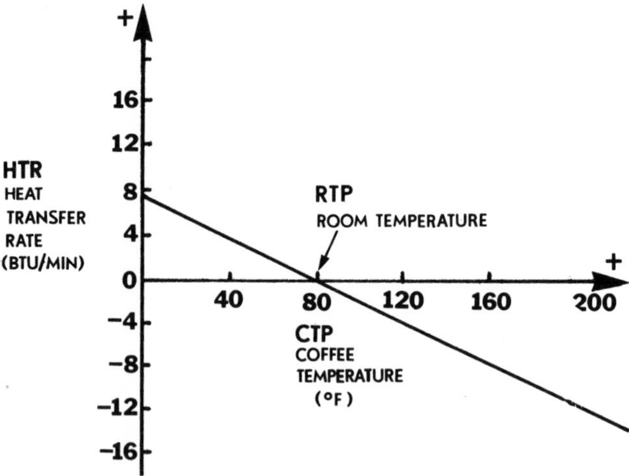

FIGURE 3-15 Rate-level graph

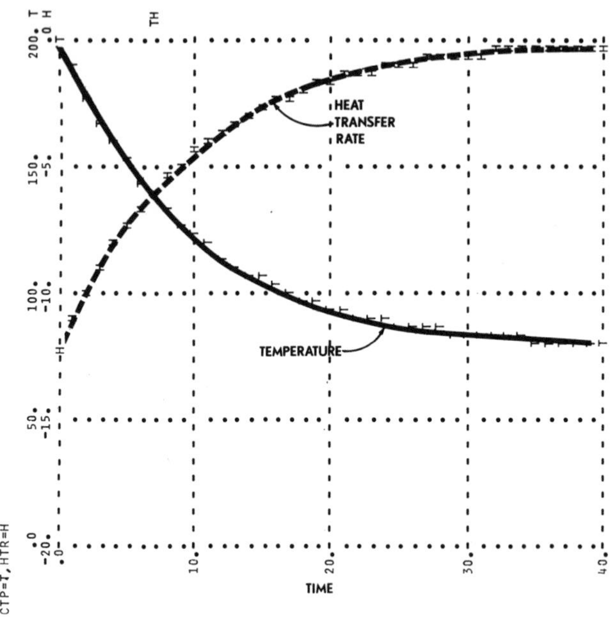

FIGURE 3-16 Coffee temperature over time

temperature is established. The time constant of the system is the reciprocal of (C2*C1) or ten minutes in this example.[10] Approximately 30 minutes later, the coffee temperature, exhibiting exponential decay, has cooled to room temperature. Figure 3-16 replicates the simulation run of the system over time.

[10] A perfectly insulated cup implies an infinite time constant. With perfect insulation, heat cannot dissipate and the initial temperature persists indefinitely.

3.12 EXAMPLE 3--POLLUTION ABSORPTION

Many simple feedback systems involve nonlinearities capable of converting a negative feedback process to a positive feedback process, or vice-versa. Chapter 4 discusses in detail the shift in dominance from positive to negative feedback. The pollution absorption example which follows exemplifies a reverse shift from negative to positive feedback.[11]

Basic Linear Pollution Model. The basic pollution system shown in Figure 3-17 (a) contains a pollution level increased by pollution generation and decreased by pollution absorption. The pollution generation rate POLGR, like the sales rate in the inventory control example, is exogenously determined. The pollution absorption rate POLAR depends upon the amount of pollution present in the environment at any given time. POLAR and POL form a negative feedback loop with a goal of zero. Figure 3-17 (b) shows the linear pollution absorption rate POLAR dependent on a fixed time constant, the pollution absorption time PAT. In the absence of pollution generation, any value of POL greater than zero yields a proportionate value of POLAR. Under these conditions, POL and POLAR decay in value until the initial amount of pollution dissipates.

Response of Basic Model to a Constant POLGR. What occurs when POLGR is constant? Figure 3-18 graphs a net pollution rate NPR (POLGR-POLAR) versus POL. The pollution level POL comes into equilibrium when POLAR equals POLGR:

POLGR = CONST = POLAR = (1/PAT)(POL)

and

POL = (PAT)(CONST)

Figure 3-19, the simulation run, illustrates that, even though the system has a goal of zero, POL asymptotically approaches a POL value of (PAT)(CONST) over time as the pollution absorption rate POLAR approaches the pollution generation rate POLGR. With any value of POLGR, regardless of magnitude, the system always achieves equilibrium.

---

[11] This example is based on the pollution sector of Forrester, World Dynamics, p. 24.

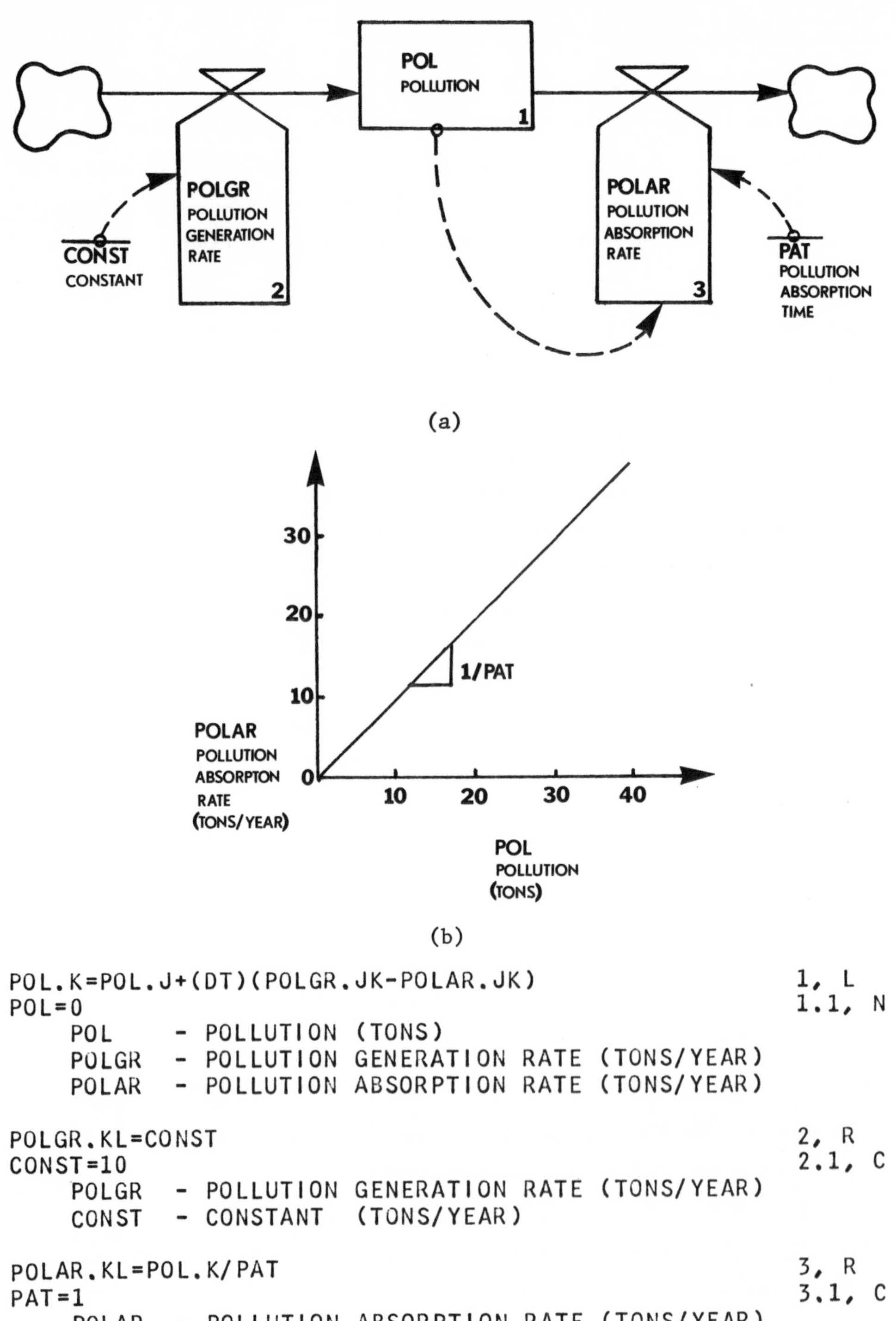


```
POL.K=POL.J+(DT)(POLGR.JK-POLAR.JK)                          1, L
POL=0                                                        1.1, N
     POL     - POLLUTION (TONS)
     POLGR   - POLLUTION GENERATION RATE (TONS/YEAR)
     POLAR   - POLLUTION ABSORPTION RATE (TONS/YEAR)

POLGR.KL=CONST                                               2, R
CONST=10                                                     2.1, C
     POLGR   - POLLUTION GENERATION RATE (TONS/YEAR)
     CONST   - CONSTANT  (TONS/YEAR)

POLAR.KL=POL.K/PAT                                           3, R
PAT=1                                                        3.1, C
     POLAR   - POLLUTION ABSORPTION RATE (TONS/YEAR)
     POL     - POLLUTION (TONS)
     PAT     - POLLUTION ABSORPTION TIME (YEARS)
```

FIGURE 3-17 Basic pollution model

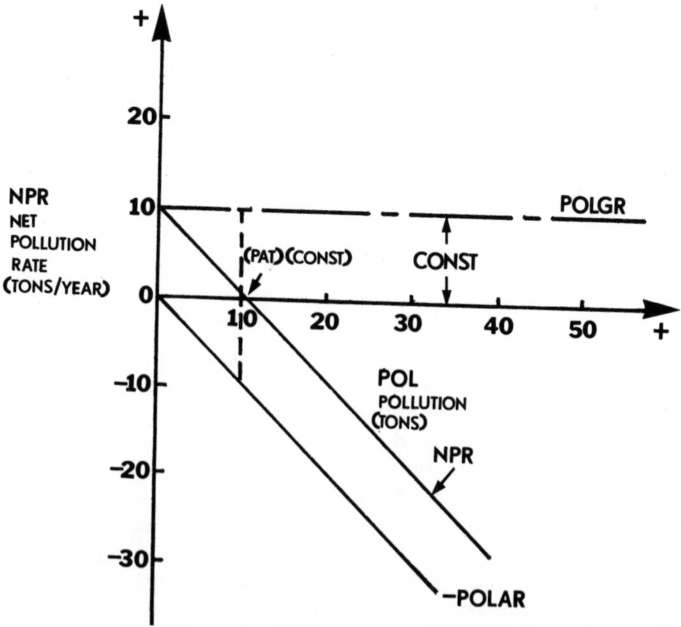

FIGURE 3-18 Net pollution rate versus pollution

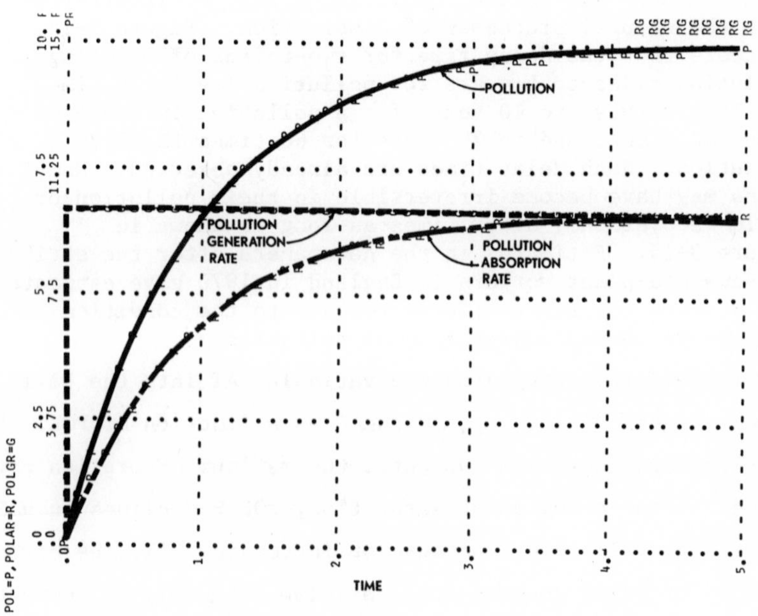

FIGURE 3-19 Response of POL to constant POLGR

Nonlinear Pollution Model. High pollution levels inhibit or destroy natural environment clean-up processes. A variable absorption time PAT captures the "overloading" effect of elevated pollution levels on the dissipative capacity of the environment. Figure 3-20 pictures a table function relating PAT to a pollution ratio POLR, the actual pollution level POL divided by a standard or definitional amount of pollution POLS (equal to one in the example). This function resembles the nonlinear relationship described in World Dynamics:

> ....A pollution ratio POLR of 1 represents the conditions existing in 1970. A value for PAT of 1 year is taken for 1970. This means an assumption that under present conditions a year would be needed to dissipate about two-thirds of the existing pollution if all new pollution generation were to stop. For some of the polluting materials, that is too slow. On the other hand, one sees estimates that 90% of all DDT that has ever been manufactured is still in the environment. Certainly many kinds of pollution, probably including the more serious kinds, take longer than a year to disappear. A year is here used as an average. But as the amount of pollution increases, the pollution-absorption time is assumed to increase. This represents the poisoning and destroying of the pollution-cleanup mechanisms. Small amounts of pollution are dissipated quickly. But large amounts can have a cumulative effect by interfering with the natural processes of dissipation. Figure 3-15 suggests that the decay time for two-thirds of existing pollution rises to 5 years for pollution levels 20 times the 1970 values, to 10 years for a pollution increase of about 40 times, and to 20 years for 60 times the 1970 pollution. Such delay times are already observed. Many lakes may have become irreversible in their pollution or would recover only after times as long as shown in Figure 3-15. Estimates in the newspapers after the strike of sewerage-plant workers in England in 1970 gave estimates of 10 years for river life to recover to the condition it had before the excessive load of pollution.[12]

Figure 3-21 (a) integrates the variable PAT into the basic structure. Figure 3-21 (b) plots POLAR as a function of POL. POLAR increases in magnitude with POL until the maximum absorption rate MAR is reached. After a region of saturation, POLAR declines in value with increasing pollution. Because of increasing PAT, the direction of the slope of POLAR changes from positive to negative, converting the negative feedback system to a positive feedback system. As PAT becomes infinitely large, POLAR approaches zero.

---

[12] See Forrester, World Dynamics, pp. 57-58.

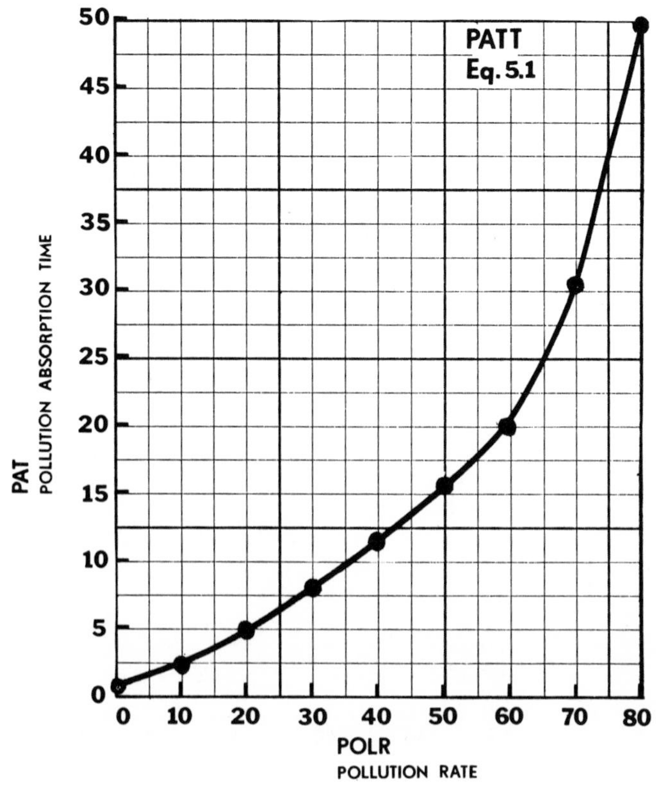

FIGURE 3-20 Pollution absorption versus pollution ratio

Response of Nonlinear Model to a Constant POLGR. How does the nonlinear pollution model behave in comparison with the linear model in the presence of constant pollution generation? As the net pollution rate NPR graph in Figure 3-22 shows, the size of the pollution generation rate POLGR is crucial. The net pollution rate NPR crosses the POL axis at P if CONST is less than the maximum absorption rate MAR. As in the linear system, POL asymptotically approaches P over time.

A value of CONST greater than MAR, however, yields a net pollution rate NPR that does not intersect the horizontal axis and therefore eliminates the possibility of system equilibrium. POL initially exhibits asymptotic, or goal-seeking, growth as POLAR reaches a maximum. An interval of linear growth follows during

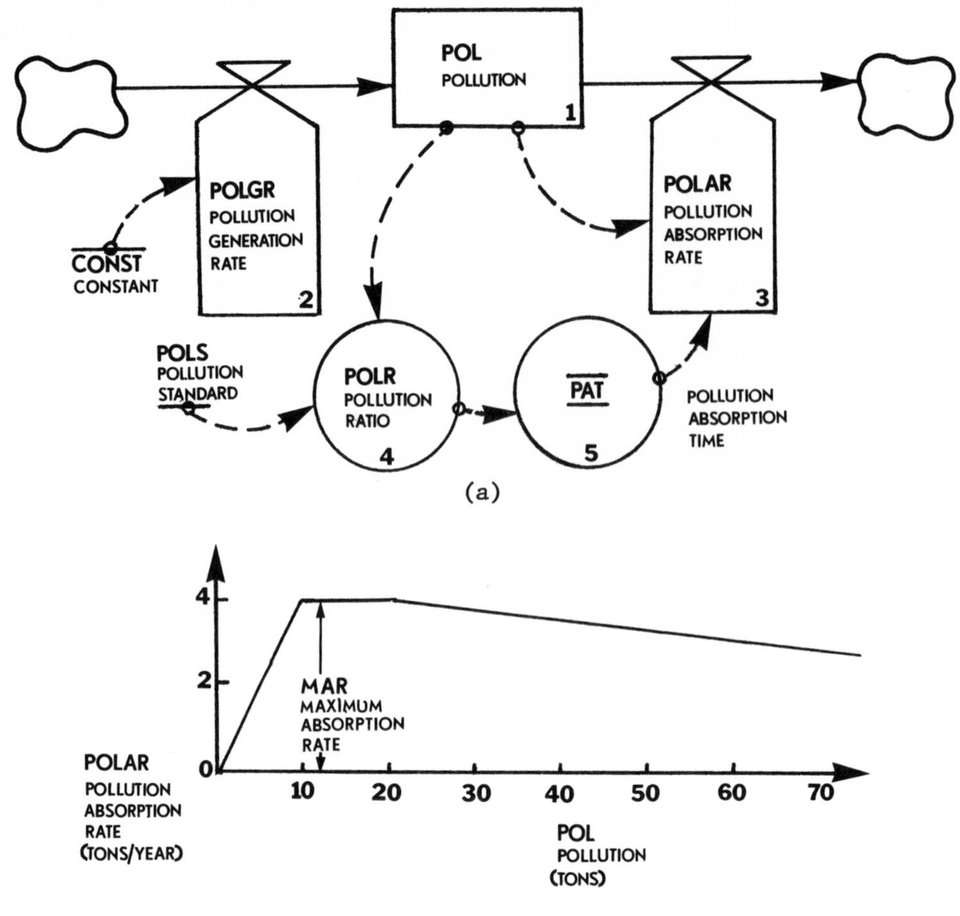

(a)

(b)

```
POL.K=POL.J+(DT)(POLGR.JK-POLAR.JK)                    1, L
POL=0                                                  1.1, N
    POL    - POLLUTION (TONS)
    POLGR  - POLLUTION GENERATION RATE (TONS/YEAR)
    POLAR  - POLLUTION ABSORPTION RATE (TONS/YEAR)

POLGR.KL=CONST                                         2, R
CONST=10                                               2.1, C
    POLGR  - POLLUTION GENERATION RATE (TONS/YEAR)
    CONST  - CONSTANT (TONS/YEAR)

POLAR.KL=POL.K/PAT.K                                   3, R
    POLAR  - POLLUTION ABSORPTION RATE (TONS/YEAR)
    POL    - POLLUTION (TONS)
    PAT    - POLLUTION ABSORPTION TIME (YEARS)
```

```
POLR.K=POL.K/POLS                                          4, A
POLS=1                                                     4.1, C
    POLR   - POLLUTION RATIO (DIMENSIONLESS)
    POL    - POLLUTION (TONS)
    POLS   - POLLUTION STANDARD (TONS)

PAT.K=TABLE(PATT,POLR,0,80,10)                             5, A
PATT=.6/2.5/5/8/11.5/15.5/20/31/50                         5.1, T
    PAT    - POLLUTION ABSORPTION TIME (YEARS)
    POLR   - POLLUTION RATIO (DIMENSIONLESS)
```

FIGURE 3-21 Complete pollution model

which POLAR maintains a maximum. When POLAR begins to decrease with increasing POL, NPR increases and produces exponential growth of POL. Eventually, POLAR approaches zero (not shown in Figure 3-22). As POLAR goes toward zero, NPR becomes constant (and equal to POLGR) and POL grows linearly without limit. The simulation runs shown in Figure 3-23 verify the rate-level graph analysis.

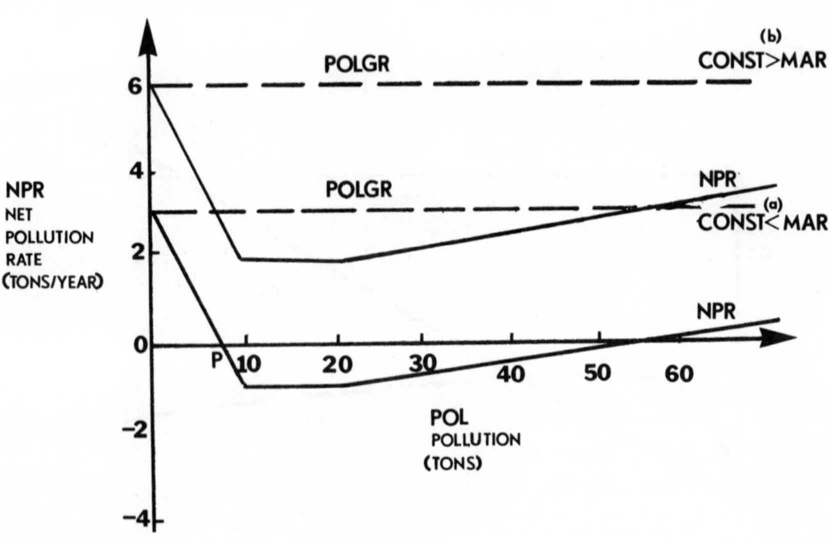

FIGURE 3-22 Net pollution rate versus pollution level

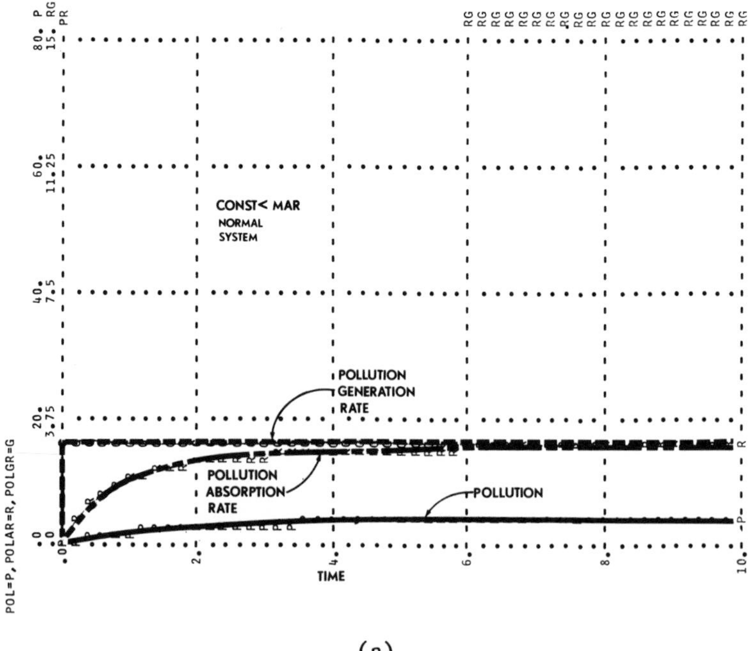

(a)

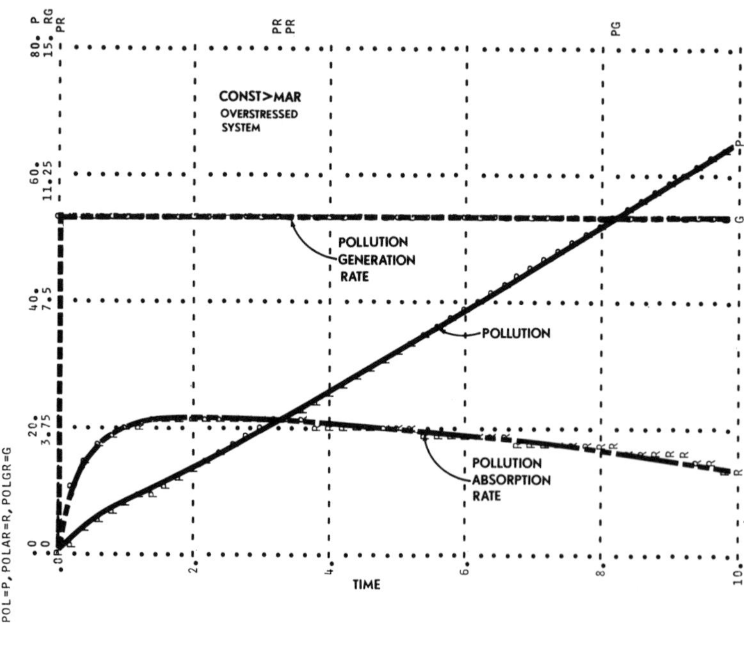

(b)

FIGURE 3-23 Constant POLGR applied to nonlinear model

The pollution dissipation system illustrates how a variable time constant (and hence a nonlinear rate-level relationship) affects the behavior of a negative feedback structure. A constant pollution absorption time PAT, regardless of the magnitude of the input POLGR, generates an equilibrium level of pollution. Pollution absorption can always match pollution generation. A variable absorption time, allowing for a diminished pollution absorption rate at high pollution levels, spawns more complex behavior. At low pollution generation rates, pollution accumulates but eventually goes to equilibrium. However, at high generation rates, the overloaded dissipative capacity of the system cannot absorb the inflow rate. Unlike the linear system, in a system with variable absorption time, accumulation of generated pollution breeds unrestrained growth of the pollution level.

Chapter 4
S-Shaped Growth Structure

Preparation: The reader should thoroughly understand linear negative and positive feedback structures and their characteristic behavior. Reader is advised to complete Exercises 4 through 7 before studying this chapter.

Purpose: Social and biological systems commonly exhibit exponential growth followed by asymptotic growth. This mode of behavior, also known as S-shaped or logistic growth, occurs in any structure where loop dominance shifts from positive feedback (discussed in Chapter 2) to negative feedback (discussed in Chapter 3). Chapter 4 employs rate-level analysis to explore the kind of nonlinear coupling which produces such a change in feedback loop dominance.

This chapter presents three examples of S-shaped growth--each example represented by a structure with only one level. The first example deals with population growth, the second analyzes the life-cycle of a contagious disease, and the third examines the behavior of a damped pendulum.

Practice: Chapter 5 and Exercise 8.

4.1 INTRODUCTION 69

4.2 S-SHAPED GROWTH STRUCTURE 71

4.3 STABLE AND UNSTABLE EQUILIBRIUM 76

4.4 SUMMARY 76

4.5 EXAMPLE 1--POPULATION GROWTH 77

Crowding and Population Growth 77

4.6 EXAMPLE 2--EPIDEMIC GROWTH 84

4.7 EXAMPLE 3--DAMPED PENDULUM 89

4.1 INTRODUCTION

Figure 4-1 illustrates S-shaped growth also known as "logistic" or sigmoid growth. It is a common type of behavior which combines both exponential and asymptotic growth. The time path includes two distinct regions. Exponential growth, the result of positive feedback, is followed by asymptotic growth, the result of negative feedback.

Population trends of various plants and animals typify S-shaped patterns. Births at first exceed deaths and the population grows exponentially. However, the excess of births over deaths decreases as environmental limitations to growth come to bear. The system enters equilibrium when the net growth rate (births minus deaths) equals zero. Many people advocate a logistic growth trend, similar to that observed in other biological populations, for human population.[1] Figure 4-2 illustrates alternative population growth patterns.

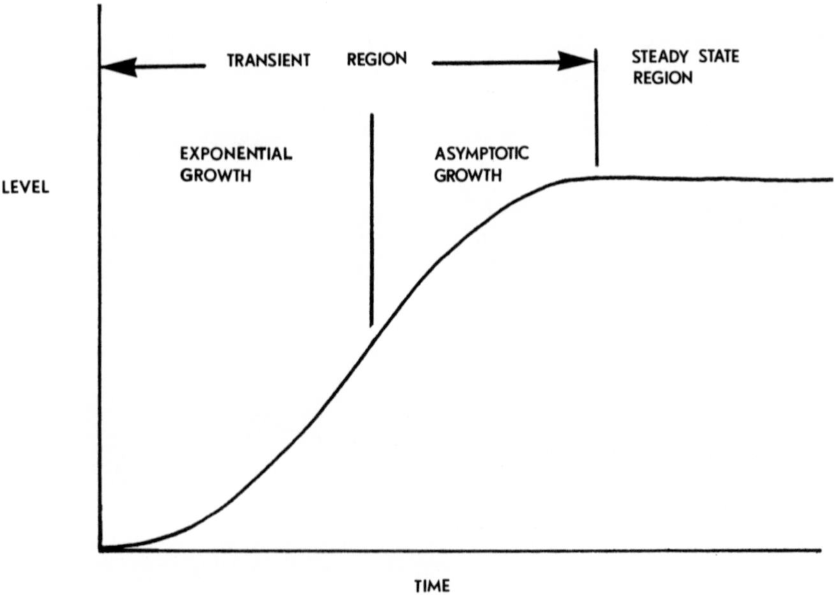

FIGURE 4-1 S-shaped growth

[1] See for example Jonas Salk, The Survival of the Wisest (New York: Harper & Row, 1973); or D.L. Meadows, et al., The Limits to Growth.

The solid line denotes past and projected growth; the dotted line suggests a desired equilibrium growth shape.

The sigmoidal trend also characterizes physical and mental development of an individual. Learning curves, for example, exhibit the S-shaped pattern. Achievement begins slowly, then gradually gains speed. Rapid achievement tapers off, however, as an individual attains his learning capacity.[2] Diffusion phenomena in a fixed population (such as the spread of riots, rumors, news, and epidemics) display S-shaped life cycles. The explosive growth of an epidemic eventually declines and levels off as the susceptible population succumbs entirely. Many socio-economic activities follow the sigmoidal pattern over time. The

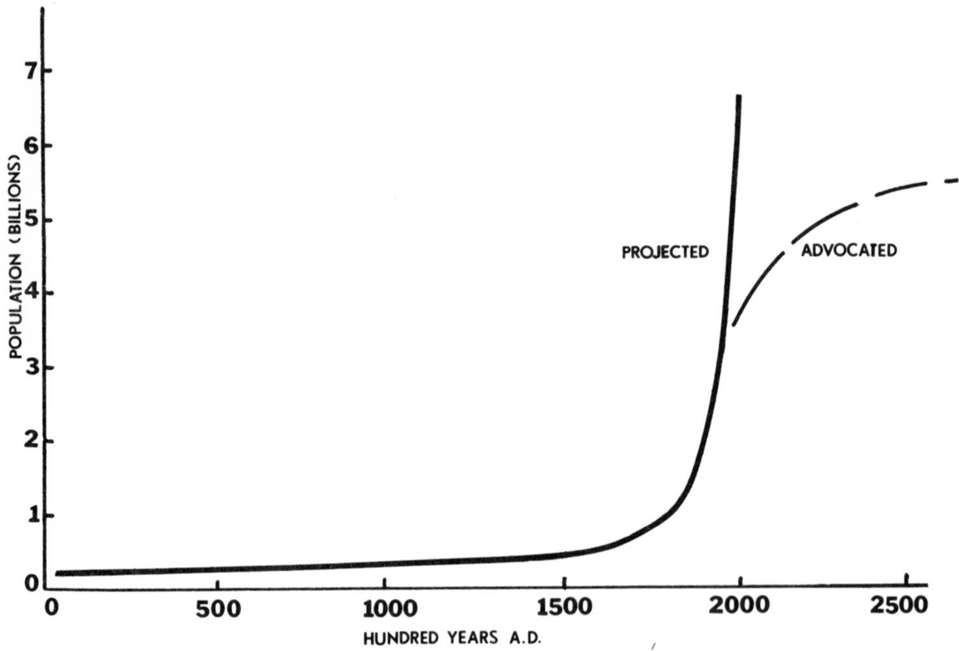

FIGURE 4-2 Human population growth

---

[2] See, for example, L. Coch and J.R.P. French, Jr., "Overcoming Resistance to Change," Human Relations (1947-48): pp. 514-15.

growth of urban structures on a tract of land exemplifies such a trend. The growth of empires may follow the sigmoidal or logistic curve.[3]

In a simple feedback system, what produces the transition from exponential to asymptotic growth? How can a single-level model produce the S-shaped trend? By answering these questions, the modeler will learn to better recognize and understand basic feedback relationships which produce sigmoidal growth in the real world.

4.2 S-SHAPED GROWTH STRUCTURE

S-shaped behavior requires that negative feedback (asymptotic growth) follow positive feedback (exponential growth). The simplified nonlinear structure pictured in Figure 4-3 contains the necessary shift in loop predominance. The rate-level relationship based on the

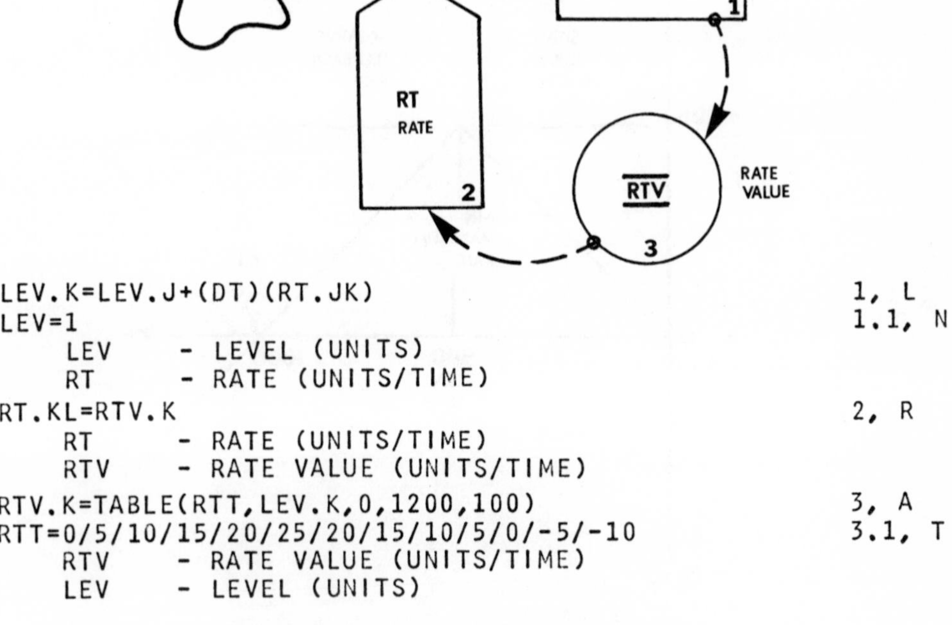


```
LEV.K=LEV.J+(DT)(RT.JK)                               1, L
LEV=1                                                 1.1, N
    LEV    - LEVEL (UNITS)
    RT     - RATE (UNITS/TIME)
RT.KL=RTV.K                                           2, R
    RT     - RATE (UNITS/TIME)
    RTV    - RATE VALUE (UNITS/TIME)
RTV.K=TABLE(RTT,LEV.K,0,1200,100)                     3, A
RTT=0/5/10/15/20/25/20/15/10/5/0/-5/-10               3.1, T
    RTV    - RATE VALUE (UNITS/TIME)
    LEV    - LEVEL (UNITS)
```

FIGURE 4-3 S-shaped growth structure

[3] Rein Taagepera, "Growth Curves of Empires," in General Systems, L.W. Bertalanffy, Anatol Rapport, Richard Meibr, eds., (Washington, D.C.: Society for General Systems Research, 1969), p. 171.

table function in Figure 4-3 appears in Figure 4-4. The rate-level graph contains two distinct curves: (1) a curve with a positive slope typical of positive feedback; (2) a curve with a negative slope typical of negative feedback.[4] At the inflection point (LEV = 500) on the graph, the rate RT attains a maximum, MR, and negative feedback begins to succeed positive feedback.

We can understand the behavior of the level over time by examining the rate-level graph. For example, an initial value of LEV slightly greater than zero yields a RT value proportionate to LEV. The RT value, accumulated over a small interval of time (DT) and added to the initial LEV value, yields a new and larger LEV value. Increasing RT values generate increasing LEV values characteristic of exponential growth until LEV reaches 500. At 500 constraints on growth begin to suppress the growth rate. The next increment to the

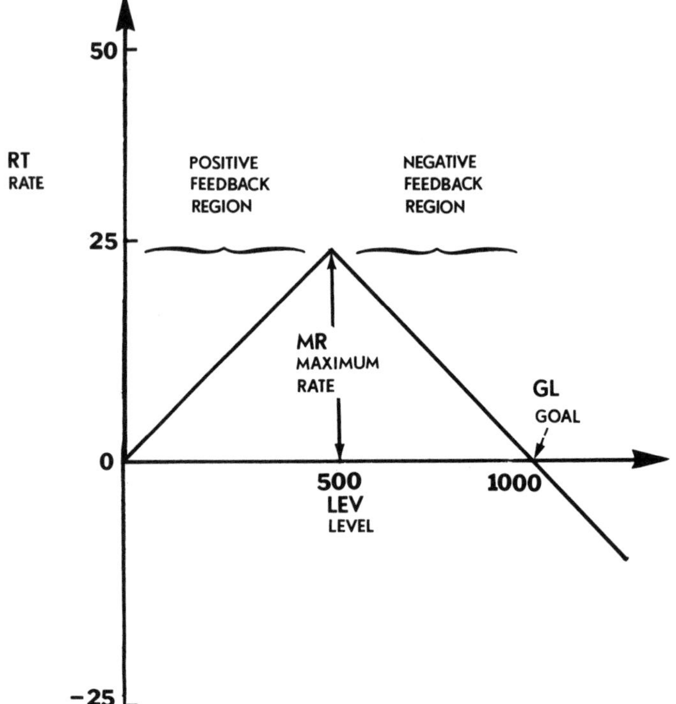

FIGURE 4-4 Nonlinear rate-level graph

[4] A single-level structure having a rate-level slope that does not change direction is only capable of one type of behavior mode as demonstrated in Chapters 2 and 3.

level LEV moves LEV into the negative feedback region of the rate-level curve. The level continues to increase, but at a decreasing rate RT. As in a purely negative feedback system, each new increment to LEV decreases as RT approaches zero. LEV asymptotically approaches the system goal GL while RT asymptotically decays to zero. Figure 4-5 shows the expected sigmoidal growth of LEV.

Any single-level structure possessing a rate-level curve of the general shape pictured in Figure 4-4 produces S-shaped growth over time. The structure in Figure 4-3 illustrates a simple rate-level relationship containing the necessary transition from positive feedback to negative feedback. However, the slope at the inflection point in Figure 4-4 need not change so abruptly. The rate-level graphs in Figure 4-6 show smoother and more typical transitions. In general, the nature of the transition depends on which phenomena we model. This chapter includes three different examples of shifts in loop dominance.

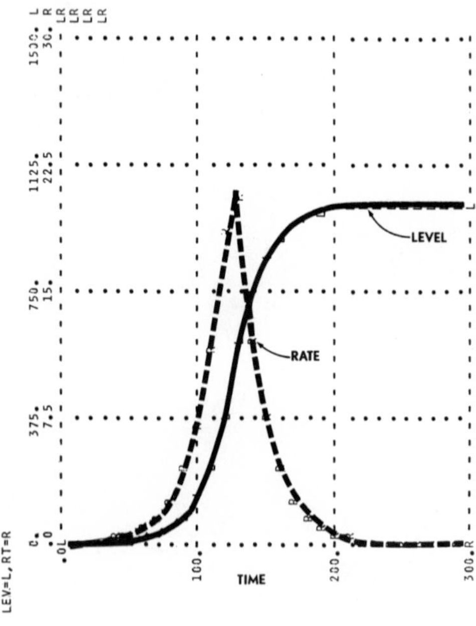

FIGURE 4-5 S-shaped simulation run

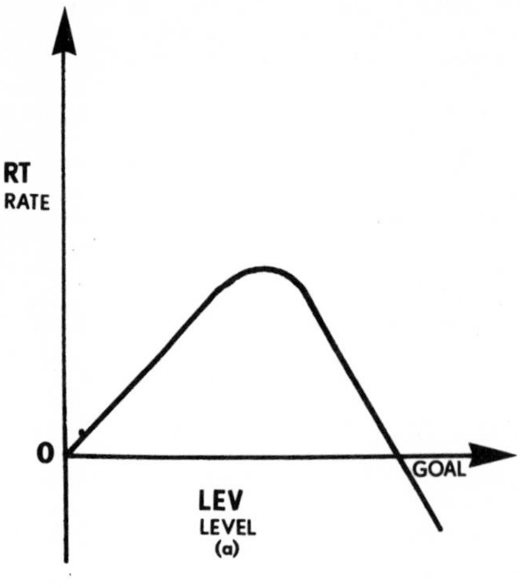

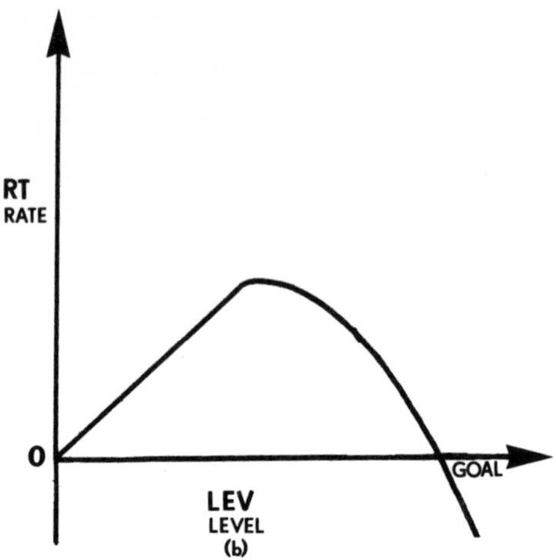

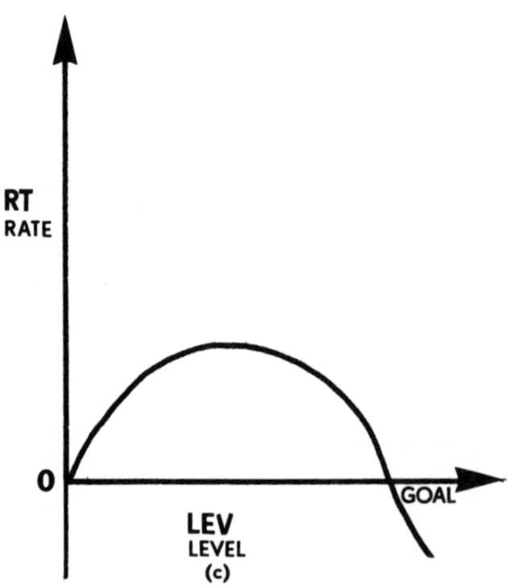

FIGURE 4-6 Rate-level graphs capable of producing sigmoidal growth

4.3 STABLE AND UNSTABLE EQUILIBRIUM

What are the stability characteristics of S-shaped growth structure? For example, how does the structure respond to varying initial conditions or exogenous influences on the level?

The goal GL, the value that LEV ultimately attains, persists indefinitely and is called stable equilibrium. No further increase or decrease in the level LEV occurs at GL because the rate RT equals zero. Figure 4-4 illustrates such a condition. If some outside influence increases LEV beyond GL, for example, a resulting negative RT returns the level LEV to stable equilibrium. Likewise, decreased LEV generates a positive RT until the system again reaches GL. Figures 4-7 (a) and 4-7 (b) contain simulations of an external change in LEV from its goal GL value by an addition of 100 units in Figure 4-7 (a) and a subtraction of 100 units in Figure 4-7 (b). The characteristic goal-directed behavior of negative feedback (Chapter 3) appears in each case.

Prior to growth when the level LEV value equals zero, the system rests at "unstable equilibrium." Unless exogenously disturbed, the system remains in unstable equilibrium. A disturbance such as minimal increase in LEV (for example, an initial condition greater than zero) produces a RT value other than zero. The system level increases in keeping with the general S-shaped growth. The system's inability to maintain equilibrium at zero in the presence of the slightest disturbance defines unstable equilibrium.

4.4 SUMMARY

Sigmoidal growth requires that dominance shift from positive feedback to negative feedback through a nonlinear relationship. A simple rate-level structure that first exhibits rate increases with rising level values and, then, rate decreases with further level growth produces the necessary shift. Stable equilibrium, characteristic of a first-order negative feedback structure, results.

Three examples of phenomena demonstrating S-shaped growth are: the growth pattern of a population of rats; the propagation of a contagious disease; and the behavior of a damped pendulum.

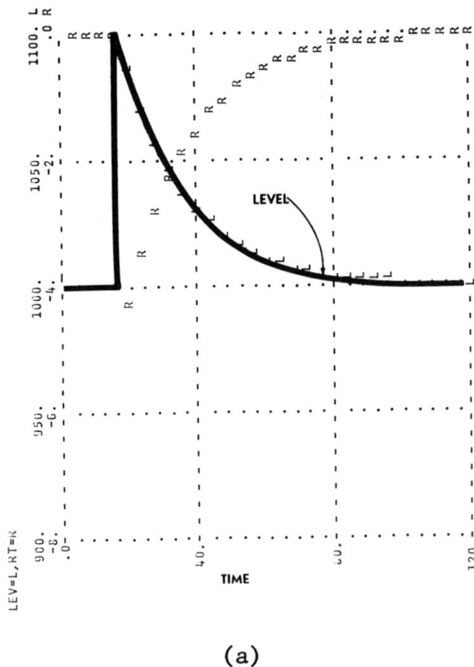

(a)

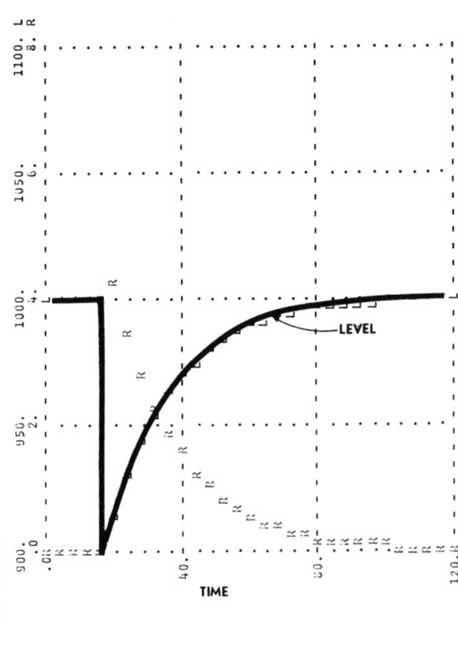

(b)

FIGURE 4-7 Level response to (a) addition and (b) subtraction of 100 units

4.5 EXAMPLE 1--POPULATION GROWTH

Population growth and stabilization trends of various organisms are familiar to population biologists. The S-shapes shown in Figure 4-8, for example, occur in experiments on fruit flies, bacteria, and sheep.

E.J. Kormondy explains:

> Although they are difficult to come by, there are enough studies on a spectrum of different kinds of plants and animals to permit the statement that most species show a sigmoidal pattern during the initial stages of their population growth. There is, in such cases, an initial slow rate of growth, in absolute numbers, followed by an increase in rate to a maximum, at which point the curve begins to be deflected downward; it terminates in a rate that gradually lessens to zero, as the population more or less stabilizes itself with respect to its environment.[5]

Biotic factors, biological relationships both within and among species, and abiotic factors, environmental characteristics, regulate population growth and stability. Kormondy observes:

> At a critical time in the life history of a given population, a physical factor such as light or a nutrient may be significant as a regulatory agent; at another time, parasitism, predation, or competition, or even some other physical factor may become the operative factor. As complex and as variable as the niche of any species is, it is unlikely that this regulation comes about by any single agency. However, there does appear to be considerable and mounting evidence, both empirical and theoretical, to suggest that populations are self-regulating through automatic feedback mechanisms. Various mechanisms and interactions appear to operate both in providing the information and in the manner of responding to it, and with the exceptional case of a catastrophe, the stimulus to do so appears to depend directly on the density of the population. The end effect is one of avoiding destruction of a population's own environment and thereby avoiding its own extinction.[6]

Crowding and Population Growth. The simple population growth model in the following example relates to crowding and infant mortality, a single density-dependent control mechanism characteristic of mammal

---

[5] E.J. Kormondy, Concepts of Ecology (Englewood Cliffs: Prentice-Hall, 1969), p. 67.

[6] Ibid., pp. 110-111.

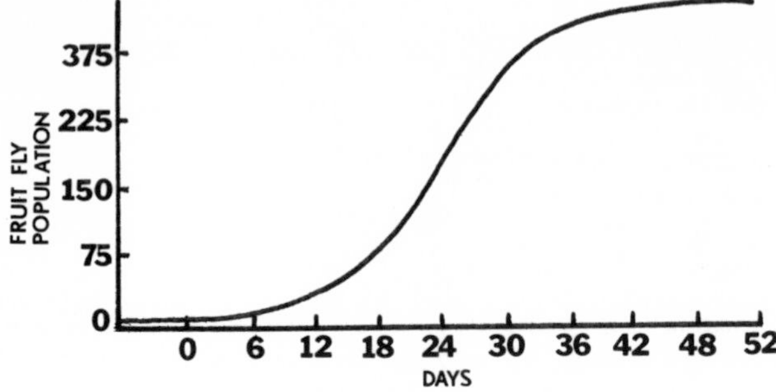

Growth of a population of Drosophila (Fruit Flies) under controlled experimental conditions.
Source: Alfred J. Lotka, Elements of Mathematical Biology (New York: Dover Publications, Inc., 1950), p. 69.

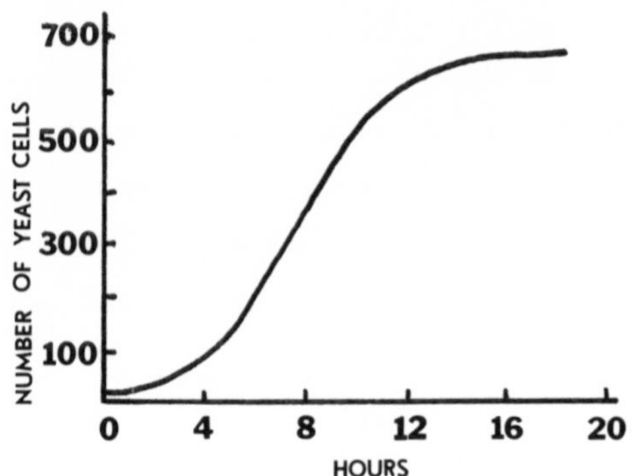

The growth curve of yeast cells in the laboratory.
Source: Kormondy, Concepts of Ecology, p. 65.

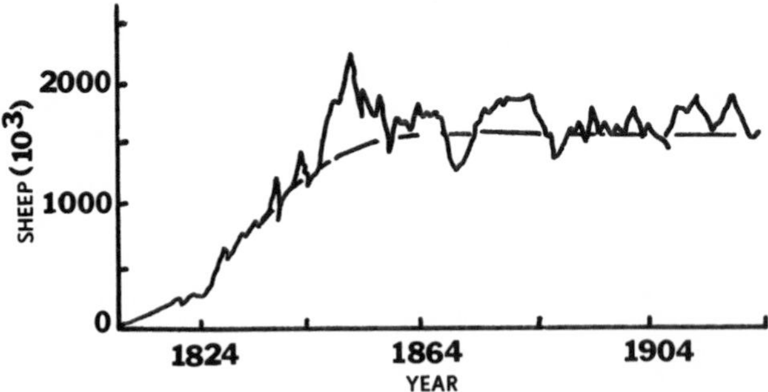

The growth curve of sheep subsequent to their introduction in Tasmania showing an initial sigmoidal pattern followed by semi-equilibrium.
Source: Kormondy, Concepts of Ecology, p. 85.

FIGURE 4-8 Population growth examples

populations. We could use any density factor or combination of factors to which Kormondy assigns a capacity for limiting population growth. Regardless of the precise density mechanism, S-shaped growth results.

B.F. Calhoun in his experiments with Norway rats observed the effect of crowding on infant mortality:

> Some years ago I attempted to submit [the question of population density on social behavior] to experimental inquiry. I confined a population of wild Norway rats in a quarter-acre enclosure. With an abundance of food and places to live and with predation and disease eliminated or minimized, only the animals' behavior with respect to one another remained as a factor that might affect the increase in their number. There could be no escape from the behavioral consequences of rising population density. By the end of 27 months the population had become stabilized at 150 adults. Yet adult mortality was so low that 5,000 adults might have been expected from the observed reproductive rate. The reason that this larger population did not materialize was that the infant morality was extremely high. Even with only 150 adults in the enclosure, stress from social interaction led to such disruption of maternal behavior that few young survived.[7]

Calhoun, using an additional indoor experiment, found that overcrowded mother rats failed to build nests or adequately nurse their young. Infant mortality rose.[8]

Calhoun's verbal description forms the basis of a model incorporating the following assumptions:

1. Confined space allows no migration and no predation.
2. Ample and sufficient food supply exists.
3. The confined space has a constant environment (e.g. no abnormal changes in weather or temperature).

For simplicity, we make two further assumptions:

4. Disregard the effects of age on reproduction capacity.
5. The sex ratio (males/females) of the population is 1.

Figure 4-9 contains a causal-loop diagram for the rat population model. Births (loop 1) increase and deaths (loop 2) decrease the rat population. The two loops can only produce exponential growth or

---

[7] B.F. Calhoun, "Population Density and Social Pathology," <u>Scientific American</u> 206 (1962), p. 139.

[8] Ibid.

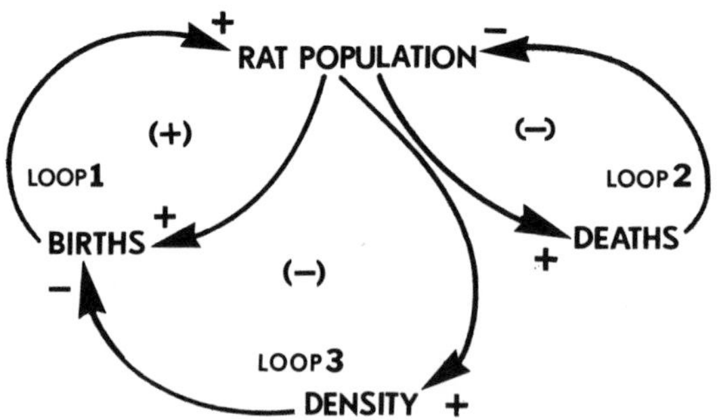

FIGURE 4-9 Causal-loop diagram--
rat population model

decay. Loop 3, an additional negative feedback loop based on Calhoun's observations, is responsible for transferring dominance from loop 1 to loop 2. As the population increases, social stresses from crowding effect a decrease in the aggregate birth rate of the population.

The flow diagram of Figure 4-10 retains the causal loops of Figure 4-9. The rat birth rate RBR is defined as the number per month of infant rats that survive to adulthood. The normal rat fertility NRF, the average number of infants per month produced by each adult female rat, equals 0.4 (rats/female/month) for a relatively low or "normal" population density. Or equivalently, every female produces approximately 5 pups per year.

The adult rat death rate ARDR is a function of the number of adult rats and average rat lifetime ARL. Average rat lifetime ARL, defined as the number of months an average rat survives during "normal" conditions, equals 22 months. Therefore, 4.5 percent of the population dies each month.

Loop 3 in the flow diagram accounts for the effect of crowding on infant mortality. The infant survival multiplier ISM makes infant survival dependent upon rat population density PD. The multiplier curve appears in Figure 4-11. At low population densities, approximately 100 percent of the pups survive. As PD increases, the percentage of surviving infant rats decreases. This depresses the normal rat fertility NRF. In extremely crowded conditions, a very low percentage of newly born pups survives and normal rat fertility NRF equals only 10 percent of its "normal" value.

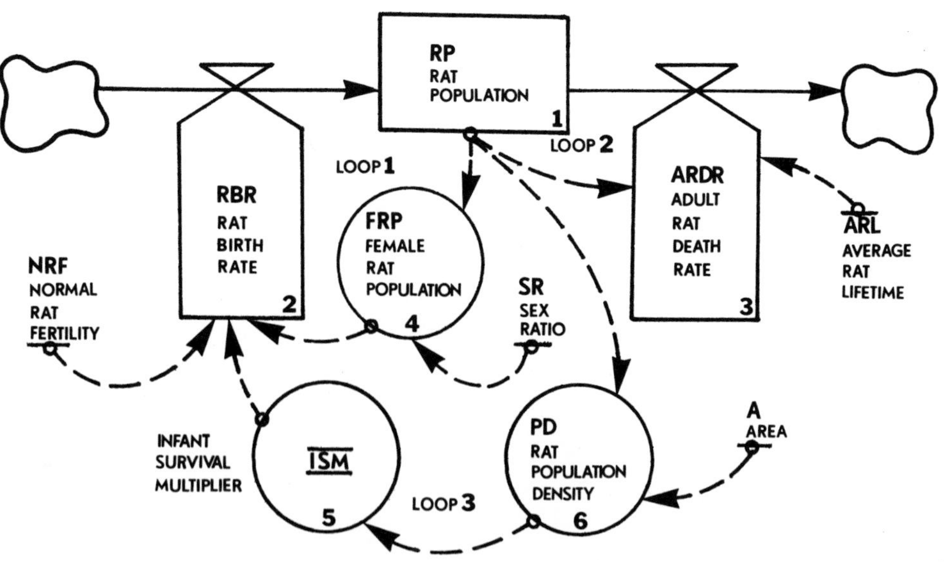

FIGURE 4-10 Rat population model

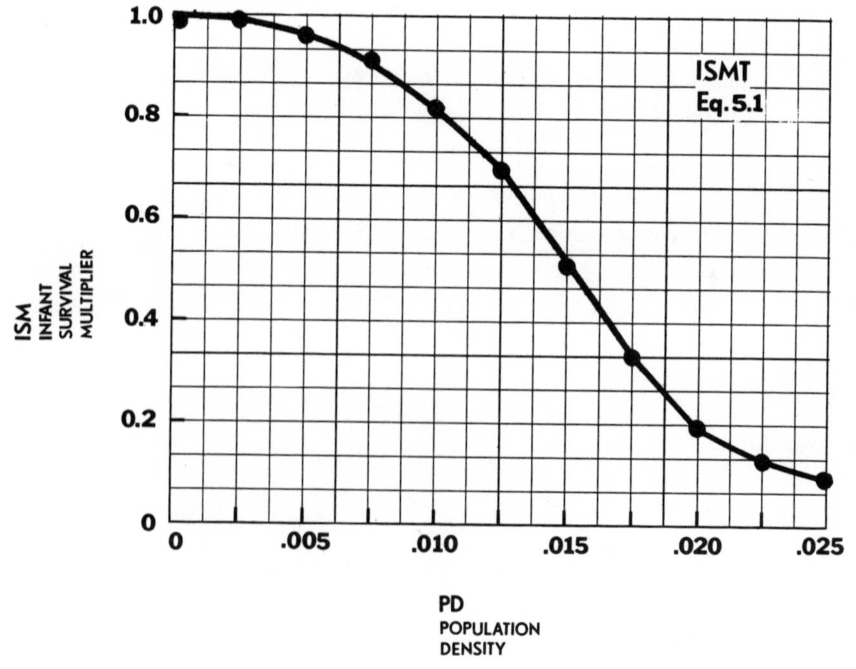

FIGURE 4-11 Infant survival multiplier table

The equations for the entire model are shown below:

```
RP.K=RP.J+(DT)(RBR.JK-ARDR.JK)                        1, L
RP=10                                                 1.1, N
    RP      - RAT POPULATION (RATS)
    RBR     - RAT BIRTH RATE (RATS/MONTH)
    ARDR    - ADULT RAT DEATH RATE (RATS/MONTH)
RBR.KL=NRF*FRP.K*ISM.K                                2, R
NRF=.4                                                2.1, C
    RBR     - RAT BIRTH RATE (RATS/MONTH)
    NRF     - NORMAL RAT FERTILITY (RATS/FEMALE/MONTH)
    FRP     - FEMALE RAT POPULATION (FEMALES)
    ISM     - INFANT SURVIVAL MULTIPLIER (DIMENSIONLESS)
ARDR.KL=RP.K/ARL                                      3, R
ARL=22                                                3.1, C
    ARDR    - ADULT RAT DEATH RATE (RATS/MONTH)
    RP      - RAT POPULATION (RATS)
    ARL     - AVERAGE RAT LIFETIME (MONTHS)
FRP.K=SR*RP.K                                         4, A
SR=.5                                                 4.1, C
    FRP     - FEMALE RAT POPULATION (FEMALES)
    SR      - SEX RATIO (DIMENSIONLESS)
    RP      - RAT POPULATION (RATS)
ISM.K=TABLE(ISMT,PD.K,0,.025,.0025)                   5, A
ISMT=1/1/.96/.92/.82/.7/.52/.34/.20/.14/.1            5.1, T
    ISM     - INFANT SURVIVAL MULTIPLIER (DIMENSIONLESS)
    PD      - RAT POPULATION DENSITY (RATS/FEET^2)
PD.K=RP.K/A                                           6, A
A=11000                                               6.1, C
    PD      - RAT POPULATION DENSITY (RATS/FEET^2)
    RP      - RAT POPULATION (RATS)
    A       - AREA (FEET^2)
```

Figure 4-12 plots net population growth rate NPGR (i.e., RBR less ARDR) as a function of rat population RP.

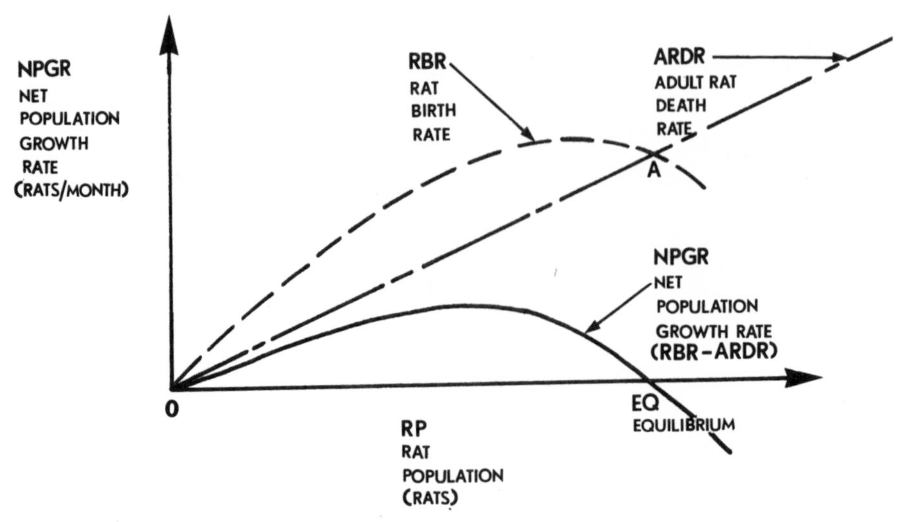

FIGURE 4-12 Net population growth rate curve

The NPGR curve in Figure 4-12 resembles the curves in Figure 4-6. In the absence of "population pressure" from crowding, the rat growth rate increases. However, the growth rate decelerates as crowding sets in. Infant mortality begins to increase and reduce the rat birth rate RBR. Further population increase reduces the net growth rate until growth stops. With births and deaths in equilibrium (point A), rat population RP enters a steady state (point EQ). The net growth rate must cross the horizontal axis to maintain a stable population. The expected sigmoid growth curve for rat population shows up in Figure 4-13, a simulation run.

The model accounts for only one density effect. What about such limiting factors on adult rat death rate ARDR as starvation, disease, and fighting? Assume, for example, an insufficient food supply (i.e., ignore assumption 2). A decrease in average rat lifetime ARL from crowding might serve as the control mechanism in this case. In Figure 4-14 a nonlinear, density-dependent death rate curve that increases with crowding intersects the birth rate curve at point B.

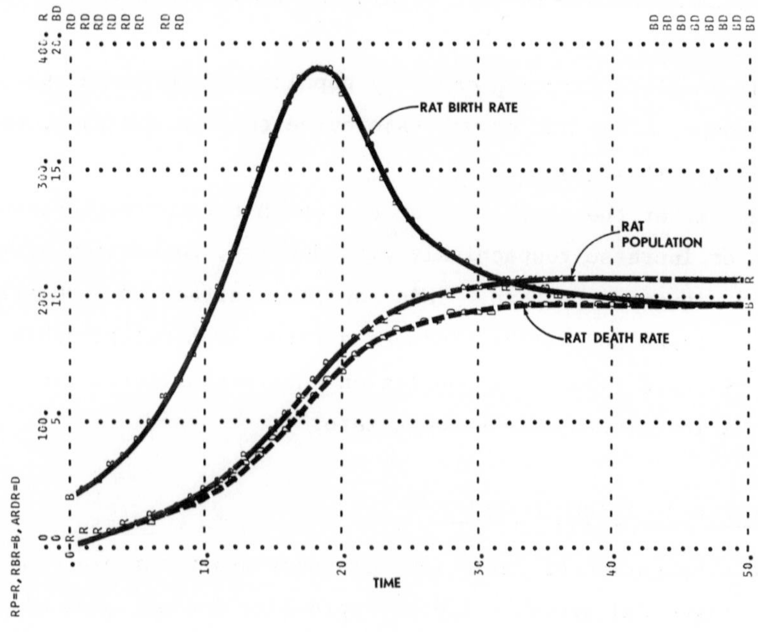

FIGURE 4-13 Rat population growth

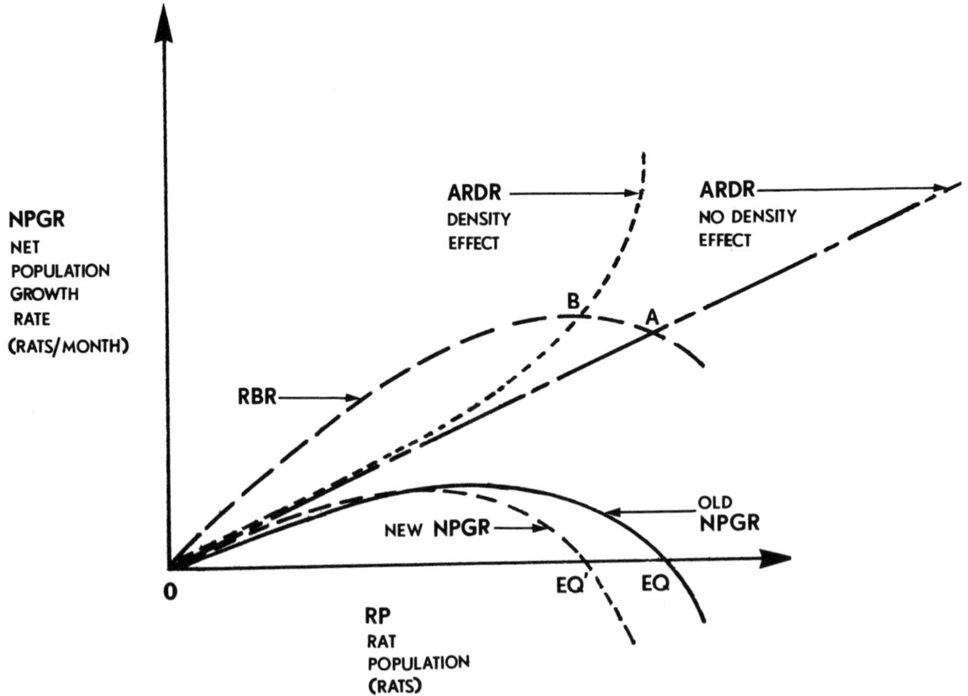

FIGURE 4-14 Net population growth rate with nonlinear death rate

A new and smaller "carrying capacity population" EQ' results. The overall shape of the net growth rate curve remains the same, however. Sigmoidal growth still occurs.

When either the birth rate or the death rate or both begin to decrease or increase respectively with density, population growth follows a sigmoidal pattern. This nonlinear effect of population growth on births and deaths produces a rate-level relationship of the general S-shaped type. The population enters equilibrium when the net population growth rate becomes zero.

4.6 EXAMPLE 2--EPIDEMIC GROWTH

The propagation of infectious diseases under certain conditions exhibits sigmoidal growth. Typical epidemics include such mild infections of the upper respiratory tract as the common cold, flu, and mild

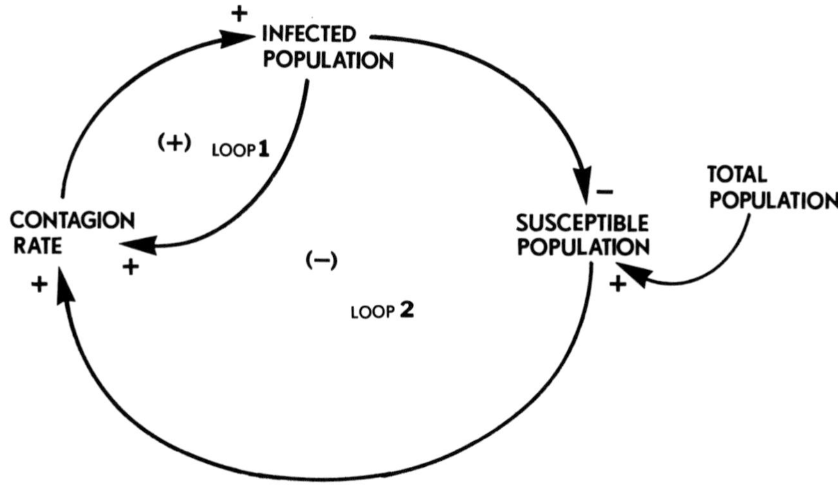

FIGURE 4-15 Causal-loop diagram of epidemic model

virus.[9]

A single-level model replicating the growth of an epidemic subsumes the following assumptions:

1. Constant population allows no migration.

2. Infectious people are never so ill that they withdraw from circulation and are not cured during the course of the epidemic (reinfection is minimized).[10]

3. Fairly homogenous mixing of the susceptible population and the infectious population occurs.

Figure 4-15 shows a causal-loop diagram which admits these three assumptions. The contagion rate depends on both the infected and

---

[9]N.T.J. Bailey, "Some Problems in Statistical Analysis of Epidemic Data," Royal Statistical Society 118, Series A, (1955), p. 39.

[10]Assumption 2 facilitates using a single level in the example. In actuality, infectious people enter either an immune pool or re-enter the susceptible pool after an infectious period. If the infectious period is very much longer than the time required to infect the entire population, then the assumption is acceptable.

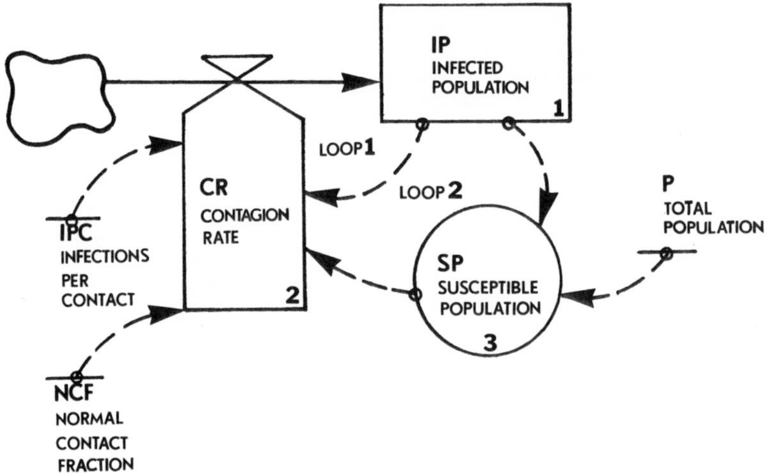

FIGURE 4-16 Epidemic model

susceptible population. In loop 1, all else being equal, an increase in the infected population generates an increase in the contagion rate. An increased contagion rate adds more people to the infected population and so forth in a positive feedback manner. Since an infinite supply of population does not exist, however, all else is not equal. Loop 2, the negative feedback loop, accounts for a finite susceptible population. As the infected population increases, the susceptible population, the difference between the total population and the infected population, decreases. The decrease in turn suppresses the contagion rate. Eventually, the contagion rate reaches zero when the entire population contracts the disease.

```
IP.K=IP.J+(DT)(CR.JK)                                  1, L
IP=10                                                  1.1, N
    IP      - INFECTED POPULATION (PEOPLE)
    CR      - CONTAGION RATE (PEOPLE/DAY)
CR.KL=IPC*NCF*IP.K*SP.K                                2, R
IPC=.1                                                 2.1, C
NCF=.02                                                2.2, C
    CR      - CONTAGION RATE (PEOPLE/DAY)
    IPC     - INFECTIONS PER CONTACT (DIMENSIONLESS)
    NCF     - NORMAL CONTACT FRACTION (FRACTION/PERSON/DAY)
    IP      - INFECTED POPULATION (PEOPLE)
    SP      - SUSCEPTIBLE POPULATION (PEOPLE)
SP.K=P-IP.K                                            3, A
P=100                                                  3.1, C
    SP      - SUSCEPTIBLE POPULATION (PEOPLE)
    P       - TOTAL POPULATION (PEOPLE)
    IP      - INFECTED POPULATION (PEOPLE)
```

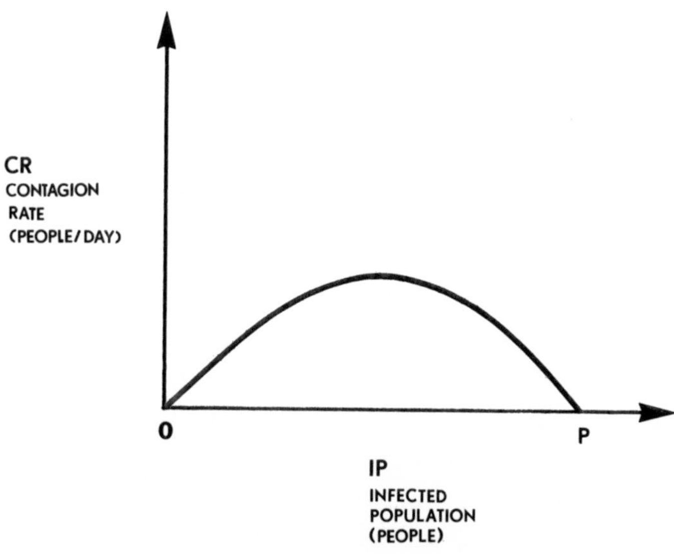

FIGURE 4-17 Contagion rate versus infected population

The flow diagram and equations of Figure 4-16 represent the single-level epidemic model. The contagion rate CR is developed from probabilistic considerations of the likelihood of infection between individuals making contact in a closed environment with uniform mixing.[11] The total number of possible contacts equals the product of IP and SP. NCF is the percentage of actual contacts per day. The number of infections per day or contagion rate CR equals contacts per day multiplied by the fraction of contacts producing infection IPC.

We can express the contagion rate CR as a function of the level of infected population by substituting auxiliaries and constants into the CR equation:

CR.KL = IPC*NCF*IP.K(P-IP.K)

Plotting the contagion rate CR versus the infected population IP (Figure 4-17) yields the rate-level relationship necessary for sigmoidal growth.[12] As the number of infected people increases, the

---

[11] J.S. Coleman, Introduction to Mathematical Sociology (New York: Free Press, 1961), p. 493.

[12] The quadratic rate equation produces the curve in Figure 4-17.

simultaneously increasing infection rate further increases the number of infected people. However, as the infected population IP increases, the uninfected population pool rapidly decreases. The likelihood of contact between infected and uninfected people diminishes despite a large infected population IP. Finally, when the entire population contracts the disease, the contagion rate CR goes to zero. Because the contagion rate CR by definition is a one-way flow and the entire susceptible population pool diminishes, the rate curve cannot extend below the horizontal axis.

Figure 4-18 contains the model simulation. The infected population IP displays sigmoidal growth. We can anticipate the inverted sigmoid curve of the susceptible population SP since the sum of infected population IP and the susceptible population SP must always equal the total population P.

While representing the spread of a mild disease in this example, the model can with definitional changes apply to a riot or a rumor. In general, epidemic processes that begin exponentially and eventually "consume" all available population must follow the sigmoid time form produced in Figure 4-18.

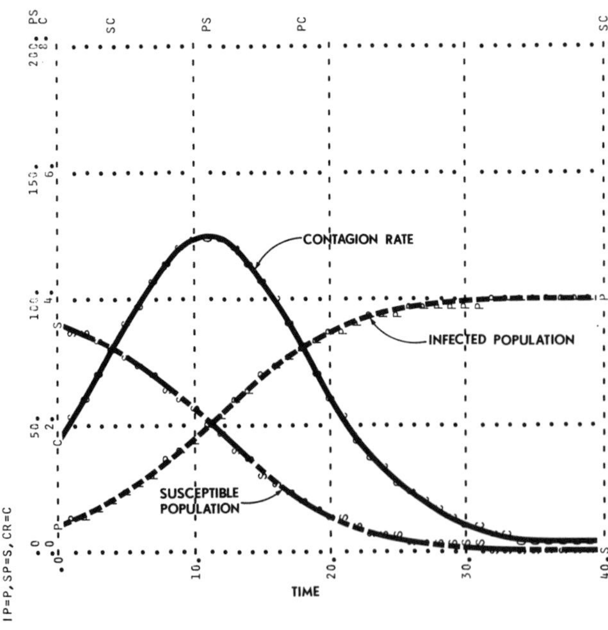

FIGURE 4-18 Epidemic growth

4.7 EXAMPLE 3--DAMPED PENDULUM

The final example of S-shaped change comes from the physical sciences. Under certain conditions, the positional behavior of a damped pendulum forms a single-level feedback system displaying sigmoidal growth.

Consider a pendulum immersed in a highly viscous fluid such as oil in the vertical plane. The diagram of Figure 4-19 pictures the pendulum with a mass at one end. When displaced from its zero degree (vertical) position, the pendulum tends to fall under the force of gravity. However, the resistive force of the viscous fluid retards the pendulum's movement.

The following equations describe the pendulum system. We first calculate the inertial torque or turning force T_I about the pivot point:

$$T_g + T_d = T_I \qquad (4.1)$$

where

T_g = Torque due to gravity,
T_d = Torque due to damping,
T_I = Inertial torque.

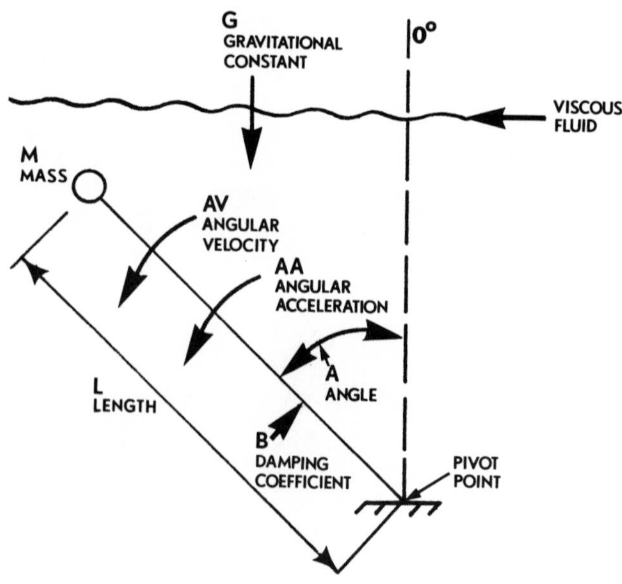

FIGURE 4-19 Damped pendulum

The torques in turn equal the following:

$$T_g = F_g{*}L{*}\sin(A) \qquad (4.2)$$

$$T_d = F_d{*}L \qquad (4.3)$$

$$T_I = M{*}L^2{*}AA,$$

and

$$F_g = \text{Force of gravity} = M{*}G,$$

$$F_d = \text{Force due to damping} = -B{*}AV.$$

Substituting equations (4.2) and (4.3) into equation (4.1):

$$M{*}G{*}L{*}\sin(A) - B{*}AV{*}L = M{*}L^2AA. \qquad (4.4)$$

If the damping or viscous torque T_d roughly equals the gravity torque T_g, the inertial torque becomes inconsequentially small or:

$$T_g + T_d \approx 0$$

and,

$$B{*}L{*}AV \approx M{*}G{*}L{*}\sin(A),$$

or,

$$AV \approx (M{*}G/B)\sin(A). \qquad (4.5)$$

The angular velocity AV becomes exclusively a function of the angle A. Figure 4-20 gives a model of the pendulum based on equation (4.5).

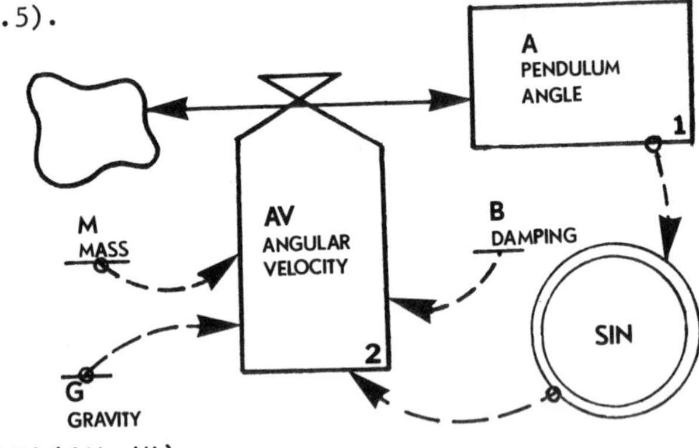


```
A.K=A.J+(DT)(AV.JK)                                  1, L
A=1                                                  1.1, N
    A       - PENDULUM ANGLE (DEGREES)
    AV      - ANGULAR VELOCITY (DEGREES/SECOND)
AV.KL=(M*G/B)(SIN(A.K*6.28/360))                     2, R
M=1                                                  2.1, C
B=20                                                 2.2, C
G=32                                                 2.3, C
    AV      - ANGULAR VELOCITY (DEGREES/SECOND)
    M       - MASS (SLUGS)
    G       - GRAVITY (FEET/SECOND^2)
    B       - DAMPING (POUNDS FORCE/DEGREES/SECOND)
    A       - PENDULUM ANGLE (DEGREES)
```

FIGURE 4-20 Pendulum model

A plot of AV for various values of A according to equation (4.5) appears in Figure 4-21. The sinusoidal rate curve has a small amplitude because of the large damping coefficient B. This rate curve is the general nonlinear form needed to produce the S-curve time shape.

The pendulum behavior over time seen in Figure 4-22 is predictable. With the pendulum displaced from its unstable equilibrium at A = 0°, velocity increases very slowly to a maximum at A = 90°. The pendulum then gradually loses speed until A = 180°, its stable equilibrium value. The pendulum does not overshoot the 180° position because the inertia (momentum) due to acceleration is absent.[13]

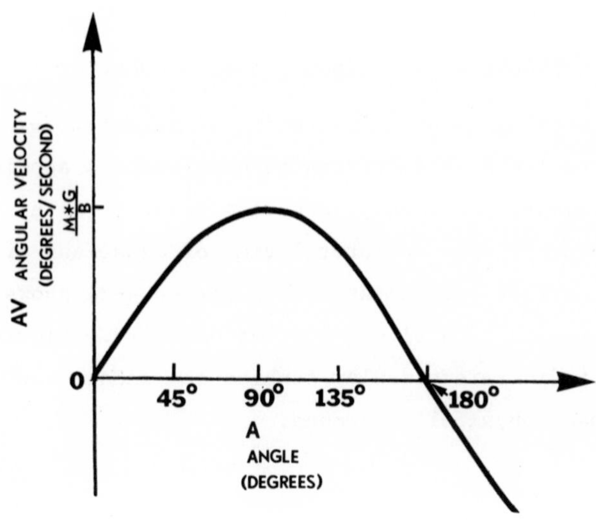

FIGURE 4-21 Angular velocity versus angle

[13] By decreasing the viscosity of the fluid (for example, by heating the oil and lowering B) we invalidate the assumption of negligible inertial torque T_I. Angular momentum appears and the model requires an additional level (velocity) to account for acceleration. The two-level model would produce damped oscillations about the equilibrium position.

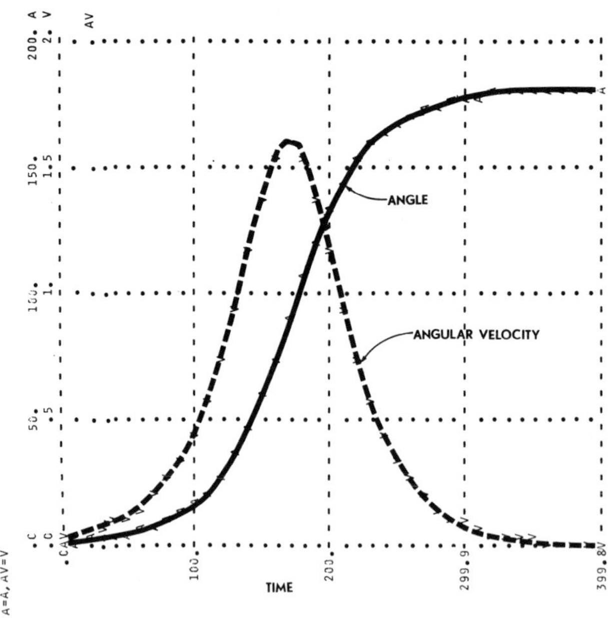

FIGURE 4-22 Damped pendulum behavior

Besides serving as an illustration of sigmoidal growth behavior, the damped pendulum also demonstrates how feedback applies to purely physical phenomena. Obviously, pendulum velocity alters the angular position. However, this unilateral view of cause and effect obscures another less obvious process at work. Figure 4-21 shows that the angular position in turn alters angular velocity. Viewing the pendulum system in this circular fashion can enhance our understanding of many physical phenomena.

Chapter 5
Review of Simple Structures: Industrial Land-Use Model

Preparation: Chapters 1-4 and Exercises 1-7 provide appropriate background for this chapter.

Purpose: Stepwise model construction and analysis provides an approach to reviewing the relationship between structure and behavior. This chapter utilizes the step-by-step approach to develop a simple, hypothetical model of industrial land use. The first section of Chapter 5 investigates the underlying growth loop of an urban area. The second section adds a nonlinear land constraint loop. The third section incorporates an aging mechanism to complete the model.

Chapter 5 also exposes the reader to various stages of model construction including causal loop diagramming, flow diagramming, equation writing, table function development, simulation, and sensitivity analysis, in preparation for the conceptualization exercises in Part III.

Practice: Complete Exercise 8, and then undertake the exercises in Part III. Exercise 12 should interest the student of more complex land-use models.

5.1 INTRODUCTION 95
5.2 GROWTH MODEL 95
Verbal Description 95
Formulation of Flow Diagram 96
Equation Writing 98
Model Behavior 101
Model Analysis 102
5.3 LAND CONSTRAINED MODEL 108
Verbal Description 108
Equation Writing 115
Model Behavior 117
Model Analysis 119
5.4 AGING AND DEMOLITION MODEL 122
Verbal Description 122
Flow Diagram 125
Equation Writing 127
Model Behavior 129
Model Analysis 131
5.5 CONCLUSIONS 132

5.1 INTRODUCTION

This chapter presents a simple model of the growth of industrial activity within a fixed land area. The chapter develops the model one step at a time through the usual format of dynamic model construction:

1. Verbally describe the phenomenon.

2. Convert the verbal description into both causal-loop and flow diagrams.

3. Write DYNAMO equations for the flow diagram.

4. Run the model on a computer.

5. Analyze the computer runs by asking the following questions:

   a) What behavior does the model show?

   b) How has behavior changed from previous computer runs?

   c) Why does the model exhibit the behavior?

   d) How can we alter the behavior?

Chapter 5 offers a self-learning format that differs from the formats in Chapters 1-4. At each stage in model development, the reader answers questions in the space provided and then compares his answers with those in the text. The reader should carefully review any material which he does not understand before going on to new material.

5.2 GROWTH MODEL

Verbal Description. The first interesting behavior is growth of industrial activity in a fixed land area. For simplicity, we assume that a certain amount of economic activity tends to attract more activity. The amount of activity attracted during any particular period of time is directly proportional to the level of existing activity. This process exemplifies positive feedback.

Draw a causal-loop diagram of this phenomenon.

Figure 5-1 diagrams the previously described positive loop. Only two variables appear: economic activity and economic growth rate. Each variable enhances the growth of the other to breed the vicious (or virtuous) circle typical of positive feedback.

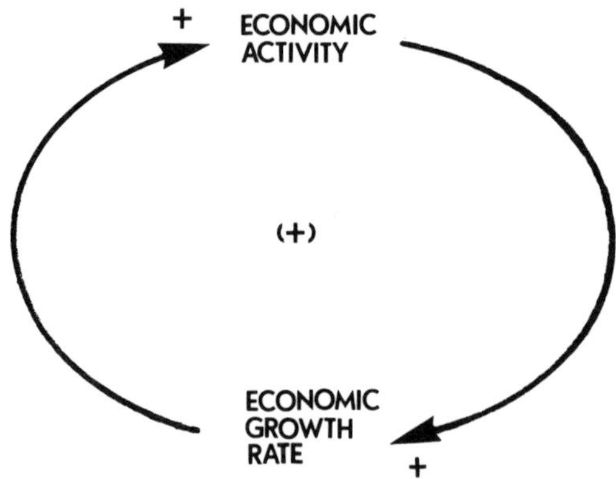

FIGURE 5-1 Causal-loop diagram of economic activity

In this exercise, "economic activity" represents the number of industrial structures (buildings) present in a fixed land area. The choice of "buildings" reflects one model purpose: to study land use. Value of goods produced could provide a measure of economic activity, but this factor yields little understanding of land use within the area. Structures, on the other hand, have meaning both in the context of taking up land and in the context of stimulating more activity (construction of buildings) because they house economic agents. At present, one structure means one industrial building housing economic activity in a balanced mix of supporting businesses which all contribute to the production of goods and services in the area. Obviously, many factors influence the construction of buildings in an area. These factors include the economic condition of the nation, local availability of labor, local tax rates, and investment opportunities. For simplicity, however, this presentation ignores those factors.

Formulation of Flow Diagram. In the preceding description, the level of economic activity is represented by the number of existing industrial structures. The rate of addition of economic activity therefore equals the construction rate of industrial structures.

Draw an appropriate flow diagram. Introduce any constants necessary to complete the diagram. Label the diagram fully.

Figure 5-2 contains the system flow diagram. The industrial construction rate IC is related to the level of industrial structures IS by the industrial growth factor IGF. IGF expresses the rate of construction for a given level of industrial structures under favorable economic conditions, for example, adequate land availability which does not restrict growth. In other words, IGF states how the rate of construction (rate of addition of economic activity) relates to the number of existing structures (level of economic activity).

The model presumes that, under "normal" conditions, the rate of growth of economic activity equals 10 percent per year of the level of economic activity. IGF is therefore 0.1. The "normal" construction factor represents an average national growth rate varying with local conditions. The choice of a "normal" growth factor is more or less arbitrary. However once defined, the "normal" factor must be made consistent with all other parameters in the model.

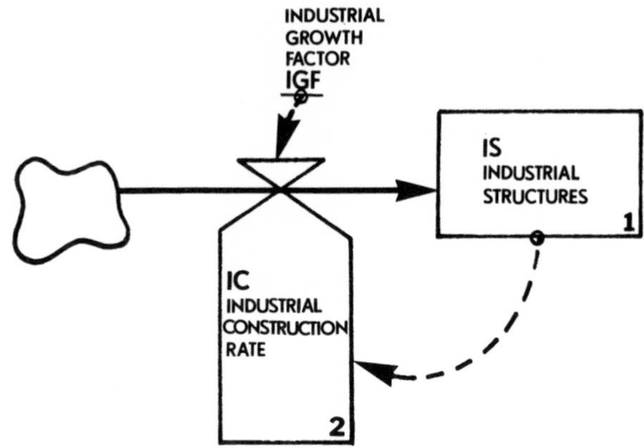

FIGURE 5-2 Flow diagram for industrial model

Keep in mind that a model simplifies reality. The simplification makes possible a meaningful analysis of the model. A model as complex as the real world permits little additional understanding of important real world processes. We add a new phenomenon only to understand a different type of behavior not represented by the previous model or to explore certain types of policy recommendations which exceed the boundary of the previous model.

Equation Writing

Write equations for the flow diagram in Figure 5-2.

The correct equations for the flow diagram appear below. Industrial structures IS represents the stock of structures, the accumulation of new structures generated by the construction rate, at any point in time. The initial number of structures in the model has an arbitrary value of 5. The industrial construction rate IC in any year equals the product of the industrial growth factor IGF and the level of industrial structures IS. The lines following the note "Control Statements" simply instruct the computer to plot model variables over time. The model is presently set to run for 10 years (LENGTH) and plot out values of IS and IC every year (PLTPER). DT sets the computational interval over which the construction rate remains constant. Hence, we calculate a new rate IC every year for this run.

```
IS.K=IS.J+(DT)(IC.JK)                                    1, L
IS=5                                                     1.1, N
    IS     - INDUSTRIAL STRUCTURES (STRUCTURES)
    IC     - INDUSTRIAL CONSTRUCTION RATE (STRUCTURES/
               YEAR)

IC.KL=IGF*IS.K                                           2, R
IGF=.1                                                   2.1, C
    IC     - INDUSTRIAL CONSTRUCTION RATE (STRUCTURES/
               YEAR)
    IGF    - INDUSTRIAL GROWTH FACTOR (FRACTION/YEAR)
    IS     - INDUSTRIAL STRUCTURES (STRUCTURES)

NOTE CONTROL STATEMENTS
PLOT IS=I(0,120)/IC=C(0,6)
C LENGTH=10
C DT=1
C PLTPER=1
```

FIGURE 5-3 DYNAMO equations

Model Behavior. Having completed the flow diagram and equations, we can have the computer trace out the performance of industrial structures IS and industrial construction IC over time.

Calculate IS and IC values for 10 years from the equations in Figure 5-3. Make computations every year and plot them on the grid below.

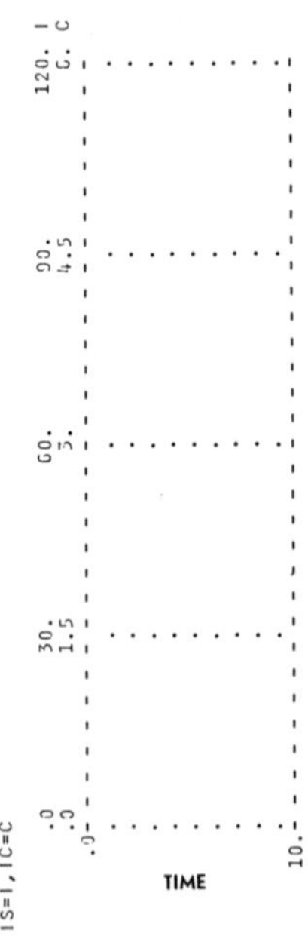

FIGURE 5-4 Sketch of industrial growth over 10 years

Figure 5-5 shows computer-generated growth of the level and rate over a 10 year period from 5 initial structures. The computer plots IS as I and IC as C.

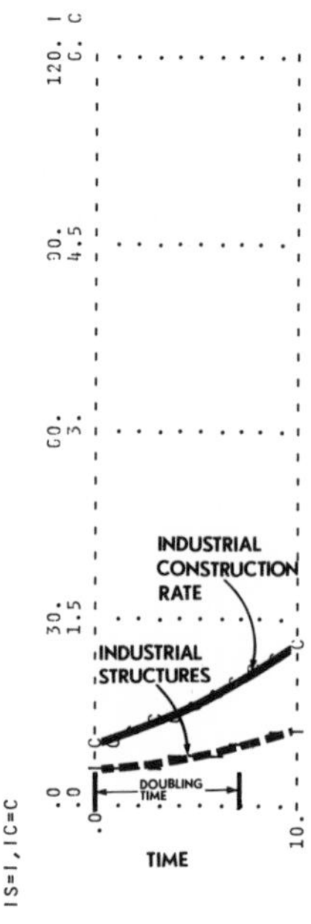

FIGURE 5-5 Industrial growth over 10 years

Model Analysis. "Doubling time" is an important aspect of exponential growth resulting from a positive feedback loop. Doubling time, explained in Chapter 2, is the amount of time required for a level (or rate) to double in value. In a simple positive loop, doubling time always remains constant. The industrial structures model doubles approximately every seven years. In a ten year time interval, only slightly more than one doubling occurs. The dramatic growth of IS and IC becomes apparent only with a longer time horizon.

Sketch the behavior of industrial structures IS and industrial construction rate IC over a period of 60 years on the grid below.

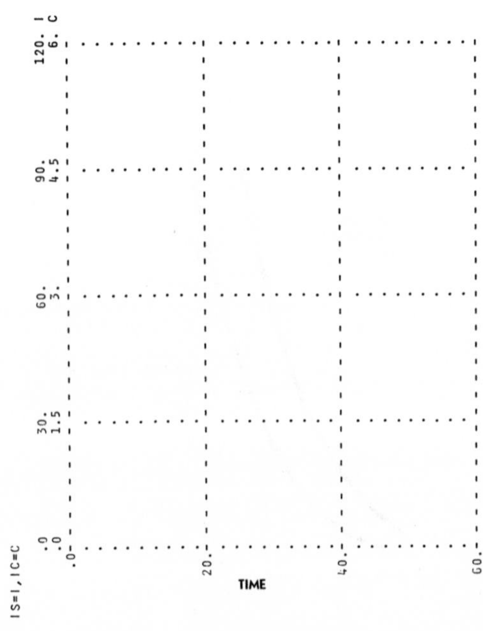

FIGURE 5-6 Sketch of industrial growth over 60 years

Figure 5-7 shows a computer run of the model over a 60 year interval. The exponential growth is explosive. In five doublings, IS has increased by a factor of 25. Figures 5-5 and 5-7 indicate the importance of time horizon in dynamic systems. Both growth trends emerged from the same system yet the short time period of Figure 5-5 gave little indication of the explosive trend in Figure 5-7.

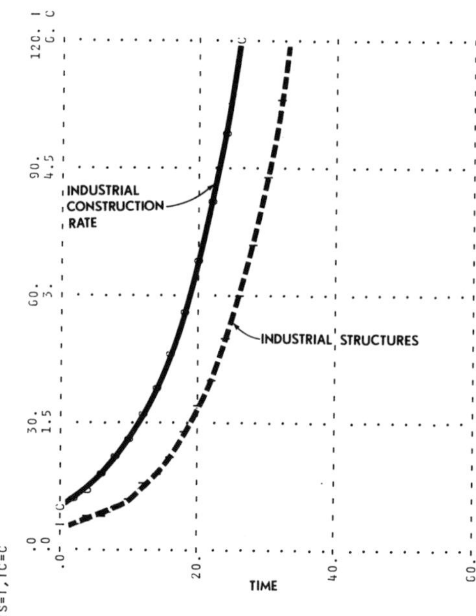

FIGURE 5-7 Industrial growth over 60 years
(Basic run)

As a check on the nature of positive feedback, test the effect of the initial values of IS and IGF on behavior. For example, what occurs with an initial value of IS equal to 2.5? Sketch the system behavior on Figure 5-7, reproduced below.

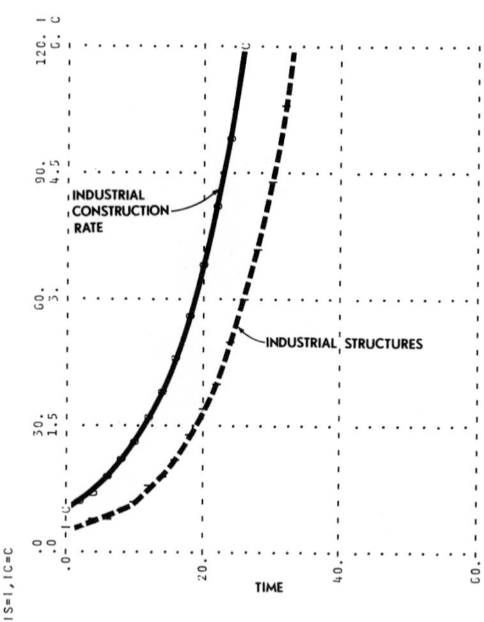

Figure 5-8 reveals the consequences of halving the number of buildings initially in the area. Comparison of Figure 5-7 and Figure 5-8 indicates that the number of structures in Figure 5-8 at any point in time equals one-half the number in Figure 5-7, the basic run. Seven years or one doubling time separates the two curves. In other words, if shifted back (left) seven years, the curves in Figure 5-7 and 5-8 would be identical. Both model runs display the same growth patterns. Varying the initial conditions does not affect qualitative behavior.

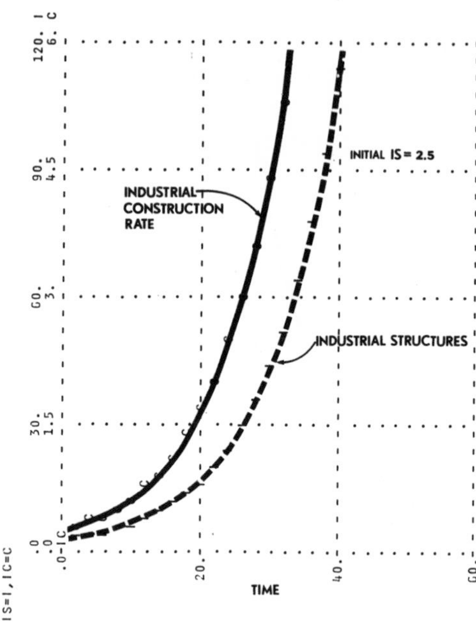

FIGURE 5-8 Industrial growth with IS initially 2.5

On Figure 5-8, reproduced below, sketch the behavior of the model with IGF set at 5 percent per year instead of 10 percent per year for an initial level value of 5.

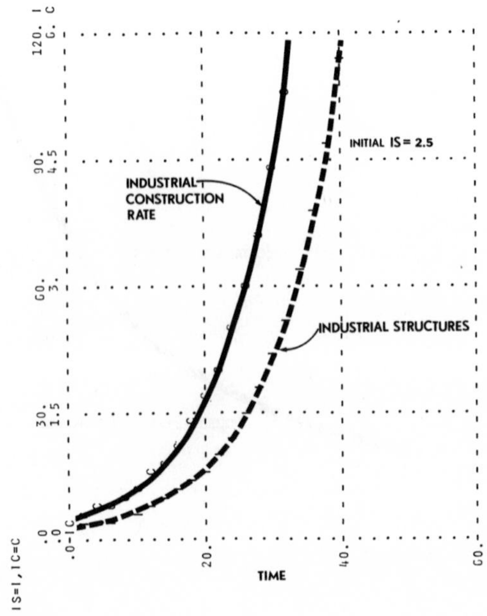

Figure 5-9 contains the growth trend for IGF equal to 0.05. The behavior basically resembles the previous run. Exponential growth results, but at a slower rate. Figure 5-9 takes two times as long as the basic run to attain a given level of industrial structures. IGF controls the time required for the system level to grow. As IGF increases, growth becomes more rapid. However, neither IGF nor the initial level value qualitatively alter the shape of the growth trends.

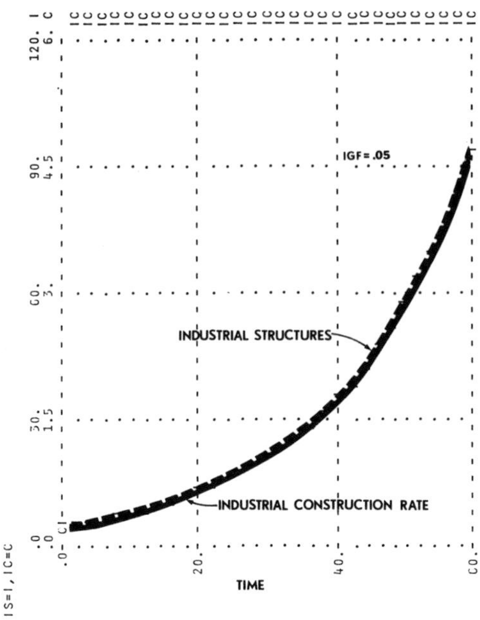

FIGURE 5-9 Industrial growth with IGF = 0.05

5.3 LAND CONSTRAINED MODEL

Verbal Description. Unlimited expansion of industrial structures cannot occur on a finite amount of land. Land availability must eventually constrain growth. The land use model should account in some way for land constraints. As the number of buildings increases, the amount of free land decreases and the rate of construction drops.

Draw a causal-loop diagram that contains both the growth loop and the land constraint loop.

Figure 5-10 displays a complete causal-loop diagram. For simplicity, we have added only one additional variable: land occupied. As the number of structures increases, the land occupied also increases. With a finite supply of available land, an increase in land occupied tends to decrease industrial construction. The negative (or inverse) relationship between land occupied and industrial construction rate creates a negative feedback loop linking industrial structures, land occupied, and the industrial construction rate. Two opposing forces operate on the industrial construction rate: a positive force from increasing structures and a negative force from increasing land occupancy.

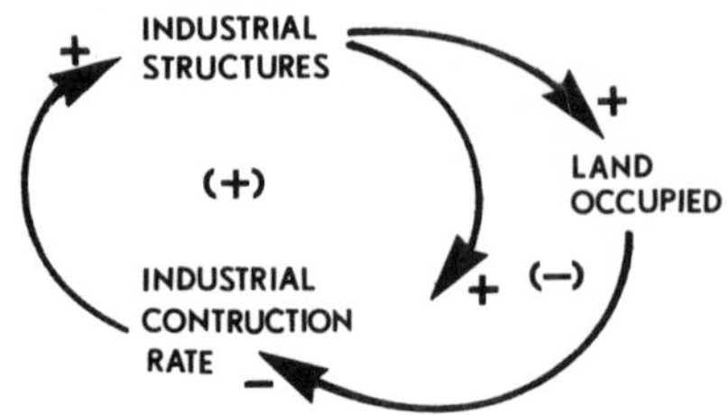

FIGURE 5-10 Causal-loop diagram of land-constrained syste

Convert the causal-loop diagram to a flow diagram. Supply any necessary variables and parameters to specify completely all relationships in the model.

Figure 5-11 shows a flow diagram with the additional land availability loop. Devising a flow diagram from a verbal description requires simplification and restatement of the phenomenon. However, we must justify the new structure and terms.

First, we explicitly specify how the construction rate is linked to the availability of land. The flow diagram in Figure 5-11 and the following discussion describe one possible connection.

We can derive the relationship between construction and land by asking the following questions:

1. When none of the land is filled, what effect does land availability have on construction?
2. What is the construction rate if 60 percent of the land fills?
3. When the land completely fills, what is the construction rate?

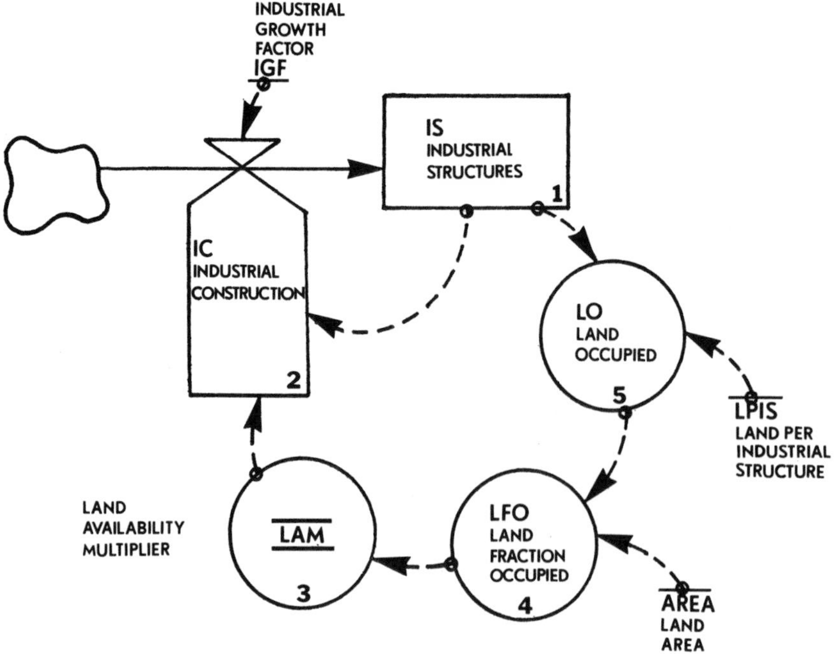

FIGURE 5-11 Flow diagram of land constrained model

These questions, although subject to discussion, might have these qualitative answers:

1. Complete land availability does not significantly affect new construction.

2. Occupation of 60 percent of the land will constrain construction.

3. Complete land occupancy means no new construction without prior demolition of older structures.

Thus, the relationship between land availability and construction has no affect on construction when a large amount of land is available, completely stops construction when no land is available, and depresses construction anywhere from 30 percent to 80 percent of its "normal" (unconstrained) value when 60 percent of the land is occupied. The curves in Figure 5-12 graphically represent plausible relationships between land availability and construction. The horizontal axis of this graph delimits the land fraction occupied LFO. The vertical axis represents the fraction of "normal" construction. For convenience the ordinate, called land availability multiplier LAM, directly modifies IGF. With little or no land occupied, the multiplier has a value of 1 and IGF remains unchanged. With all the land filled and LFO equal to 1, the multiplier must have a value of zero. All new construction stops. Intermediate values of land occupancy have a partial influence on construction. For example, when the land fraction occupied is 0.5, then the multiplier or the percent of normal construction falls somewhere between 0.5 (50 percent) and 0.9 (90 percent). The exact multiplier value depends upon available data and educated guesses about the particular tract of land modeled. Therefore, the curve labeled "Alternative 2" might represent a more uniform piece of land than the curve labeled "Alternative 1." Alternative 1 more often contains swamp lands, marshes, or a stream running through its middle. With only a few buildings constructed, the remaining land acts as a depressant on new construction. Initially, there are only a few good sites. In either case, as a smaller fraction of total land becomes available, the probability increases that remaining land has the wrong area or shape, is too rocky or too wet, needs special and expensive foundations, or has other undesirable features. The declining marginal quality of land supports the assumption of a declining multiplier value before land completely fills.

The model can, of course, test the effect of each curve. For example, in order to design policies <u>insensitive</u> to precise curve values, we might need to know model sensitivity to extreme cases. For the moment, we need only include a mechanism to represent the most plausible influence(s) of land availability on construction.

Having established the link between construction and land availability, we can easily add the remaining variables. Land fraction occupied LFO represents the ratio of land occupied LO to total land area AREA. The product of IS and the land per industrial structure LPIS yields land occupied LO. For simplicity, assume a total land area AREA of 100 acres. Imagine a 100-acre tract of land zoned for industrial development within a city for example. Assume also that an industrial building occupies one acre (including road access and parking). Potentially, 100 structures can fill the 100 acres. Both AREA and LPIS are constants. The new feedback loop is now complete.

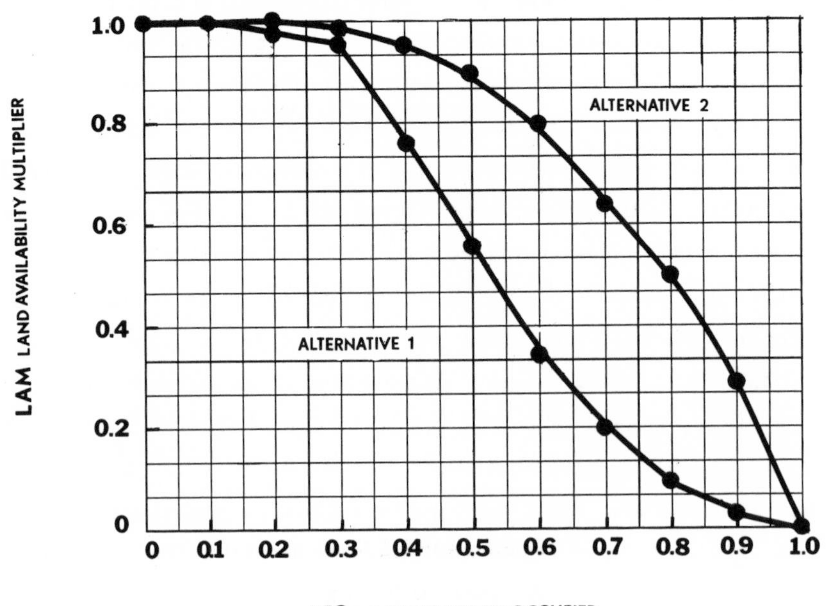

FIGURE 5-12 Land availability table function

Equation Writing

Write equations for the new flow diagram. Use the values of previously defined constants.

Equations for the appropriate flow diagram appear below. The reader should take special care to learn how to represent the table function for computer use. Notice the use of the curve labeled "Alternative 1" in Figure 5-12 in the table function.

```
IS.K=IS.J+(DT)(IC.JK)                                          1, L
IS=5                                                           1.1, N
    IS      - INDUSTRIAL STRUCTURES (STRUCTURES)
    IC      - INDUSTRIAL CONSTRUCTION (STRUCTURES/YEAR)

IC.KL=IGF*IS.K*LAM.K                                           2, R
IGF=.1                                                         2.1, C
    IC      - INDUSTRIAL CONSTRUCTION (STRUCTURES/YEAR)
    IGF     - INDUSTRIAL GROWTH FACTOR (FRACTION/YEAR)
    IS      - INDUSTRIAL STRUCTURES (STRUCTURES)
    LAM     - LAND AVAILABILITY MULTIPLIER
                (DIMENSIONLESS)

LAM.K=TABLE(LAMT,LFO.K,0,1,.1)                                 3, A
LAMT=1/1/.98/.95/.75/.55/.35/.2/.1/.05/0                       3.1, T
    LAM     - LAND AVAILABILITY MULTIPLIER
                (DIMENSIONLESS)
    LFO     - LAND FRACTION OCCUPIED (DIMENSIONLESS)

LFO.K=LO.K/AREA                                                4, A
AREA=100                                                       4.1, C
    LFO     - LAND FRACTION OCCUPIED (DIMENSIONLESS)
    LO      - LAND OCCUPIED (ACRES)
    AREA    - LAND AREA (ACRES)

LO.K=IS.K*LPIS                                                 5, A
LPIS=1                                                         5.1, C
    LO      - LAND OCCUPIED (ACRES)
    IS      - INDUSTRIAL STRUCTURES (STRUCTURES)
    LPIS    - LAND PER INDUSTRIAL STRUCTURE (ACRE/
                STRUCTURE)

NOTE CONTROL STATEMENTS
PLOT IS=I(0,120)/IC=C(0,6)/
X LFO=F(0,1)
C LENGTH=60
C DT=1
C PLTPER=2
```

FIGURE 5-13 DYNAMO equations

Model Behavior

Using the grid provided in Figure 5-14, sketch LFO, IS, and IC over 60 years.

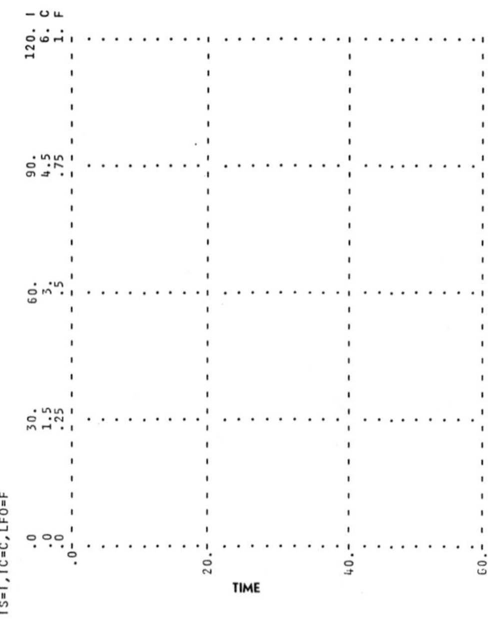

FIGURE 5-14 Sketch of industrial growth with land constraint

Figure 5-15 shows the computer run for the new model. Initially, the industrial construction rate IC rises as if unrestrained. Construction falls after 20 years, however, when the prime land fills. Eventually, new construction stops as all the land is occupied. Since the model does not permit demolition, no new construction can occur. Notice that the system does not quite equilibrate over the 60 year period. Again, we observe the importance of the time horizon in simulation analysis. The simulation requires an additional 20 years to reach equilibrium.

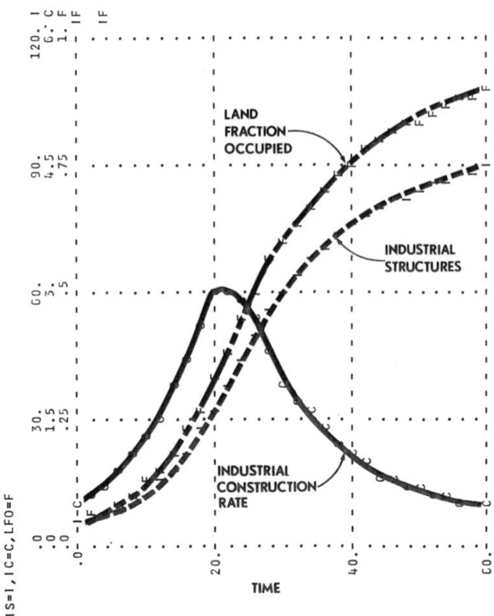

FIGURE 5-15 Industrial growth with land constraint "Alternative 1"

Model Analysis. Finite land availability converts the exponential growth mode of industrial structures IS into an "S-shaped" or "logistic" growth mode.[1] Construction rises exponentially, peaks, and then falls as structures accumulate. For the first 22 years, the positive growth loop dominates. When the land fraction occupied LFO reaches 40 percent, the negative loop begins to force construction downward. The positive loop succumbs to the negative loop. Construction continues to occur, but at a slower and slower rate. Once 100 structures have been built, construction falls to zero and the system enters equilibrium. The addition of the land occupancy mechanism significantly alters model behavior.

What might be expected from a simulation using the "Alternative 2" curve? Sketch your answer on the run in Figure 5-15, reproduced below.

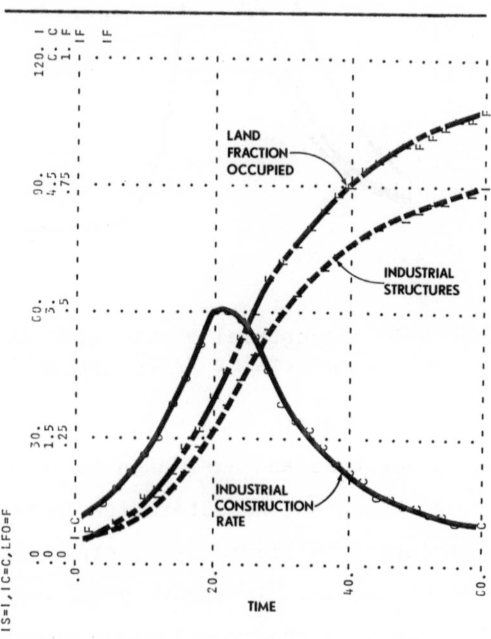

[1]The rate-level graph of the model displays a parabolic shape. As Chapter 4 explained, this shape is typical of a single level S-shaped growth structure.

Figure 5-16 shows the effect of changing the table function in the model to the upper curve described in Figure 5-12. The qualitative behavior of IS and IC remain the same for the runs in Figure 5-15 and Figure 5-16. In the former run, construction rises for 20 years. Construction then falls for more than 40 years before ceasing in Figure 5-15. Construction rises for 26 years and ceases in less than 20 years in Figure 5-16. The equilibrium values of both systems are identical:

LFO = 1, IC = 0, IS = 100.

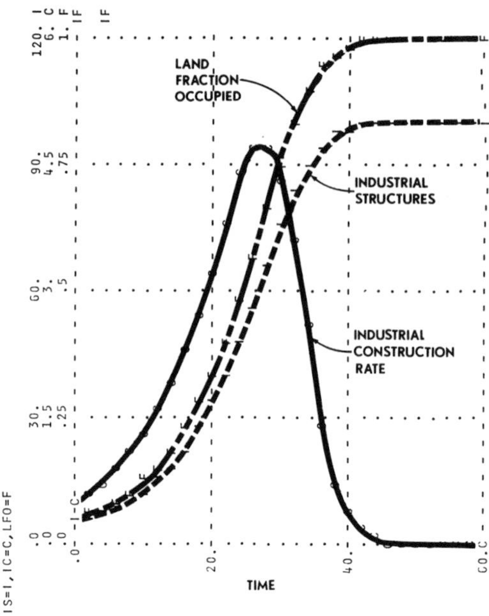

FIGURE 5-16 Industrial growth with land constraint "Alternative 2"

Figure 5-17 and 5-18 contain runs with the industrial growth factor IGF doubled in one case and area AREA increased by 50 percent in the other case. As expected, no alteration in the basic behavior mode of the system occurs. In Figure 5-17, since growth is faster, equilibrium comes about sooner. In Figure 5-18, equilibrium is delayed slightly because more construction can occur before the land fills, but the same transitional behavior results. The level in Figure 5-18 tends toward a value of 150 in correspondence with increased area.

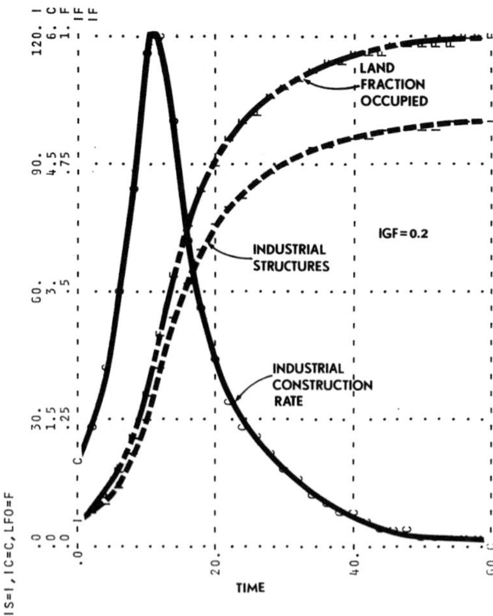

FIGURE 5-17 Industrial growth with IGF = 0.2

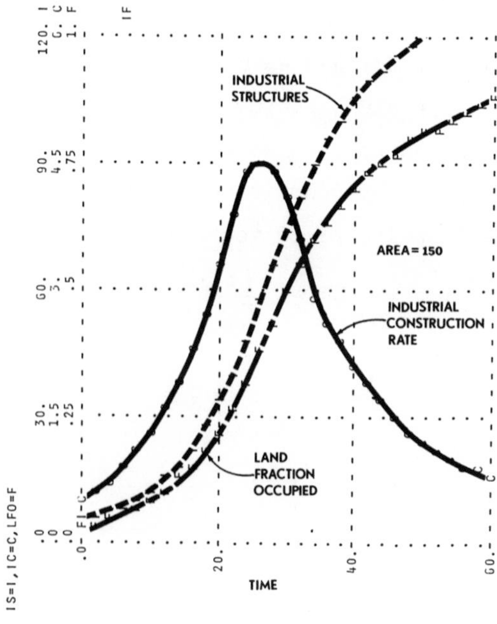

FIGURE 5-18 Industrial growth with AREA = 150

Introducing a feedback loop to account for the influence of land availability on construction demonstrates one limit to industrial growth. With more uniform land ("Alternative 2"), growth stops more abruptly and equilibrium occurs earlier than with less uniform land ("Alternative 1"). Altering other parameters in the system (i.e., IGF and AREA) generates little effect on the qualitative behavior. We now turn to the effect of aging and demolition on the behavior mode.

5.4 AGING AND DEMOLITION MODEL

Verbal Description. Our simple model has assumed that a building, once erected, stands forever. Over a 60-year time horizon this assumption loses validity in a land-use model. Therefore, we now add an aging process to show the life cycle of industrial structures.

Ideally, we should distinguish between at least two kinds of industrial structures. The first type, relatively new, encourages the construction of more new buildings and has high employment density relative to older buildings. These new structures eventually age into older buildings of the second type which differ in employment density and attractiveness for further economic expansion. For simplicity, this model ignores the old structures which inevitably accumulate. We assume that aging structures which no longer stimulate industrial growth leave the system by demolition and no longer occupy land. We assume an average lifetime of 20 years as a first approximation.

Draw a new causal-loop diagram that adds aging and demolition to the model.

The new causal-loop diagram for the system appears below. The addition of aging causes an outflow from the stock of structures. As the number of structures increases, eventually the outflow from aging (buildings per year) must increase. This process, in turn, reduces the number of structures and eventually decreases aging. A typical negative feedback loop develops. The model now contains a positive growth loop, a negative loop from land availability, and a negative loop from aging. These three loops represent fundamental assumptions of the model.

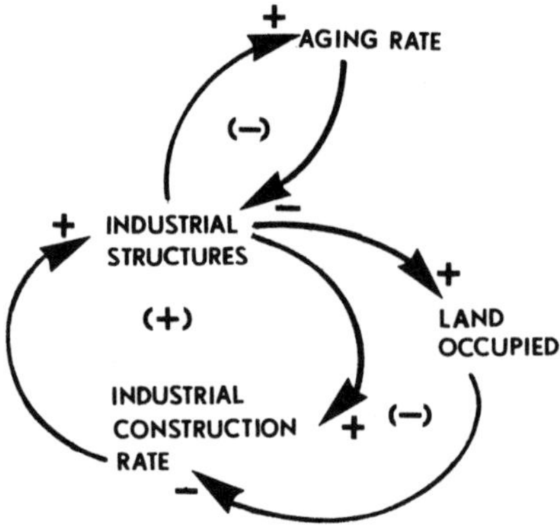

FIGURE 5-19 Aging and demolition diagram

Flow Diagram

Using 20 years as the time constant or average life time of the aging loop, construct a complete flow diagram of the system.

Figure 5-20 shows the new flow diagram for the land-use model. The aging of structures, an outflow rate, continuously decreases the level of structures. LTIS is defined as the industrial structure lifetime, the amount of time an average structure stays in the level.[2]

The flow diagram now embodies three major model assumptions:

1. Industrial structures tend to stimulate further economic expansion by generating and/or attracting additional industrial activity.

2. Land availability plays an important role in controlling the construction of industrial structures.

3. As buildings age and no longer attract additional structures, they leave the system.

These three assumptions comprise the <u>theory</u> or <u>structure</u> of the model. Such model parameters as IGF, LTIS, AREA, and LPIS and the values of the table function only detail the assumptions. Section 5.3 proved that adding the second loop to the structure substantially altered model behavior. On the other hand, changes in parameter values did little to change basic behavior. In general, altering or adding assumptions can have a greater effect on dynamic behavior than changes in parameter values. The addition of a third assumption illustrates this principle once again.

---

[2]The time constant of a negative-feedback loop equals the average amount of time a unit resides in a level. Appendix A in Exercise 9 proves this theorem. For a review of the time constant, refer to Chapter 3 of this text.

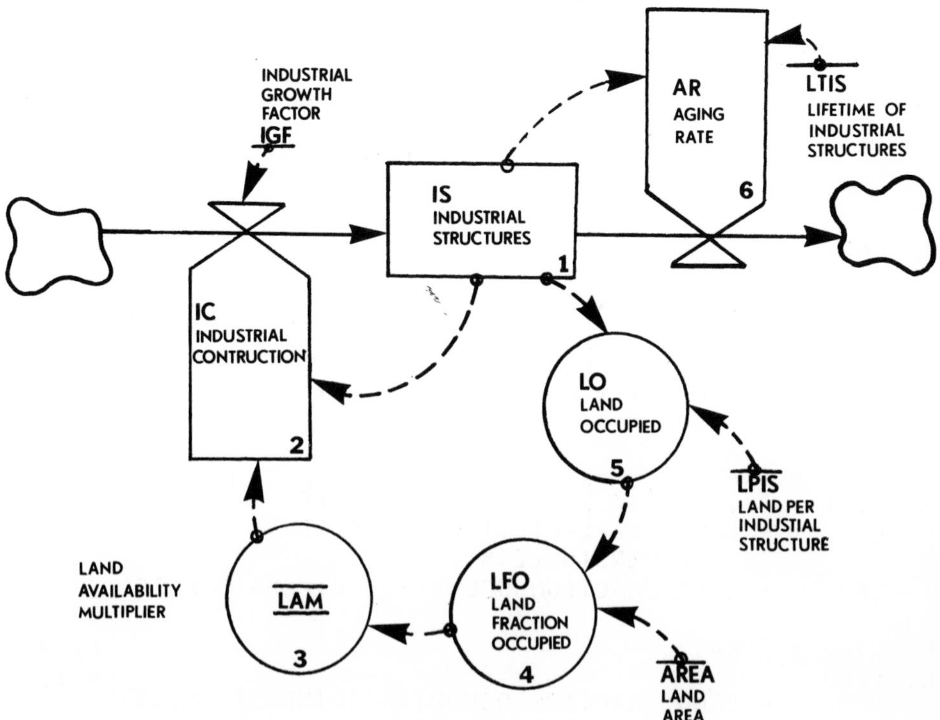

FIGURE 5-20 Complete flow diagram

Equation Writing

Write equations for the extended industrial system. Check the dimensions of variables and constants.

DYNAMO equations corresponding to the flow diagram in Figure 5-20 appear below.

```
IS.K=IS.J+(DT)(IC.JK-AR.JK)                            1, L
IS=5                                                   1.1, N
    IS      - INDUSTRIAL STRUCTURES (STRUCTURES)
    IC      - INDUSTRIAL CONSTRUCTION (STRUCTURES/YEAR)
    AR      - AGING RATE (STRUCTURES/YEAR)

IC.KL=IGF*IS.K*LAM.K                                   2, R
IGF=.1                                                 2.1, C
    IC      - INDUSTRIAL CONSTRUCTION (STRUCTURES/YEAR)
    IGF     - INDUSTRIAL GROWTH FACTOR (FRACTION/YEAR)
    IS      - INDUSTRIAL STRUCTURES (STRUCTURES)
    LAM     - LAND AVAILABILITY MULTIPLIER
                (DIMENSIONLESS)

LAM.K=TABLE(LAMT,LFO.K,0,1,.1)                         3, A
LAMT=1/1/.98/.95/.75/.55/.35/.2/.1/.05/0               3.1, T
    LAM     - LAND AVAILABILITY MULTIPLIER
                (DIMENSIONLESS)
    LFO     - LAND FRACTION OCCUPIED (DIMENSIONLESS)

LFO.K=LO.K/AREA                                        4, A
AREA=100                                               4.1, C
    LFO     - LAND FRACTION OCCUPIED (DIMENSIONLESS)
    LO      - LAND OCCUPIED (ACRES)
    AREA    - LAND AREA (ACRES)

LO.K=IS.K*LPIS                                         5, A
LPIS=1                                                 5.1, C
    LO      - LAND OCCUPIED (ACRES)
    IS      - INDUSTRIAL STRUCTURES (STRUCTURES)
    LPIS    - LAND PER INDUSTRIAL STRUCTURE (ACRES/
                STRUCTURE)

AR.KL=IS.K/LTIS                                        6, R
LTIS=20                                                6.1, C
    AR      - AGING RATE (STRUCTURES/YEAR)
    IS      - INDUSTRIAL STRUCTURES (STRUCTURES)
    LTIS    - LIFETIME OF INDUSTRIAL STRUCTURES (YEARS)

NOTE CONTROL STATEMENTS
PLOT IS=I(0,120)/IC=C,AR=A(0,6)/
X LFO=F(0,1)
C LENGTH=60
C DT=1
C PLTPER=2
```

FIGURE 5-21 DYNAMO equations

Model Behavior

Given the new structural addition, sketch on the grid in Figure 5-22 the behavior of IS, IC, AR, and LFO over a period of 60 years. What equilibrium values of IS, IC, AR, and LFO do you expect?

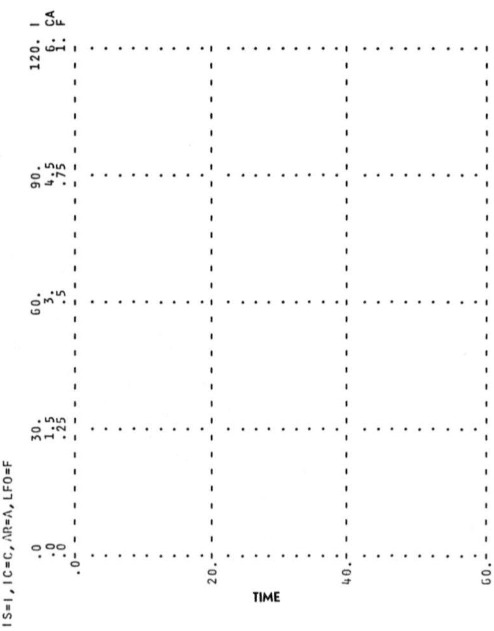

FIGURE 5-22 Sketch of growth with complete model

Figure 5-23 displays the model run. Since IS, IC, and AR seem to continue changing after 60 years, the system has not yet entered equilibrium. However, the behavior in Figure 5-23 seems reasonable. Aging retards the growth of IS. Outflow from the level accompanies inflow from construction. The net growth rate (inflow less outflow) is less than the growth rate observed in previous runs. The slower growth rate prevents the system from equilibrating in the 60-year time interval. A 10-year perspective, as in Figure 5-5, would completely miss the important long-term processes in this relatively simple model. The long time horizon necessary for dealing with feedback systems once more seems verified.

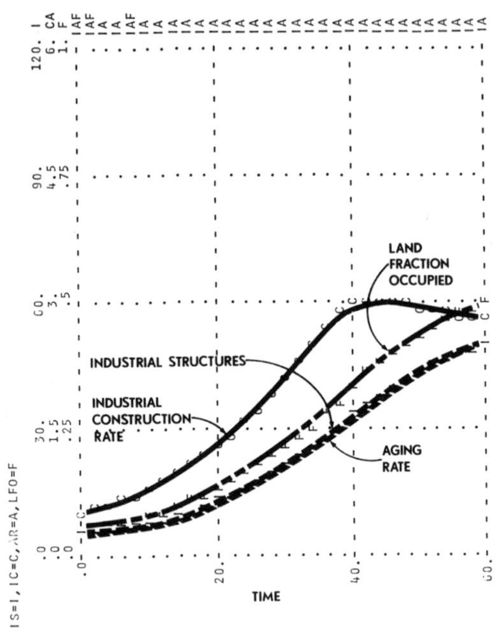

FIGURE 5-23 Growth with complete model over 60 years

Figure 5-24 contains the simulation run for a 100 year interval. Equilibrium seems to occur in year 80.

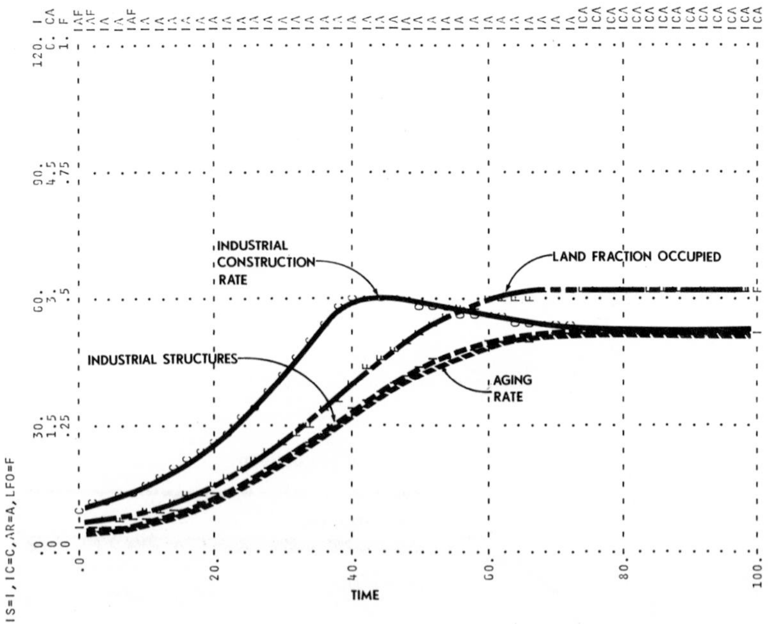

FIGURE 5-24 Growth with complete model over 100 years

Model Analysis. In equilibrium the area does not completely fill. LFO in fact reaches a maximum of 0.52. Construction proceeds while the number of industrial structures remains constant. The net growth rate must equal zero. Therefore, aging just equals construction when LFO equals 0.52. Equilibrium in Figure 5-24 occurs when the inflow rate equals the outflow rate. Prior to equilibrium, construction exceeds aging. The level grows. Eventually, land constraints begin to suppress construction. When construction and aging equilibrate, growth in IS ceases. Buildings are constructed and removed continuously without changing the total stock of buildings.

What effect would doubling IGF have on the system? Sketch your answer on Figure 5-24.

Figure 5-25 shows the impact of a larger IGF. As expected from the run in Figure 5-17, faster growth drives the system to equilibrium within 30 years, only 1/3 of the time required with IGF equal to 0.10. LFO in equilibrium has only a slightly higher value (0.66 versus 0.52) in the last run despite IGF doubled. The equilibrium in Figure 5-25 also contains a higher set of construction and aging rates than the previous run. The change in IGF does not substantially alter qualitative model behavior.

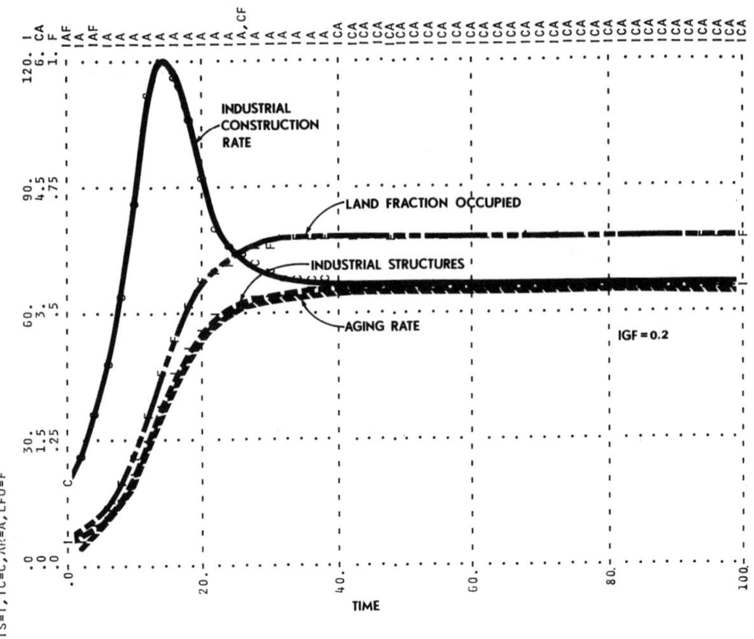

FIGURE 5-25 Complete model with IGF = 0.2

5.5 CONCLUSIONS

The complete land-use model is structurally analogous to the population growth model in Chapter 4, section 4.5. The population model contains three loops--one positive, two negative--that correspond closely to the loops developed in this chapter. The sigmoidal behavior mode of the land-use model should not be unexpected. The level's lower equilibrium value, caused by the aging loop in section 5.4, also seems predictable. The rate-level analysis performed in section 4.5 (especially Figure 4-14) would anticipate sigmoidal behavior and equilibrium conditions in similar systems. This exercise in model development and analysis tries to clearly emphasize the importance of structure in dynamic feedback models.[3]

[3]For a complete step-by-step analysis of a land-constrained urban system, see Louis Edward Alfeld, Introduction to Urban Dynamics (Cambridge: Wright-Allen Press, 1975). Alfeld's textbook begins with the same simple models presented here but continues through 10 successive steps to evolve a nine-level urban model.

Part Two
Exercises in Simple Structures

Exercise 1
Causal-Loop Diagramming

This exercise gives the reader experience in identifying causal hypotheses and converting simplified verbal descriptions of feedback processes into causal-loop diagrams. E1.1 asks the reader to fully develop the two causal-loop diagrams suggested in brief verbal descriptions. E1.2 asks the reader to identify and diagram feedback processes contained in descriptions from four fields. Review Chapter 1 before attempting this exercise.

Time Required: Two hours.

E1.1

This part of the exercise introduces two basic mechanisms underlying the life cycle of growth and stagnation in our urban areas--the population and economic expansion interactions characteristic of the growth phase of the urban life cycle and the population and land-use interactions characteristic of the growth suppression (stagnation) phase of the cycle.

Develop causal-loop diagrams from the following verbal descriptions. Identify the polarity between each pair of variables as well as the polarity of each loop.

Population and Economic Growth Loop. As employment opportunities increase in a city, people are attracted into the urban area. However, in-migrants do not immediately swarm to employment opportunities in the area. Since migrants react to perceived opportunity, the lag in acquiring information may cause 5- to 20-year delay in response. Population growth from the influx of migrants tends to encourage business expansion in the growing urban area. The additional economic expansion creates demand for additional labor. This demand further increases employment opportunities in the area.

Population and Land-Use Loop. While tending to reinforce economic growth, population growth also tends to drive housing construction at a greater pace to match population growth. Assuming only a fixed amount of land available for industrial and housing use, increasing the housing stock makes less land available for business expansion. As the unavailability of more land begins to suppress business expansion in the area, the demand for labor decreases. Consequently, local employment opportunities decline. Once potential migrants perceive the lack of opportunities, declining in-migration generates a reduction in the population growth of the area.

E1.2 Sketch causal-loop diagrams that represent the feedback processes described in the excerpts taken from the literature in the following fields:

Ecology
Economics
Sociology
Psychology.

Ecology--Population Growth and Regulation.

> ....At a critical time in the life history of a given population, a physical factor such as light or a nutrient may be significant as a regulatory agent; at another time, parasitism, predation, or competition, or even some other physical factor may become the operative factor. As complex and as variable as the niche of any species is, it is unlikely that the regulation comes about by any single agency. However, there does appear to be considerable and mounting evidence, both empirical and theoretical, to suggest that populations are self-regulating through automatic feedback mechanisms. Various mechanisms and interactions appear to operate both in providing the information and in the manner of responding to it, and with the exceptional case of a catastrophe, the stimulus to do so appears to depend on the density of the population. The end effect is one of avoiding destruction of a population's own environment and thereby avoiding its own extinction.

Source: E.J. Kormondy, *Concepts of Ecology* (Englewood Cliffs: Prentice-Hall, 1969), p. 110.

Economics--Economic Development.

....The principal source of development--that is, increase in the "productive powers of labor," as Adam Smith also saw--is the specialization in the production of human knowledge.... Here we suppose that I_1 represents the input of human labor in terms of hours of productive activity, O_1 represents the output of consumable, enjoyable goods, O_2 represents the output of capital goods, which may include knowledge in the human head and also may include machines, equipment, and so on which represents knowledge imposed on the material world. O_2 in the process F_2 then produces another inputs, I_2, of aids to labor, and the larger I_2 is, the larger will be both O_1 and O_2 for any given input of I_1. The real secret of this process is that I_2 is a function of the total capital stock rather than that of the additions to it, that is, we can think of O_2 as implying a gross investment. As long as this is greater than the consumption or depreciation of the capital stock, whether of things or of knowledge, the capital stock in F_2 will increase and I_2 will increase accordingly. As I_2 increases, however, so does O_1 and O_2, and we have a characteristic process of disequilibrating feedback. The accumulation of physical capital makes it easier to produce more and to accumulate more, and similarly the accumulation of knowledge also enables us to accumulate it and to teach it more easily and more rapidly.

Source: Kenneth E. Boulding, "Business and Economic Systems," Positive Feedback, ed. John H. Milsum (New York: Pergamon Press, 1968), p. 110.

Sociology--General Group Processes.

(1) The intensity of interaction depends upon, and increases with, the level of friendliness and the amount of activity carried on within the group. Stated otherwise, we postulate that interaction is produced, on the one hand, by friendliness, on the other, by the requirements of the activity pattern; and that these two causes of communication are additive in their effect. We will postulate, further, that the level of interaction adjusts itself rapidly--almost instantaneously--to the two variables on which it depends.

(2) The level of group friendliness will increase if the actual level of interaction is higher than that "appropriate" to the existing level of friendliness. That is, if a group of persons with little friendliness are induced to interact a great deal, the friendliness will grow; while, if a group with a great deal of friendliness interact seldom, the friendliness will weaken. We will postulate that the adjustment of friendliness to the level of interaction requires time to be consummated.

> (3) The amount of activity carried on by the group will tend to increase if the actual level of friendliness is higher than that "appropriate" to the existing amount of activity, and if the amount of activity imposed externally on the group is higher than the existing amount of activity. We will postulate that the adjustment to the activity level to the "imposed" activity level and to the actual level of friendliness both require time for their consummation.
>
> Source: Herbert A. Simon, Models of Man (New York: John Wiley & Sons, 1957) p. 101.

Psychology--Managerial Goal Setting. Three components of the operational goals of any organization include: the absolute goal attributed to an enduring long-term component of top-management attitude; the traditional performance which the organization has been able to achieve; and a weighting factor which depends on the management ability to project its absolute goals and on the effective size of the organization.

> ....The weighting factor determines the relative influence of the traditional performance and the absolute goal in creating the operational goal. To trace the behavior, suppose for simplification that the operational goal is determined entirely by the traditional performance. This means that the organization lacks absolute goals and simply tries to equal previous record. The organization acts on the discrepancy between observed conditions and the operational goal which here equals traditional performance. The organization strives to equal the goal of past accomplishment. Because of system time delays, conflicting pressures, and opposing demands for resources, the actual performance falls slightly below the goal. But today's actual performance becomes the tradition or history for tomorrow. As actual performance persists below traditional performance, the traditional performance declines accordingly. The organizational goals decline, and the stresses and pressures continue to depress actual performances below the goal. A downward spiral occurs in this positive feedback loop where tradition determines goals, accomplishment fails to meet the goal, and the lowered accomplishment sets lower traditions and goals for the future.
>
> Source: Jay W. Forrester, "Modeling the Dynamic Processes of Corporate Growth," Proceedings of the IBM Scientific Computing Symposium, (Yorktown Heights, N.Y., 1964) pp. 37-38.

Solution 1

The causal-loop diagrams developed in this solution are not necessarily the only correct diagrams. But, differences in formulations are limited to alternative variable names or degree of detail (disaggregation) rather than the nature of the respective feedback process.

Population and Economic Growth Loop. S1.1

As employment opportunities increase in a city, people are attracted into the urban area.

A positive relationship exists between the number of employment opportunities in the city and the amount of migration to the city:

We assume that employment opportunities include available jobs in the area.

However, in-migrants do not immediately swarm to employment opportunities in the area. Since migrants react to perceived opportunity, the lag in acquiring information may cause a 5- to 20-year delay in response.

An information lag or delay occurs between the time a change in opportunities takes place and the time potential migrants recognize and act upon the change. Some migrants having close contact with the city respond to a change very quickly, while others, more remote from the city, may require many years to respond. We simply note the perception delay on the link between opportunities and in-migration:

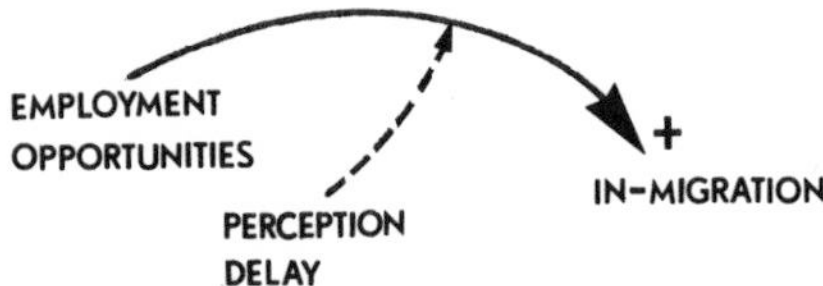

The addition of the delay in the diagram does not imply this is the only delay in the system. The significant length of the delay, however, justifies its explicit inclusion.

Population growth from the influx of migrants tends to encourage business expansion in the growing urban area.

Two new variables appear: population and business expansion. Since in-migration must expand the existing resident population, a positive link connects in-migration and population.

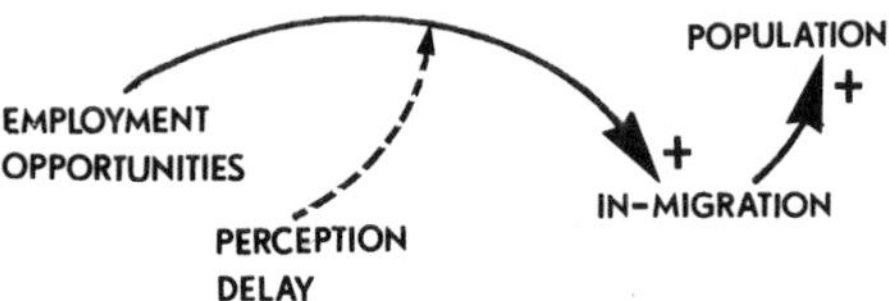

The assumed relationship between the size of population and business expansion is also positive. The expanded population not only increases demand for city service industries, for example, but also makes the area attractive to business enterprise. Population growth also allows economic activity to expand by providing a readily available labor supply.

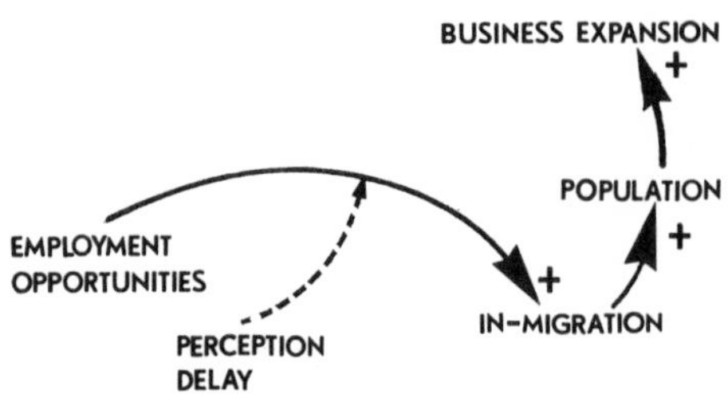

The additional economic expansion creates demand for additional labor. This demand further increases employment opportunities in the area.

An additional positive link from business expansion to demand for labor enters the diagram to form a closed loop:

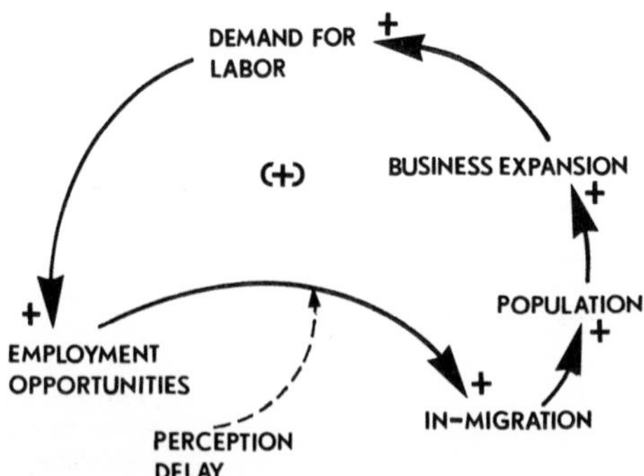

Population-economic growth loop

The complete population-economic growth loop has a positive sign since all the variables are mutually enhancing. Increase in any one variable eventually increases all the others, which in turn, produce even further increases. Of course, continual business expansion could not persist for very long. Resource constraints must, in the long term, control economic growth in a city. We consider one such constraint, available land, in the next causal-loop diagram.

Population and Land-Use Loop.

While tending to reinforce economic growth, population growth also tends to drive housing construction at a greater pace to match population growth.

A positive relationship connects population growth and housing construction. As population increases, the pressure (demand) for housing rises. The housing industry then accelerates development and construction of new housing units. For simplicity, we omit intervening market mechanisms, such as price, involved in translating excess demand into new housing stock. We assume that an increase in population produces an increase in housing construction:

Assuming only a fixed amount of land available for industrial and housing use, increasing the housing stock makes less land available for business expansion.

An increase in housing stock from accelerated housing construction simply exhausts the land available for construction of additional business enterprises. The increase in housing stock eventually inhibits business expansion:

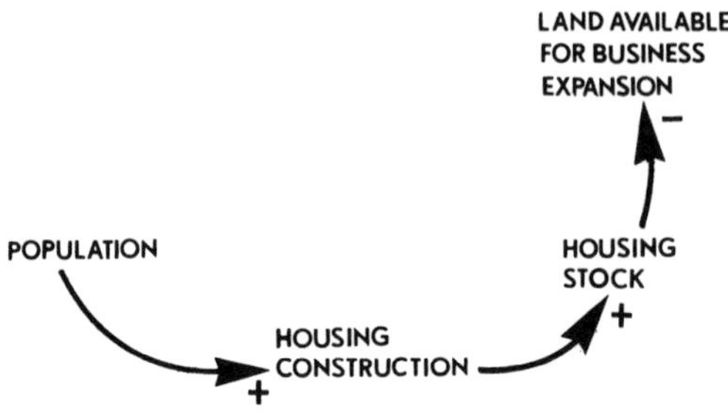

As the unavailability of more land begins to suppress business expansion in the area, the demand for labor decreases. Consequently, local employment opportunities decline. Once potential migrants perceive the lack of opportunities, declining in-migration generates a reduction in the population growth of the area.

The negative causal string developed above ties into the population-economic growth loop. The lack of available land adversely affects business expansion and causes a decline in demand for labor. A drop in labor demand leads to eventual decline of in-migration as employment opportunities also decline. The negative population and land-use loop appears below together with the positive population and economic growth loop. The outer land-use loop acts as a brake on the positive inner loop. In a very aggregated and simplified way, the multi-loop structure can describe the life cycle of a city with a limited resource such as land. During early years, emerging

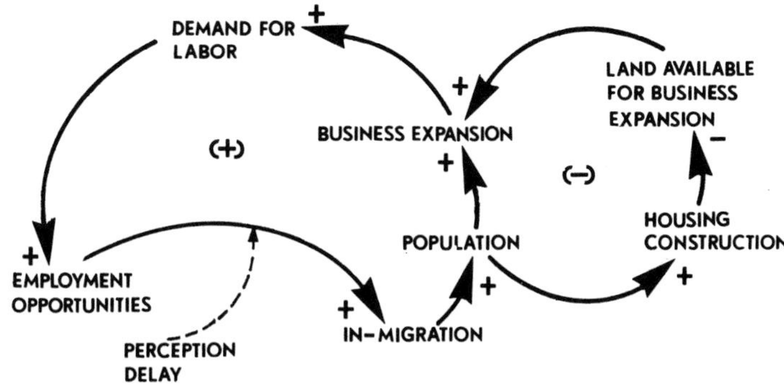

Combined loops

business enterprises create a demand for labor that attracts migrants to the area. The growing population encourages and permits accelerating economic expansion. However, the rapidly growing population also requires accelerated housing construction. Less available land for industrial use restricts continued physical growth of business enterprise. The land constraint gradually supresses further business expansion in the following transition period. In-migration drops off as employment opportunities decline. Finally, all the land fills with industrial and housing structures. Growth ceases.

Ecology--Population Growth and Regulation. **S1.2**

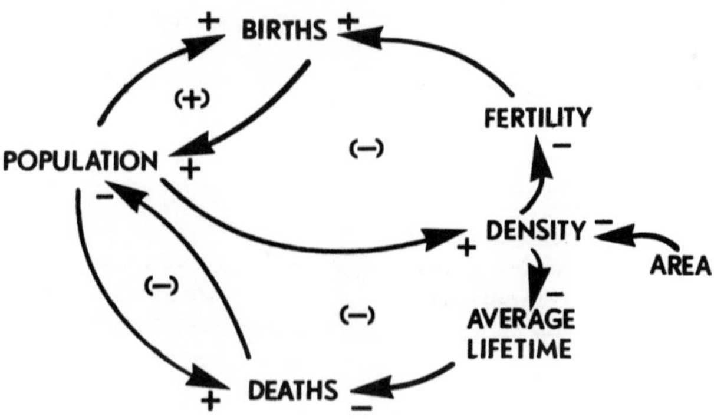

Population growth diagram

Economics--Economic Development.

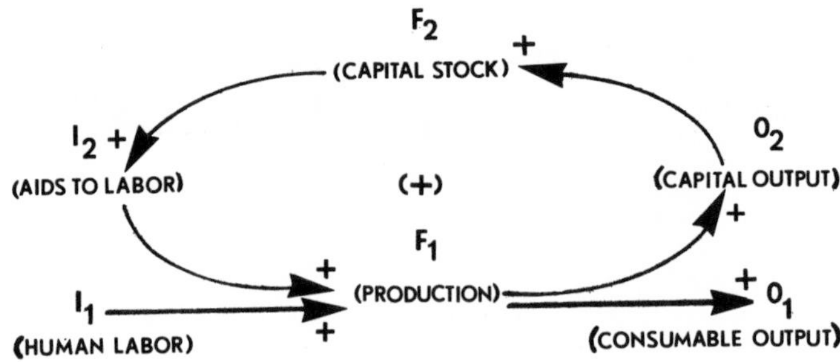

Economic development diagram

Sociology--General Group Processes.

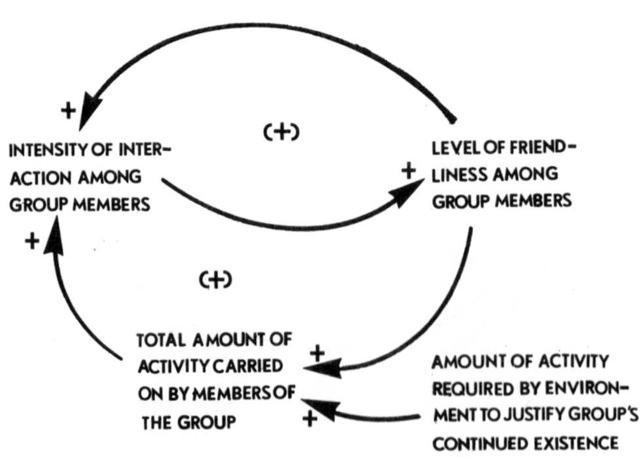

Group process diagram

Psychology--Managerial Goal Setting.

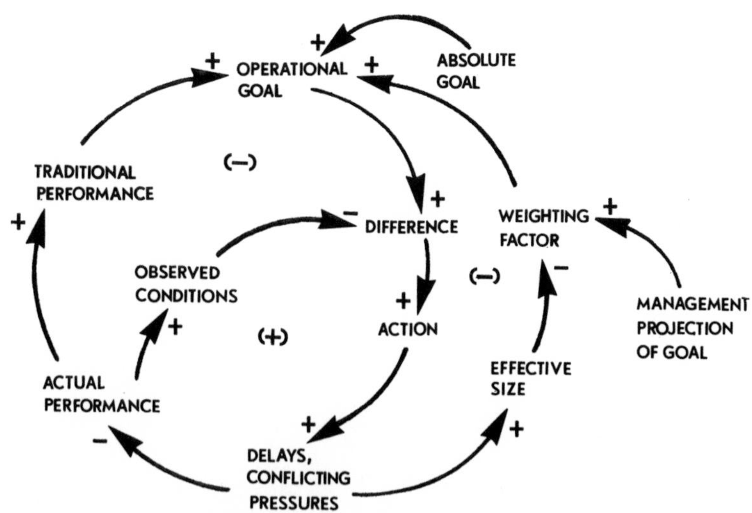

Goal setting diagram

Source: Forrester, "Modeling the Dynamic Processes of Corporate Growth," p. 38.

Exercise 2
Graphical Integration

This exercise tests the reader's understanding of integration (or accumulation), the basis of the rate and level structure in system dynamics. The exercise demonstrates how integration converts a time-varying rate into a time-varying level. This introduction to integration prepares the reader for other work in this volume. Workbook Chapter W5 (especially sections W5.3 and W5.5) in <u>*Principles of Systems*</u> *provides background for this exercise.*

Time Required: One hour.

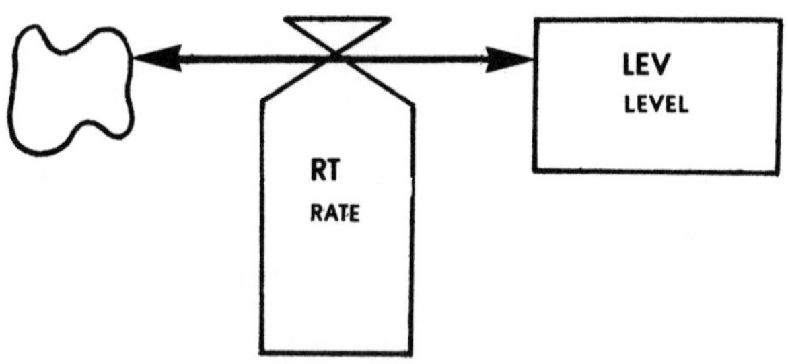

Single rate structure

Each graph below contains the exogenous rate RT as a function of time for the system shown above. No information flows from the level LEV to rate RT; therefore we call this structure an open-loop system. The level LEV has an initial value of zero. **E2.1**

Draw the values of LEV as a function of time on each of the three RT graphs. The left hand scale indicates the RT value, the right hand scale indicates the LEV value. (Use piecewise integration by calculus to verify your graphical results if you desire.)

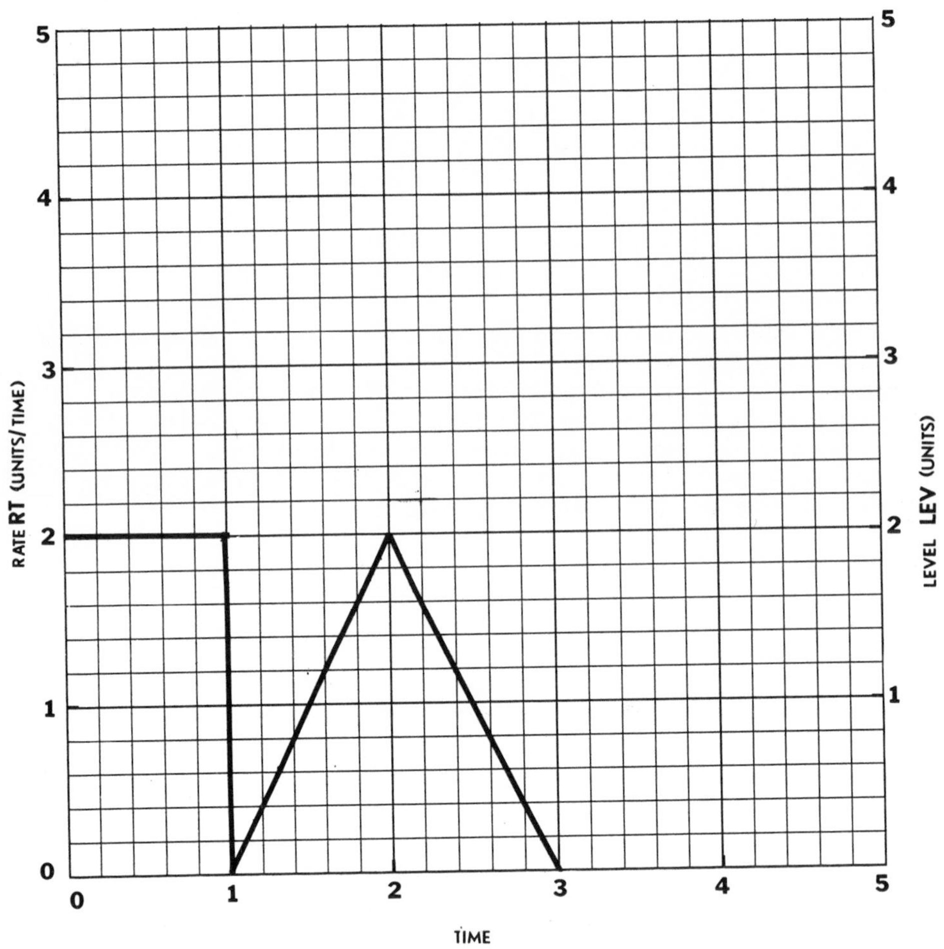

(a)

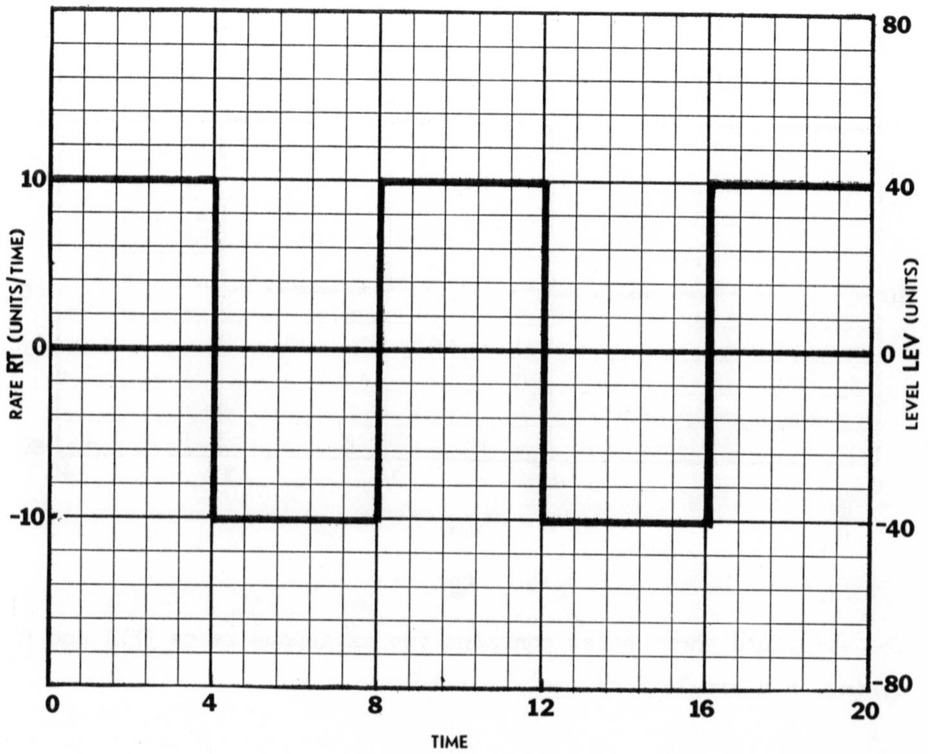
10
0
-10
RATE RT (UNITS/TIME)
80
40
0
-40
-80
LEVEL LEV (UNITS)
0
4
8
12
16
20
TIME

(b)

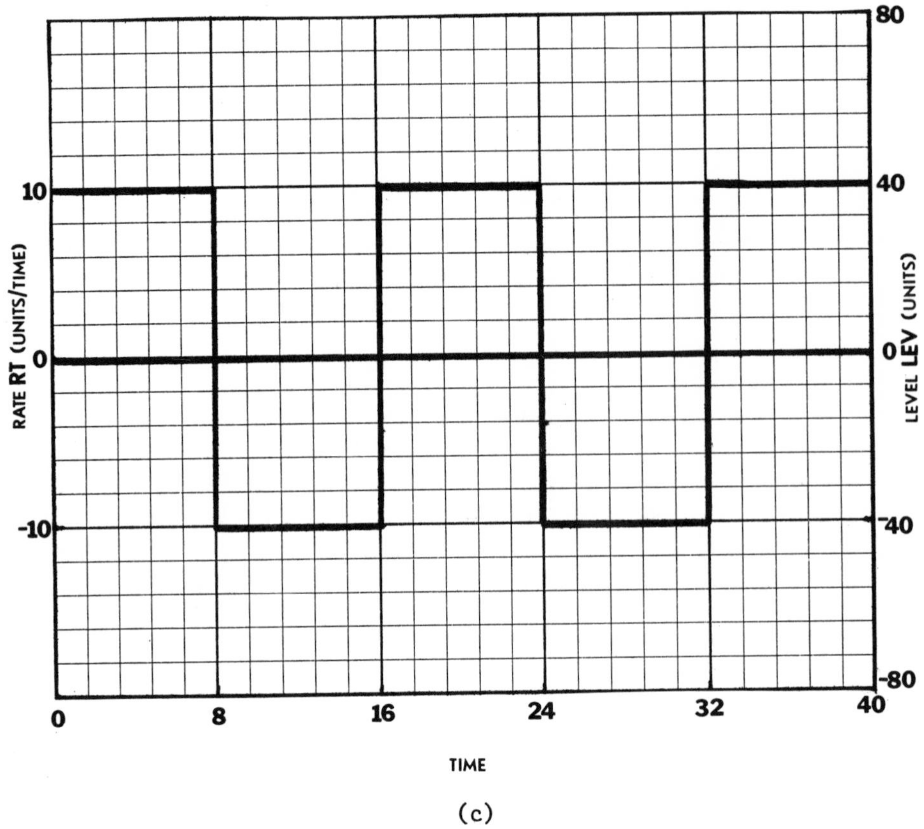

(c)

E2.2 The structure shown below contains two exogenous rates (RT1 and RT2) acting on a single level.

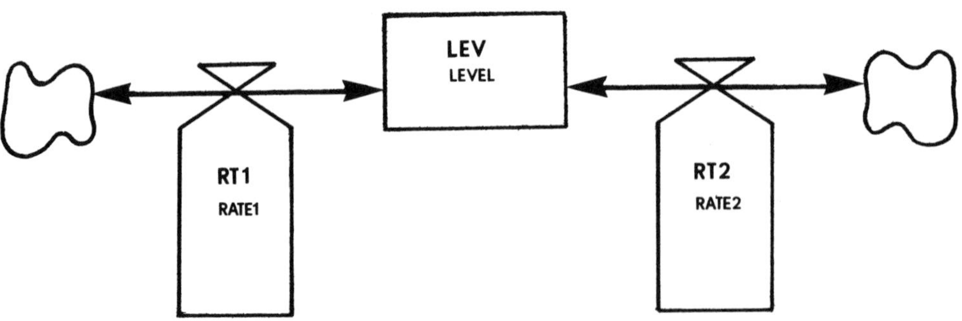

Multiple-rate structure

Indicate the behavior of the level LEV over time on the graphs.

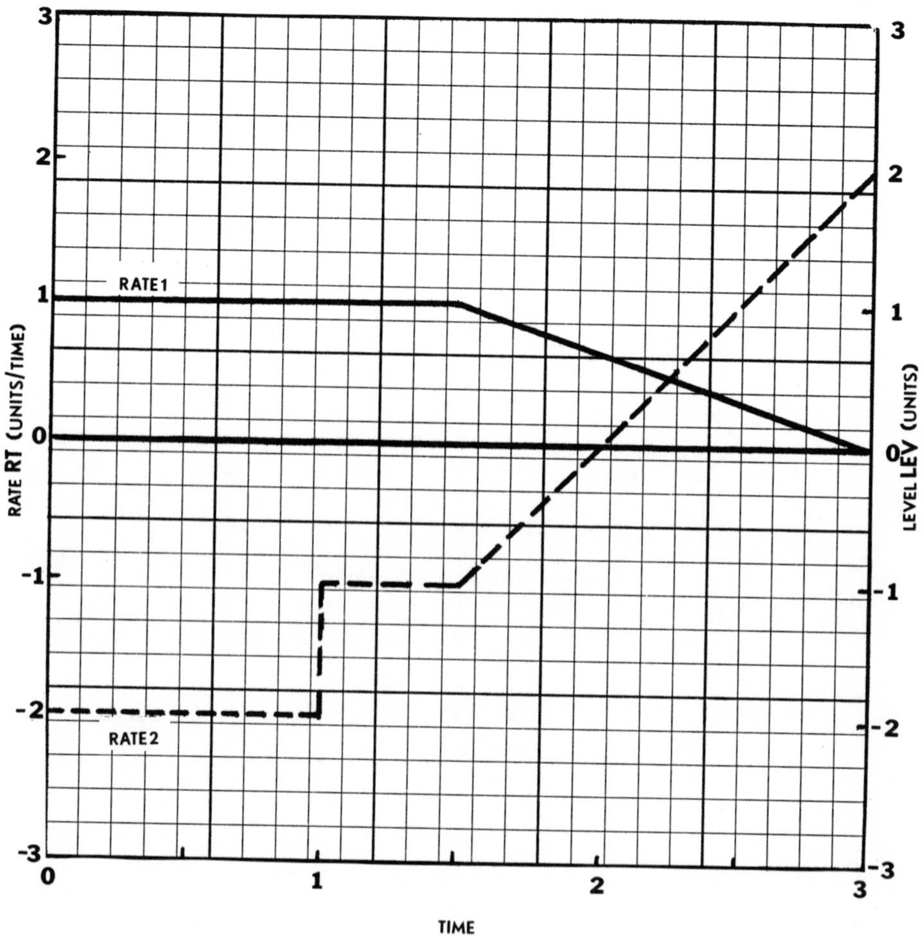

(a)

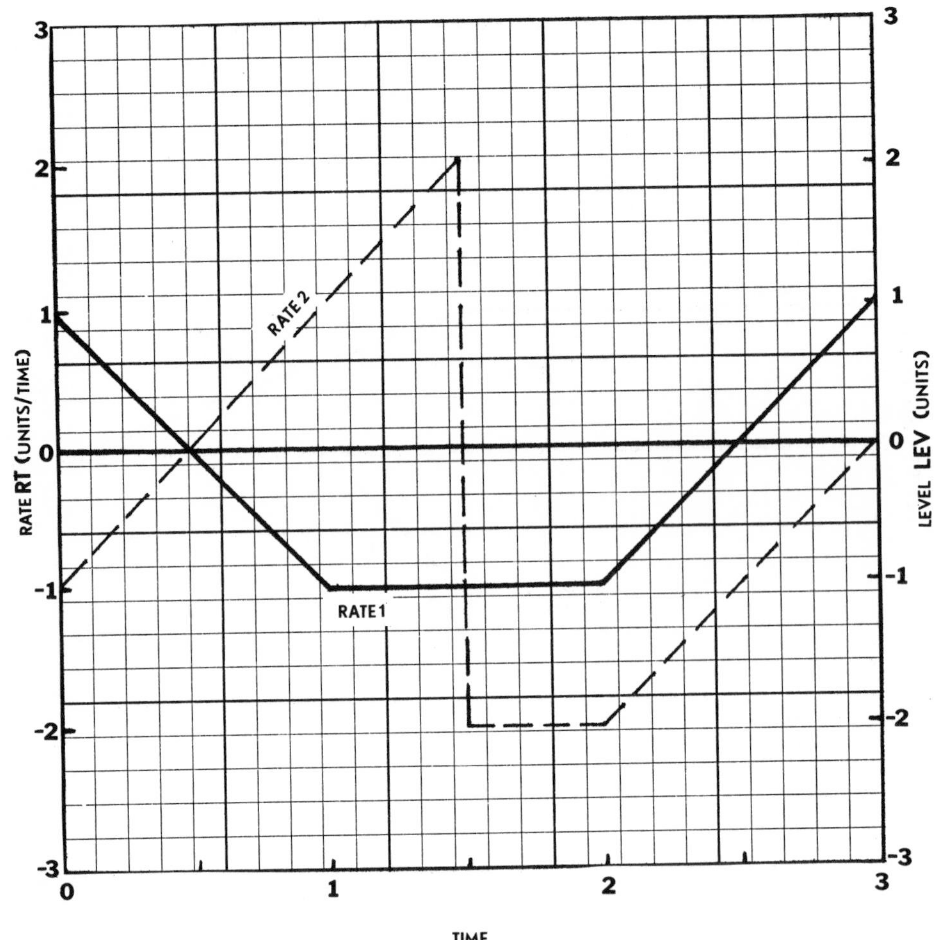

3
2
1
0
-1
-2
-3
RATE RT (UNITS/TIME)
LEVEL LEV (UNITS)
RATE 2
RATE 1
0
1
2
3
TIME

(b)

Solution 2

S2.1

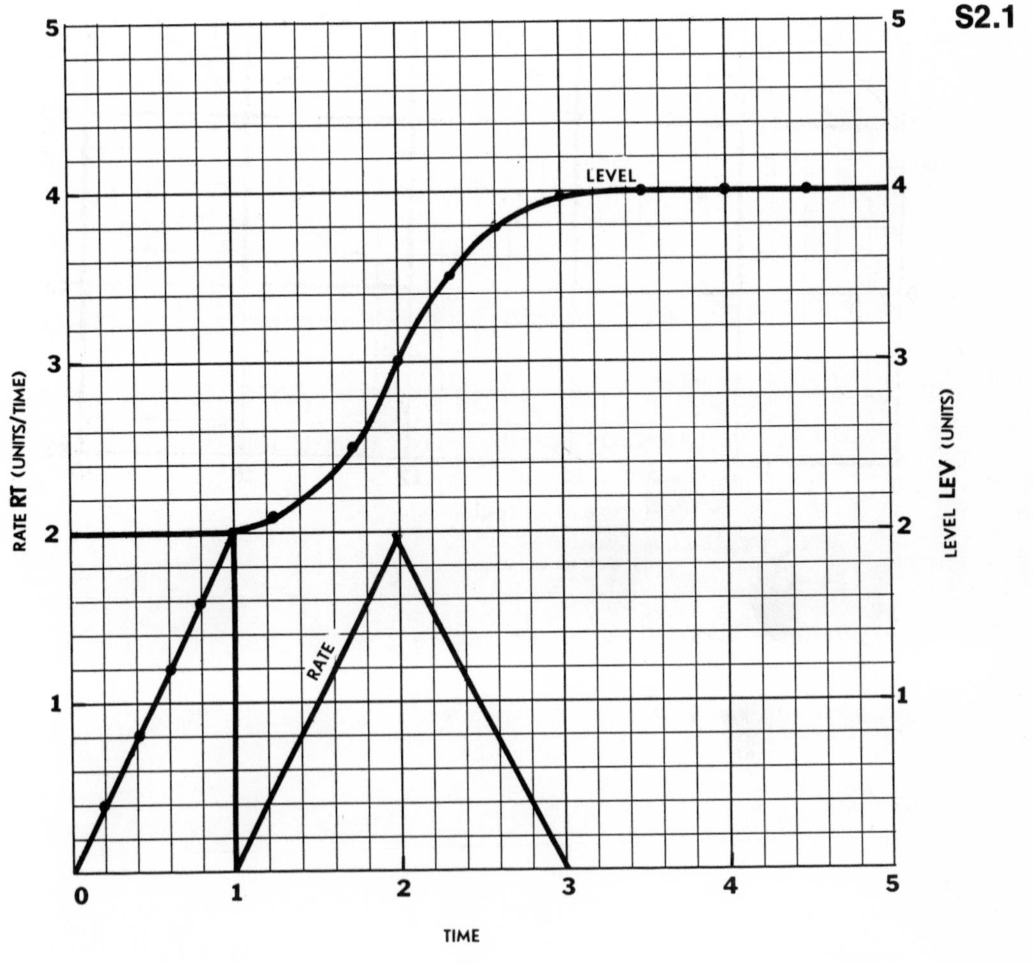

(a)

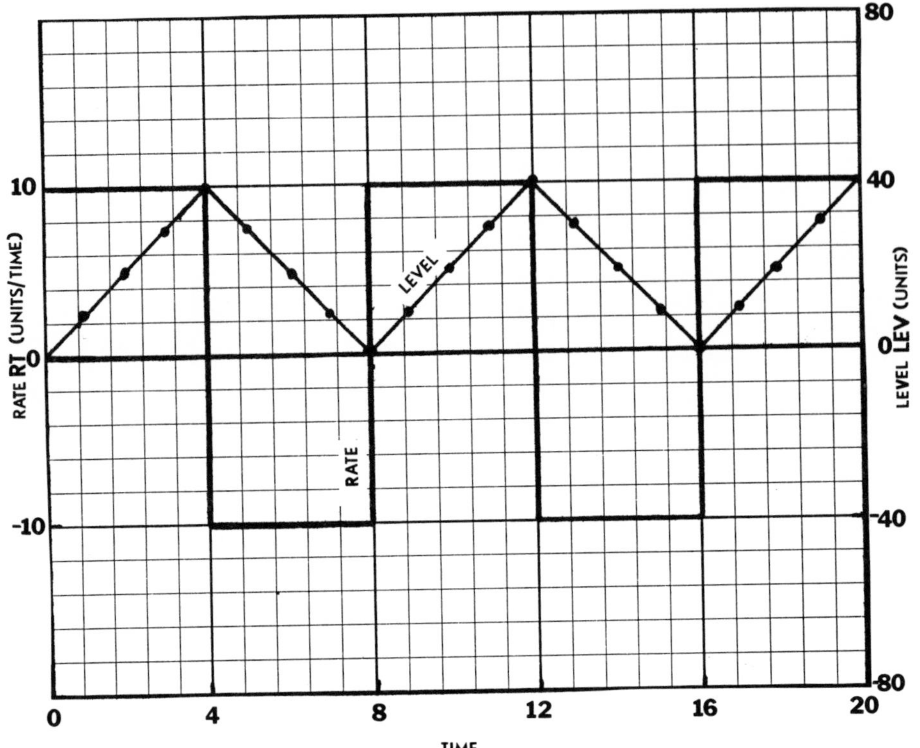
10
0
-10
RATE RT (UNITS/TIME)
80
40
0
-40
-80
LEVEL LEV (UNITS)
LEVEL
RATE
0
4
8
12
16
20
TIME

(b)

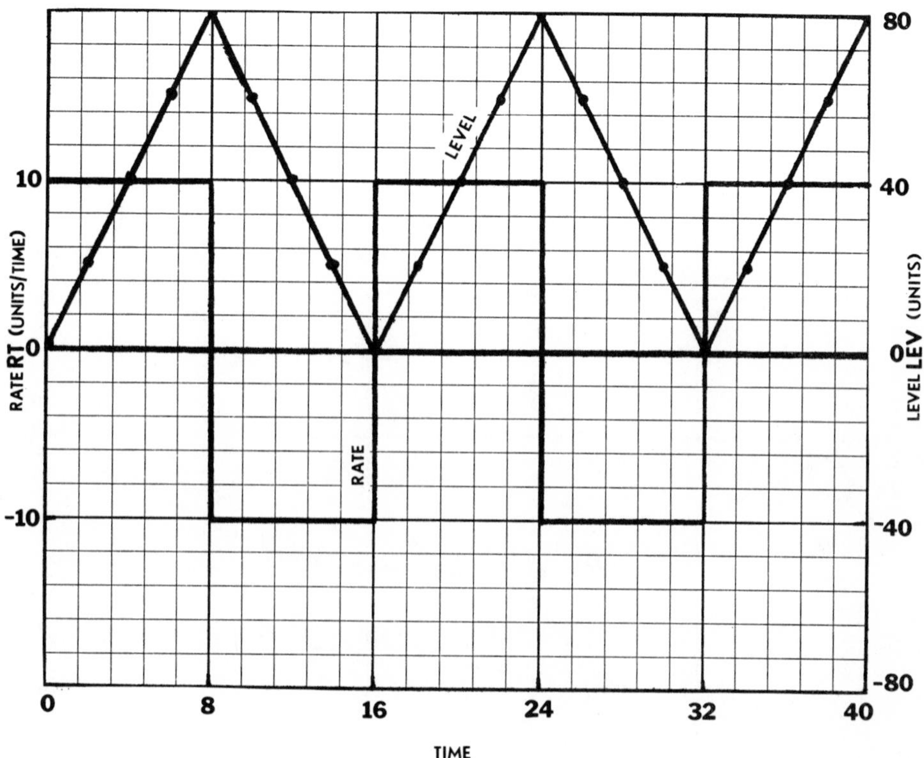
10
0
-10
RATE RT (UNITS/TIME)
80
40
0
-40
-80
LEVEL LEV (UNITS)
LEVEL
RATE
0
8
16
24
32
40
TIME

(c)

S2.2

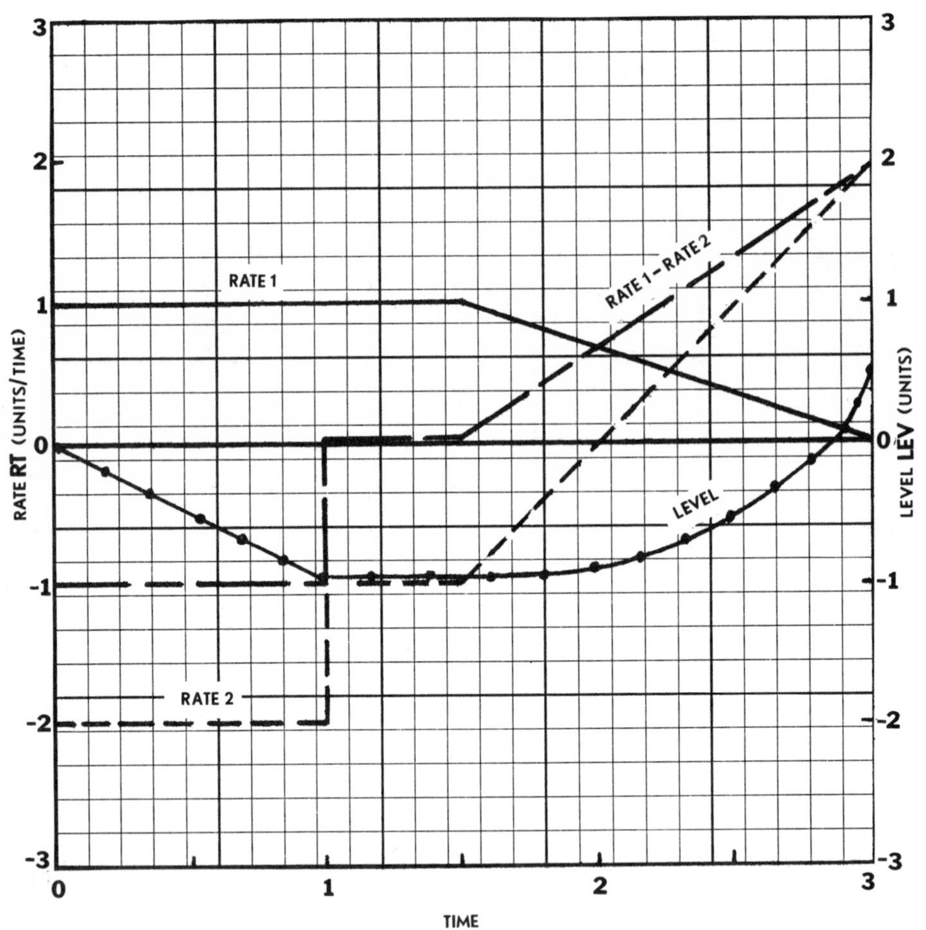
3
2
1
0
-1
-2
-3
RATE RT (UNITS/TIME)
RATE 1
RATE 2
RATE 1 – RATE 2
LEVEL
LEVEL LEV (UNITS)
TIME
0
1
2
3

(a)

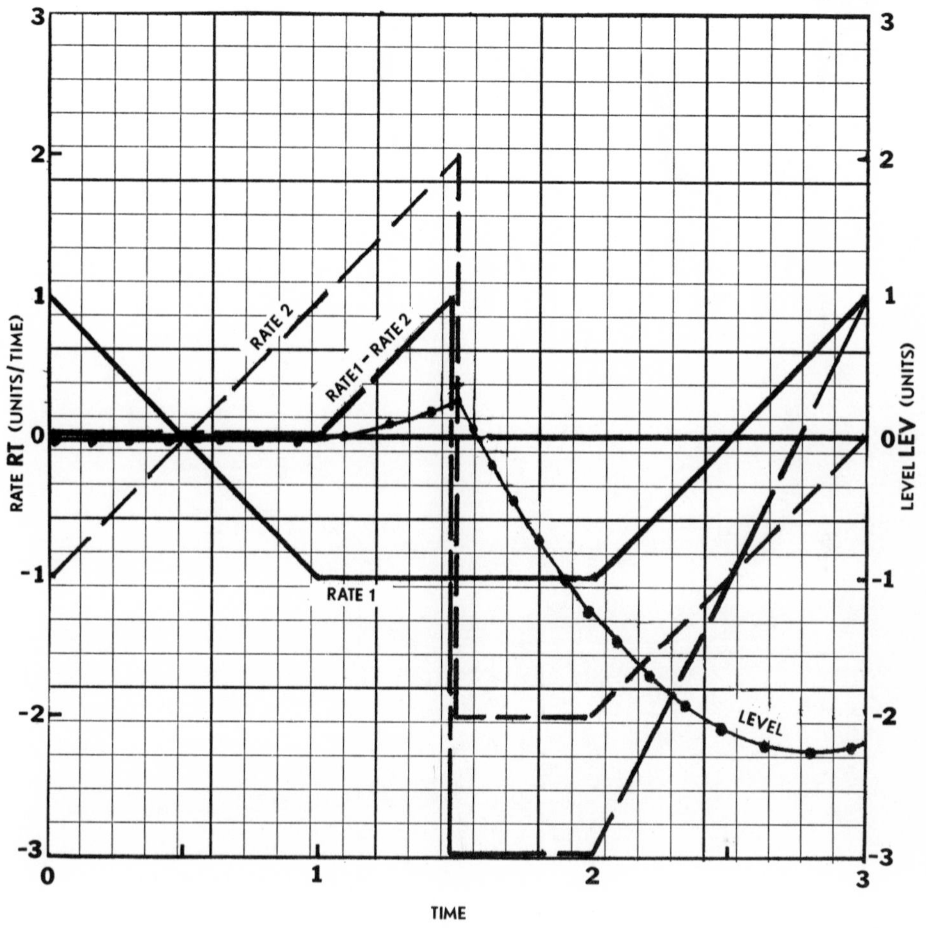
RATE RT (UNITS/TIME)
LEVEL LEV (UNITS)
RATE 2
RATE1 - RATE 2
RATE 1
LEVEL
TIME
3
2
1
0
-1
-2
-3
0
1
2
3

(b)

Exercise 3
Flow Diagramming

This exercise tests the reader's ability to convert a set of DYNAMO equations into an appropriate flow diagram and vice-versa. Chapters 5, 7, and 8 (through page 8-13) of Principles of Systems or Chapters 5, 6, 7, 8, and 12 in Industrial Dynamics provide appropriate background information. Fluency in DYNAMO and flow diagramming is essential for fully benefiting from Study Notes.

Time Required: One hour.

E3.1

Construct a flow diagram of the system described by the following program listing.

```
HP.K=HP.J+(DT)(HPC.JK)                                  1, L
HP=HPI                                                  1.1, N
HPI=196                                                 1.2, C
    HP     - HUMAN POPULATION (HUMANS)
    DT     - COMPUTATION INTERVAL (YEARS)
    HPC    - HUMAN POPULATION CHANGE (HUMANS PER YEAR)
    HPI    - HUMAN POPULATION INITIAL (HUMANS)

HPC.KL=NHPCR.K*HP.K                                     2, R
    HPC    - HUMAN POPULATION CHANGE (HUMANS PER YEAR)
    NHPCR  - NET HUMAN POPULATION CHANGE RATE (FRACTION
               PER YEAR)
    HP     - HUMAN POPULATION (HUMANS)

NHPCR.K=TABHL(NHPCRT,FPC.K,0,1113E3,371E3)              3, A
NHPCRT=-.3/-.07/.011/.013                               3.1, T
    NHPCR  - NET HUMAN POPULATION CHANGE RATE (FRACTION
               PER YEAR)
    NHPCRT - TABLE FOR NHPCR
    FPC    - FOOD PER CAPITA (CALORIES PER PERSON PER
               YEAR)

FPC.K=F.K/HP.K                                          4, A
    FPC    - FOOD PER CAPITA (CALORIES PER PERSON PER
               YEAR)
    F      - FOOD (CALORIES PER YEAR)
    HP     - HUMAN POPULATION (HUMANS)

F.K=AL*YPA.K*I.K                                        5, A
AL=972                                                  5.1, C
    F      - FOOD (CALORIES PER YEAR)
    AL     - ARABLE LAND (ACRES)
    YPA    - YIELD PER ACRE (CALORIES PER ACRE PER YEAR)
    I      - INTENSITY (DIMENSIONLESS)
```

```
I.K=TABHL(IT,PF.K/FN.K,0,24,4)                              6, A
IT=1/.25/.125/.0834/.0625/.05/.0417                         6.1, T
    I       - INTENSITY (DIMENSIONLESS)
    IT      - TABLE FOR I
    PF      - POTENTIAL FOOD (CALORIES PER YEAR)
    FN      - FOOD NEEDED (CALORIES PER YEAR)

PF.K=PYPA.K*AL                                              7, A
    PF      - POTENTIAL FOOD (CALORIES PER YEAR)
    PYPA    - PERCEIVED YIELD PER ACRE (CALORIES PER ACRE
                PER YEAR)
    AL      - ARABLE LAND (ACRES)

PYPA.K=SMOOTH(YPA.K,YPT)                                    8, A
YPT=5                                                       8.1, C
    PYPA    - PERCEIVED YIELD PER ACRE (CALORIES PER ACRE
                PER YEAR)
    YPA     - YIELD PER ACRE (CALORIES PER ACRE PER YEAR)
    YPT     - YIELD PERCEPTION TIME (YEARS)

FN.K=HP.K*DFPC                                              9, A
DFPC=742E3                                                  9.1, C
    FN      - FOOD NEEDED (CALORIES PER YEAR)
    HP      - HUMAN POPULATION (HUMANS)
    DFPC    - DESIRED FOOD PER CAPITA (CALORIES PER
                PERSON PER YEAR)

YPA.K=YPA.J+(DT)(YR.JK-YD.JK)                               10, L
YPA=YPAI                                                    10.1, N
YPAI=4.4E6                                                  10.2, C
    YPA     - YIELD PER ACRE (CALORIES PER ACRE PER YEAR)
    DT      - COMPUTATION INTERVAL (YEARS)
    YR      - YIELD REGENERATION (CALORIES PER ACRE PER
                YEAR PER YEAR)
    YD      - YIELD DEGRADATION (CALORIES PER ACRE PER
                YEAR PER YEAR)
    YPAI    - YIELD PER ACRE INITIAL (CALORIES PER ACRE
                PER YEAR)

YR.KL=(IYPA-YPA.K)/YRT.K                                    11, R
IYPA=4.4E6                                                  11.1, C
    YR      - YIELD REGENERATION (CALORIES PER ACRE PER
                YEAR PER YEAR)
    IYPA    - INHERENT YIELD PER ACRE (CALORIES PER ACRE
                PER YEAR)
    YPA     - YIELD PER ACRE (CALORIES PER ACRE PER YEAR)
    YRT     - YIELD REGENERATION TIME (YEARS)

YRT.K=TABHL(YRTT,YPA.K/IYPA,0,1,.1)                         12, A
YRTT=400/140/105/80/60/45/34/25/20/17/16                    12.1, T
    YRT     - YIELD REGENERATION TIME (YEARS)
    YRTT    - TABLE FOR YRT
    YPA     - YIELD PER ACRE (CALORIES PER ACRE PER YEAR)
    IYPA    - INHERENT YIELD PER ACRE (CALORIES PER ACRE
                PER YEAR)

YD.KL=DR.K*YPA.K                                            13, R
    YD      - YIELD DEGRADATION (CALORIES PER ACRE PER
                YEAR PER YEAR)
    DR      - DEGRADATION RATE (FRACTION PER YEAR)
    YPA     - YIELD PER ACRE (CALORIES PER ACRE PER YEAR)

DR.K=TABHL(DRT,1.0/I.K,0,24,4)                              14, A
DRT=.5/.15/.06/.02/0/0/0                                    14.1, T
    DR      - DEGRADATION RATE (FRACTION PER YEAR)
    DRT     - TABLE FOR DR
    I       - INTENSITY (DIMENSIONLESS)

CC.K=AL*I.K*YPA.K/DFPC                                      15, S
DT=.5                                                       15.1, C
LENGTH=200                                                  15.2, C
PLTPER=5                                                    15.3, C
PRTPER=0                                                    15.4, C
    CC        CARRYING CAPACITY (HUMANS)
    AL      - ARABLE LAND (ACRES)
    I       - INTENSITY (DIMENSIONLESS)
```

```
    YPA     - YIELD PER ACRE (CALORIES PER ACRE PER YEAR)
    DFPC    - DESIRED FOOD PER CAPITA (CALORIES PER
                PERSON PER YEAR)
    DT      - COMPUTATION INTERVAL (YEARS)

PLOT  HP=H,CC=C(0,500)/FPC=F(0,1E6)/I=N(0,1)/YPA=*(0,5E6)
RUN
```

Source: Steven B. Shantzis and William W. Behrens III, "Population Control Mechanisms in a Primitive Agricultural Society," ed. Dennis L. Meadows and Donella H. Meadows, Towards Global Equilibrium, p. 265.

E3.2 Write DYNAMO equations corresponding to the flow diagram provided below. The variable dimensions are as follows:

Variable	Dimension
NUT, BIO	food units
FR, FOOD, GRO, DIE, FRI	food units/time
FTIME, LT	time
GT	food units*time

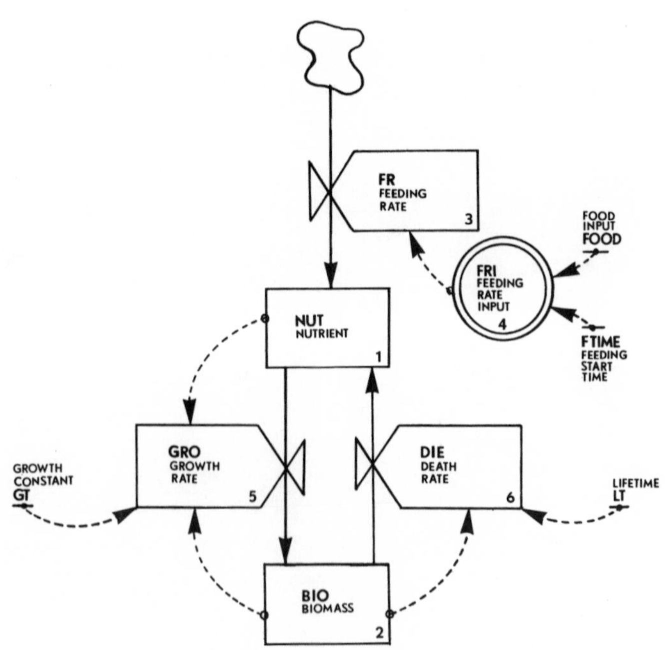

Solution 3

S3.1

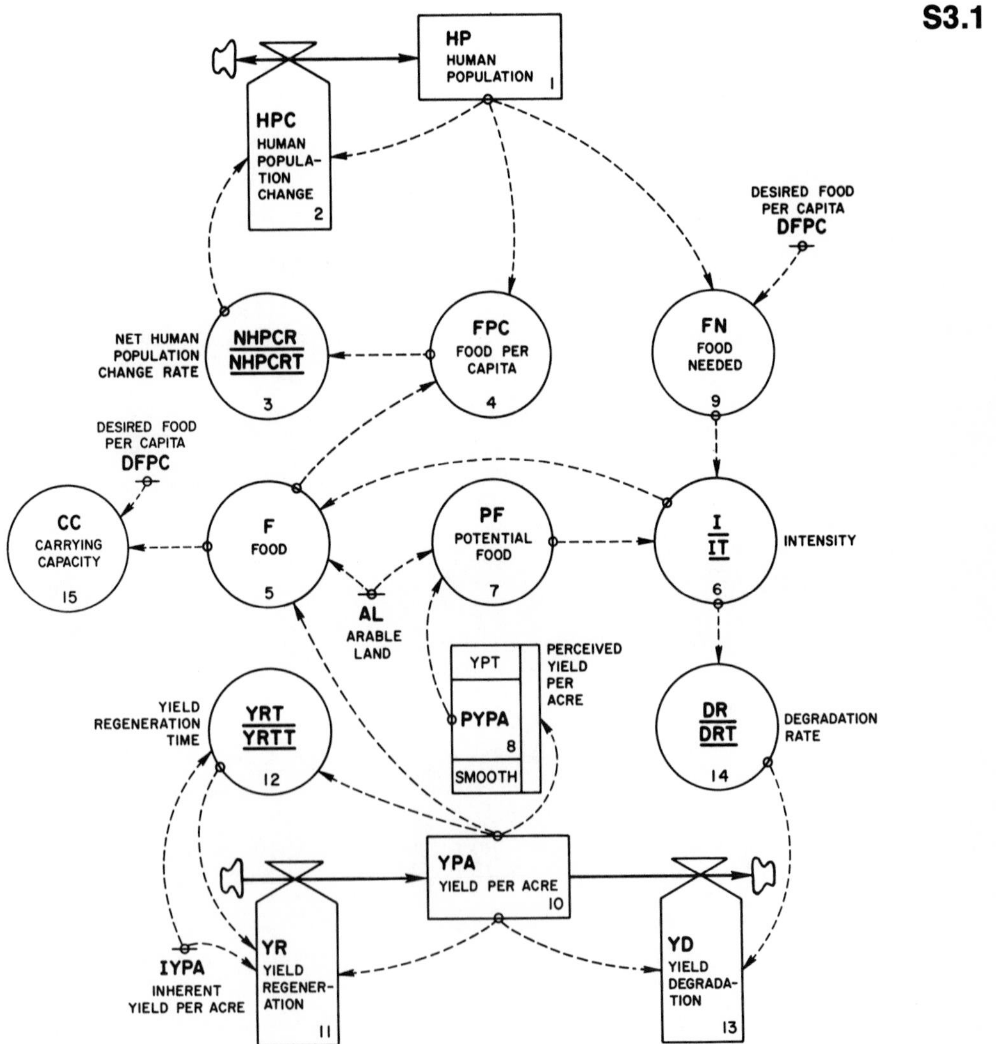

S3.2 (Numerical values of constants are left blank.)

```
*  SINGLE NUTRIENT IN EQUILIBRIUM WITH BIOMASS
L  NUT.K=NUT.J+(DT)(FR.JK-GRO.JK+DIE.JK)  (Food units)        1
N  NUT=NUT1                               (Food units)        1.1
C  NUT1=                                  (Food units)        1.2
L  BIO.K=BIO.J+(DT)(GRO.JK-DIE.JK)        (Food units)        2
N  BIO=BIO1                               (Food units)        2.1
C  BIO1=                                  (Food units)        2.2
R  FR.KL=FRI.K                            (Food units/time)   3
A  FRI.K=STEP(FOOD,FTIME)                 (Food units/time)   4
C  FOOD=                                  (Food units)        4.1
C  FTIME=                                 (Time)              4.2
R  GRO.KL=BIO.K*NUT.K/GT                  (Food units/time)   5
C  GT=                                    (Food units*time)   5.1
R  DIE.KL=BIO.K/LT                        (Food units/time)   6
C  LT=                                    (Time)              6.1
```

Exercise 4
Positive Feedback

This exercise introduces the first-order positive loop by developing a simple population growth model. Attempt this exercise before reading Chapter 2.

Time Required: One hour.

POPULATION GROWTH MODEL

Consider the positive feedback loop shown below. As births per year increase, population increases and causes an additional increase in births, and so forth. Assume no deaths or migration and that the birth rate BR is proportional to the population POP. The population grows by 7 percent per year. The birth rate equation without subscripts is:

```
BR=BRF*POP                    (People/year)
BRF=0.07                      (Fraction/year)
```

We denote the 7 percent per year growth constant as the birth rate fraction BRF. The relevant causal-loop diagram appears below.

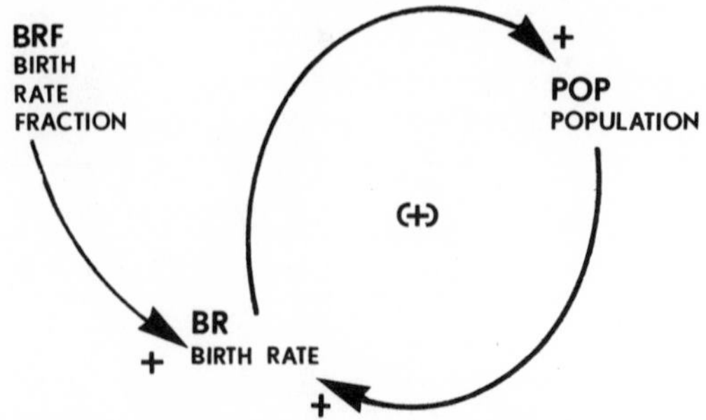

FIGURE E4-1 Population-birth rate loop

E4.1 Convert the causal-loop diagram to a flow diagram and DYNAMO equations. POP initially equals 10 people.

Offer intuitive responses to questions 2 through 8 before running the model on the computer.

E4.2 How long does POP take to double its initial value?

E4.3 What value does POP have in year 50?

E4.4 Sketch the time shape of POP and BR.

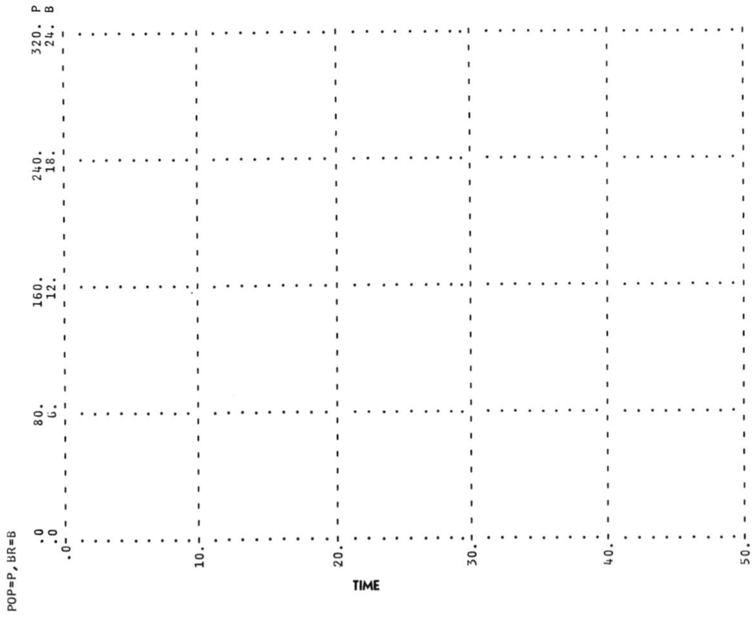

Let POP have an initial value of 5. How does system behavior change? Sketch the new time shape of POP and BR below.

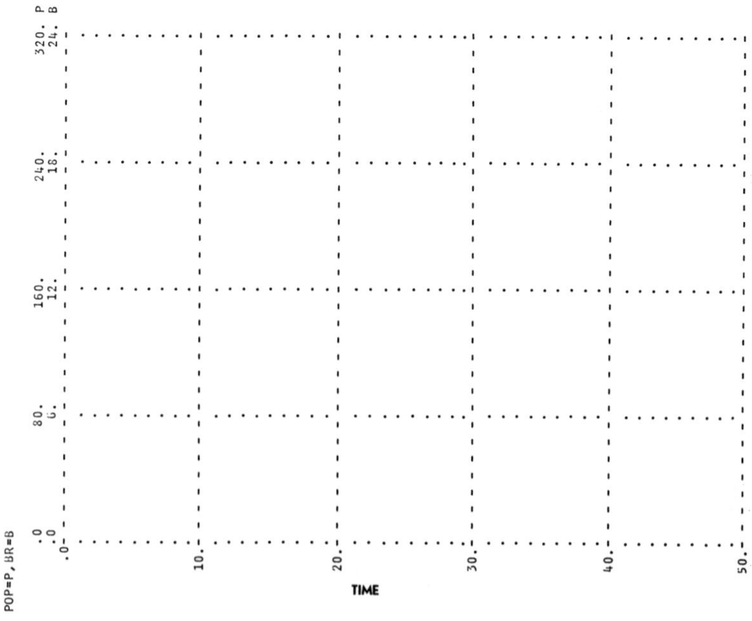

E4.6

If BRF equals 3.5 percent per year (0.035), how does system behavior change? Sketch the appropriate time shape of POP and BR below. Use an initial POP of 10.

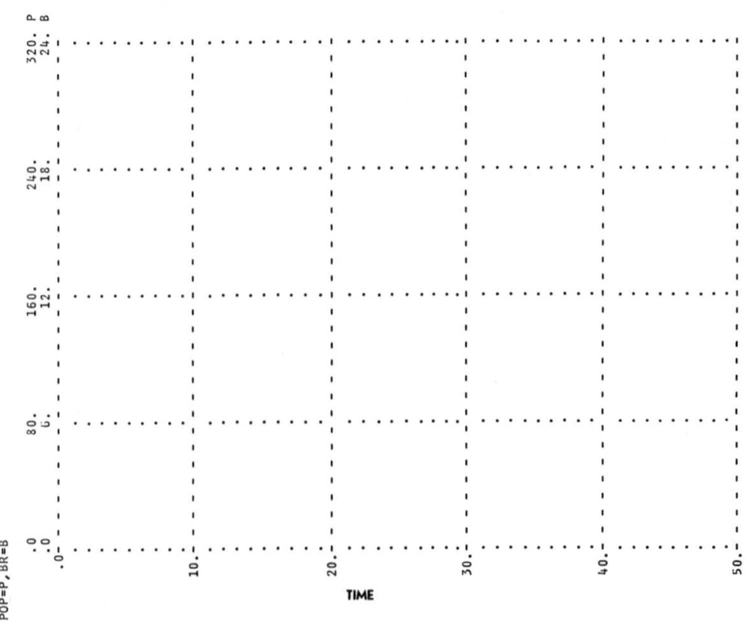

E4.7 If BRF equals 14 percent per year (0.14), how does system behavior change? Sketch the appropriate time shape of POP and BR below. Use an initial POP of 10.

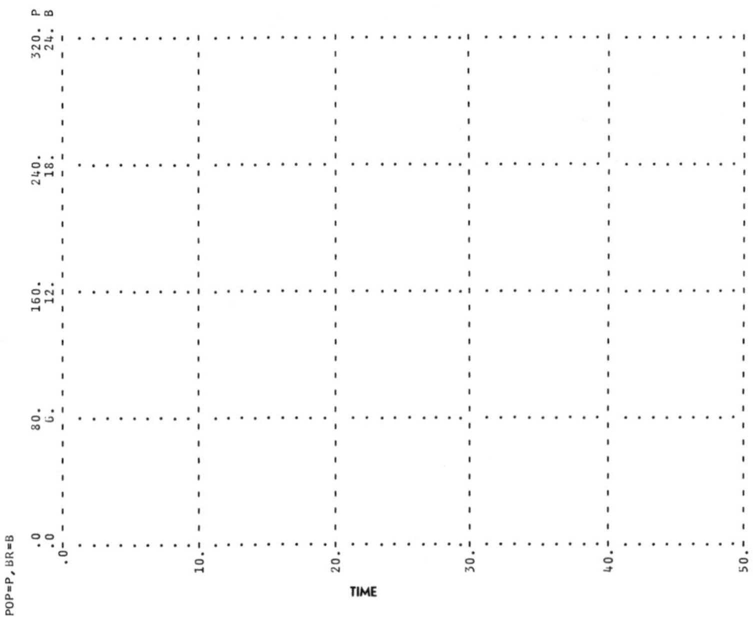

E4.8 Imagine a pond on which a water lily is growing. The lily plant doubles in size each day. If the lily grows unchecked, it will completely cover the pond in 30 days and choke off the other forms of life in the water. For a long time the lily plant seems small. We do not worry about cutting back the lily until it covers half the pond. How much time will we have left to cut back the lily? (This riddle comes from D.H. Meadows, et al., The Limits to Growth (New York: Universe Books, 1972), p. 29.)

E4.9 Check your answers to the preceding questions on the computer. Assume a DT of 0.1, a LENGTH of 50, and a PLTPER of 1.

Solution 4

POPULATION GROWTH MODEL

Flow diagram: **S4.1**

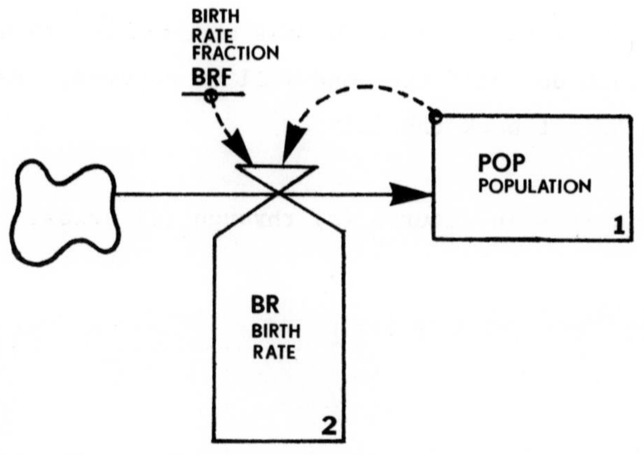

DYNAMO equations:

```
POP.K=POP.J+(DT)(BR.JK)                                   1, L
POP=10                                                    1.1, N
     POP     - POPULATION (PEOPLE)
     BR      - BIRTH RATE (PEOPLE/YEAR)
BR.KL=BRF*POP.K                                           2, R
BRF=.07                                                   2.1, C
     BR      - BIRTH RATE (PEOPLE/YEAR)
     BRF     - BIRTH RATE FRACTION (FRACTION/YEAR)
     POP     - POPULATION (PEOPLE)
DT=.1                                                     2.2, C
PLTPER=1                                                  2.3, C
LENGTH=50                                                 2.4, C
PLOT POP=P(0,320)/BR=B(0,24)
```

S4.2 Doubling time equals approximately 10 years since the time constant equals 1/BRF or roughly 14 years.

S4.3 Approximately 320.

S4.4 See Figure (a) in S4.9.

S4.5 See Figure (b) in S4.9.

S4.6 See Figure (c) in S4.9.

S4.7 See Figure (d) in S4.9.

S4.8 If the lily covers the pond in 30 days and doubles in size every day, then on the 29th day half the pond will be covered. We will have one day left to cut back the lily.

S4.9 See simulation runs in figures (a) through (d) below.

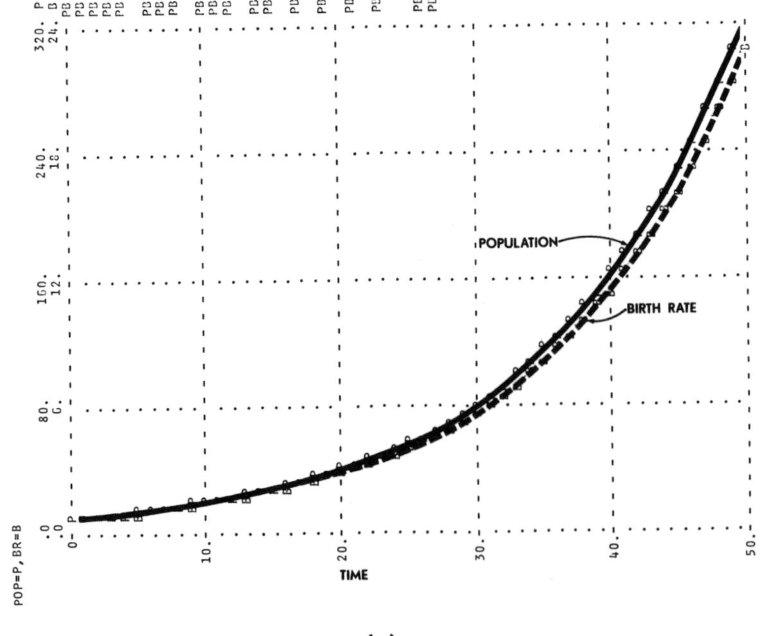

(a)

Base run of population model

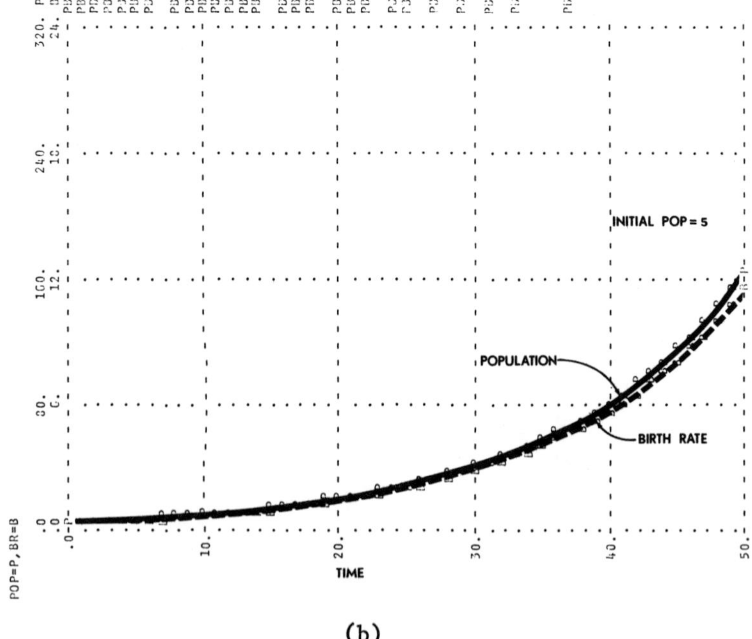

(b)
Initial POP = 5

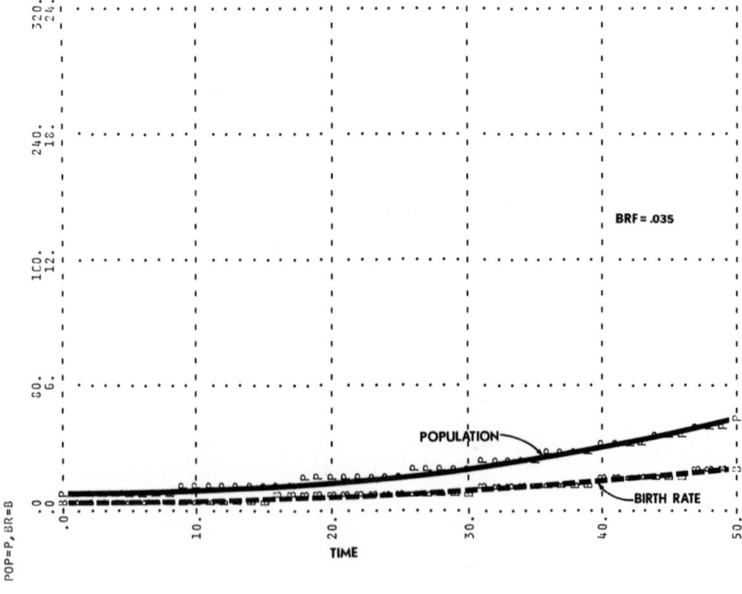

(c)
BRF = 0.035

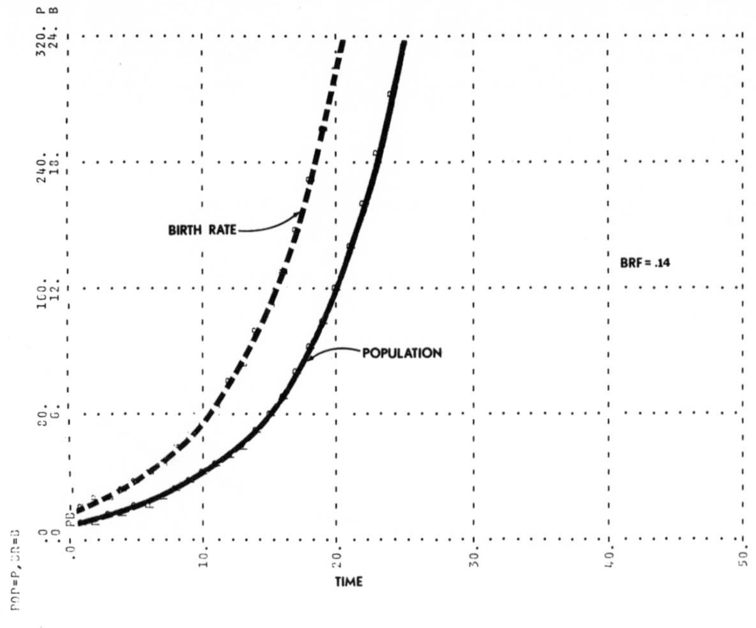

(d)

BRF = 0.14

Exercise 5
Negative Feedback: Application to Population Decay

This exercise introduces the negative feedback loop: the analog of the single-level positive loop of Exercise 4. A simple population decline model provides an example of negative feedback for Exercise 5. Attempt this exercise before reading Chapter 3.

Time Required: One hour.

POPULATION DECAY MODEL

Consider the simple population-death rate loop in Figure E5-1. For simplicity assume a fixed percentage of the population POP dies each year. This fixed percentage is the death rate fraction DRF.

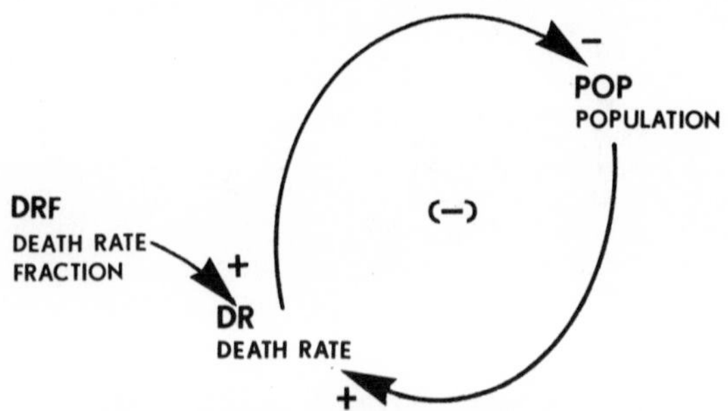

FIGURE E5-1 Population-death rate loop

Answer the following questions before running the model on the computer. Intuitive responses to questions 2 through 7 are sufficient.

E5.1 Convert the causal-loop diagram to a flow diagram and DYNAMO equations. Use an initial POP value of 320. The DRF value is 7 percent per year (0.07).

E5.2 How long will POP take to reach half its initial value?

E5.3 What value has POP in year 50?

E5.4 Sketch the time shape of POP and DR.

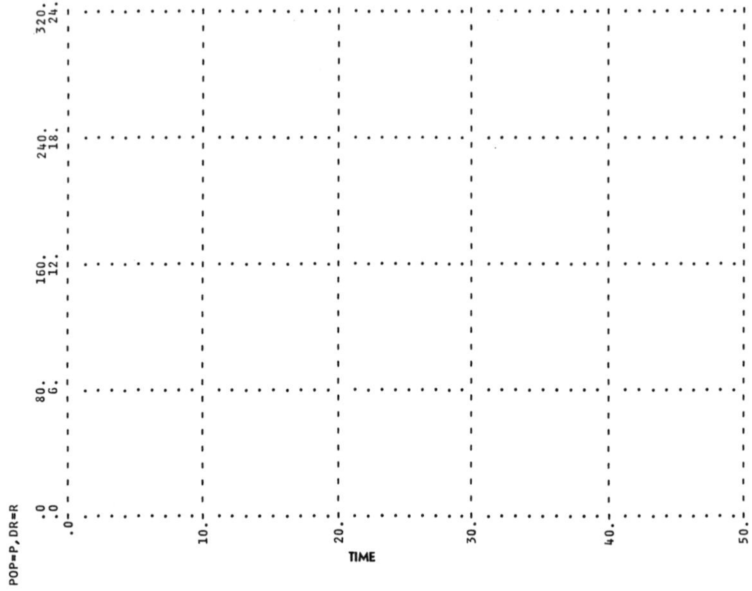

Let POP have an initial value of 160. How does system behavior change? Sketch the time shape of POP and DR below.

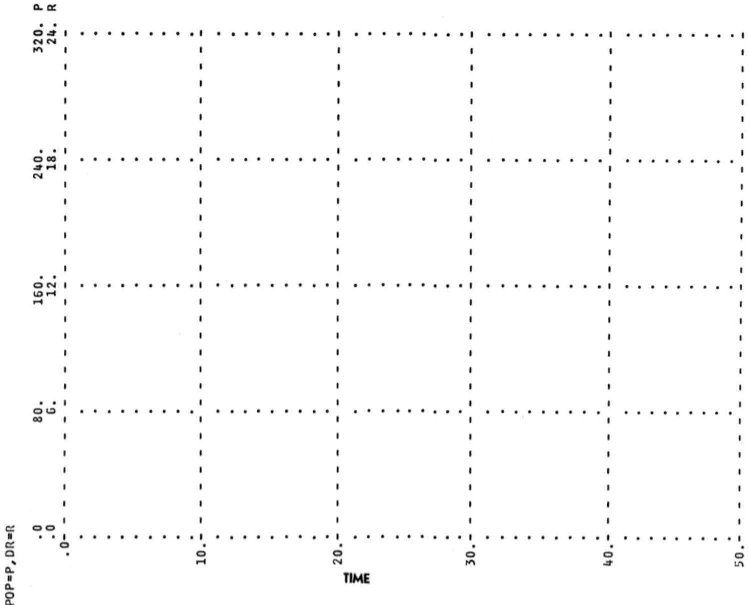

E5.6

If DRF equals 3.5 percent per year (0.035), how does system behavior change? Sketch the appropriate time shape of POP and DR below. Use an initial POP of 320.

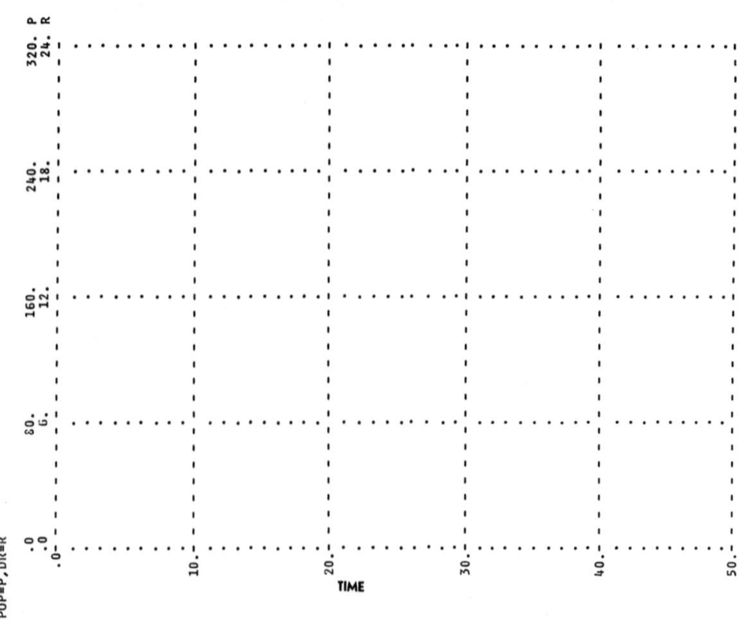

E5.7 If DRF equals 14 percent per year (0.14), how will system behavior change? Sketch the time shape of POP and DR below. Use an initial POP of 320.

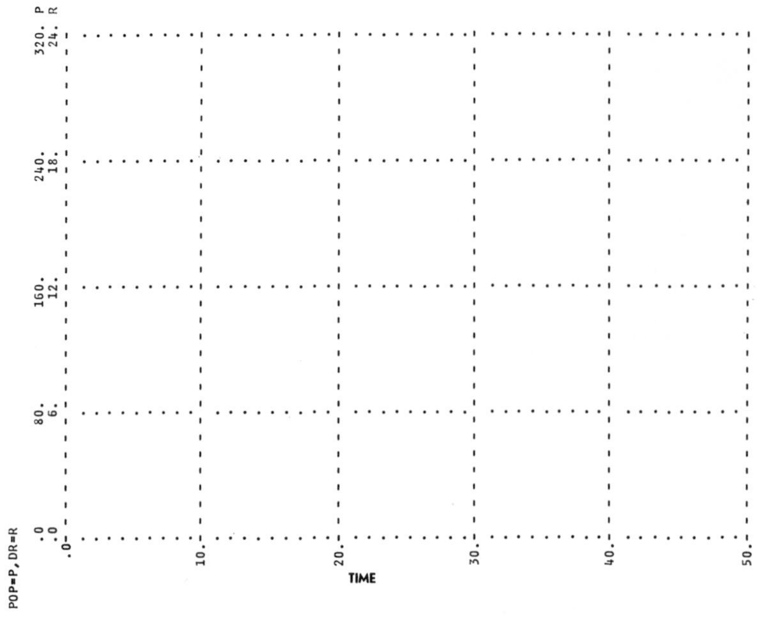

E5.8 Using the computer, check your answers to the preceding questions. Use a DT of 0.1, a LENGTH of 50, and a PLTPER of 1.

Solution 5

POPULATION DECAY MODEL

Flow diagram: **S5.1**

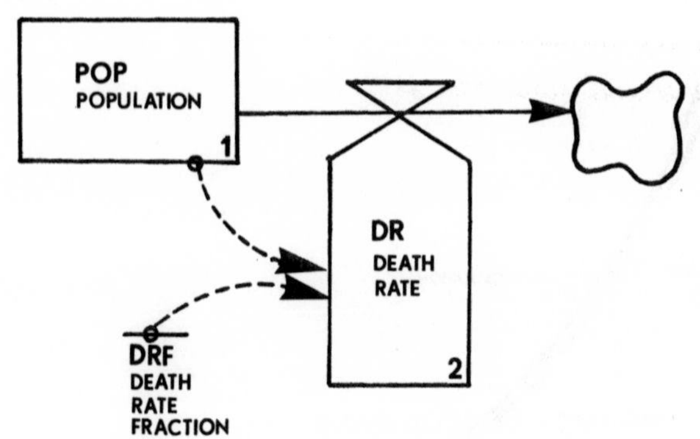

DYNAMO equations:

```
POP.K=POP.J+(DT)(-DR.JK)                                1, L
POP=320                                                 1.1, N
     POP     - POPULATION (PEOPLE)
     DR      - DEATH RATE (PEOPLE/YEAR)

DR.KL=DRF*POP.K                                         2, R
DRF=.07                                                 2.1, C
     DR      - DEATH RATE (PEOPLE/YEAR)
     DRF     - DEATH RATE FRACTION (FRACTION/YEAR)
     POP     - POPULATION (PEOPLE)
DT=.1                                                   2.2, C
PLTPER=1                                                2.3, C
LENGTH=50                                               2.4, C
PLOT POP=P(0,320)/DR=R(0,24)
```

S5.2 Half-life equals approximately 10 years since the time constant equals 1/DRF or roughly 14 years.

S5.3 Approximately zero.

S5.4 See Figure (a) in S5.8.

S5.5 See Figure (b) in S5.8.

S5.6 See Figure (c) in S5.8.

S5.7 See Figure (d) in S5.8.

S5.8

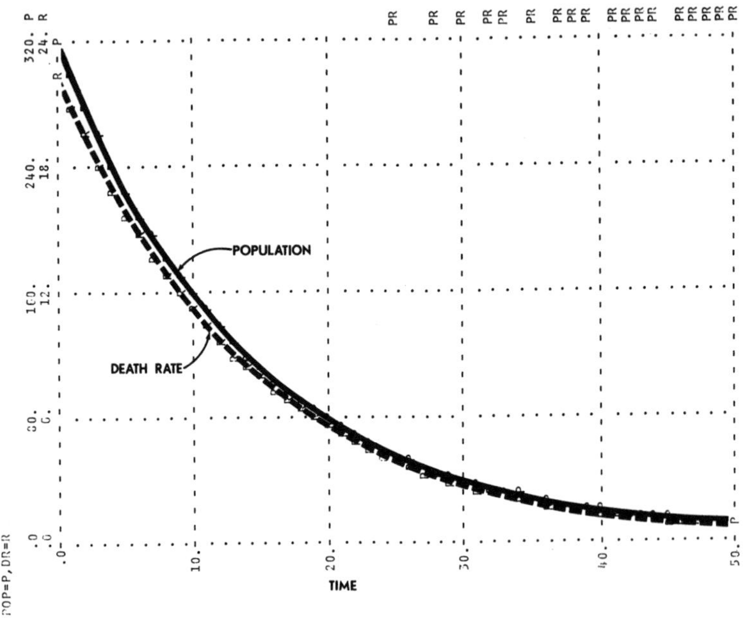

(a)

Base run of population model

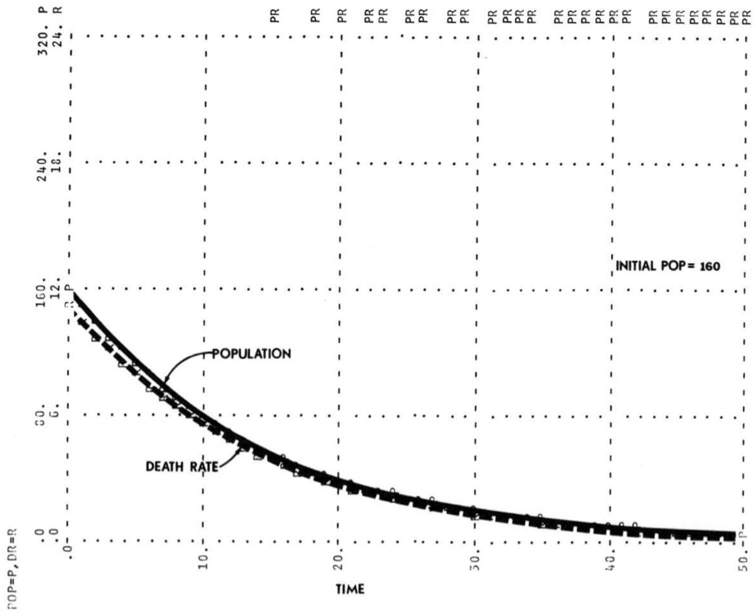

(b)
Initial POP = 160

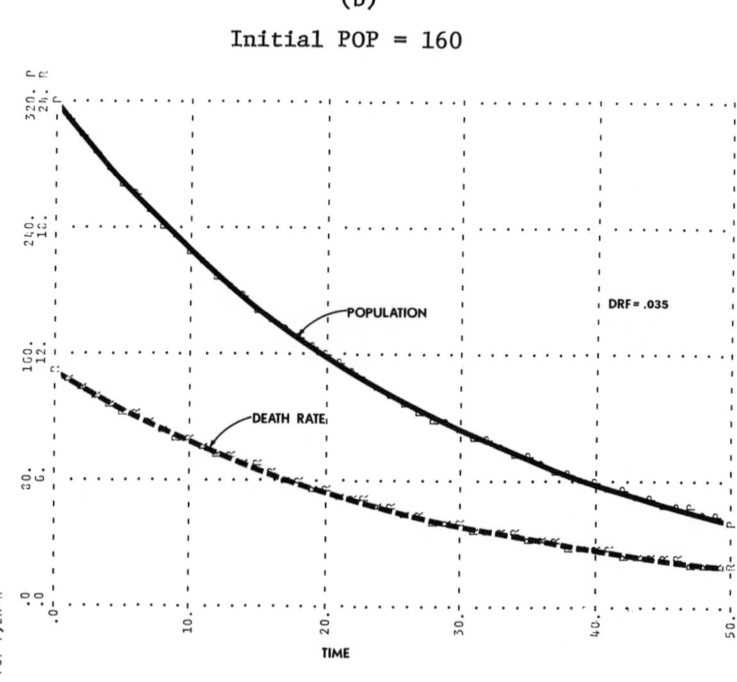

(c)
DRF = 0.035

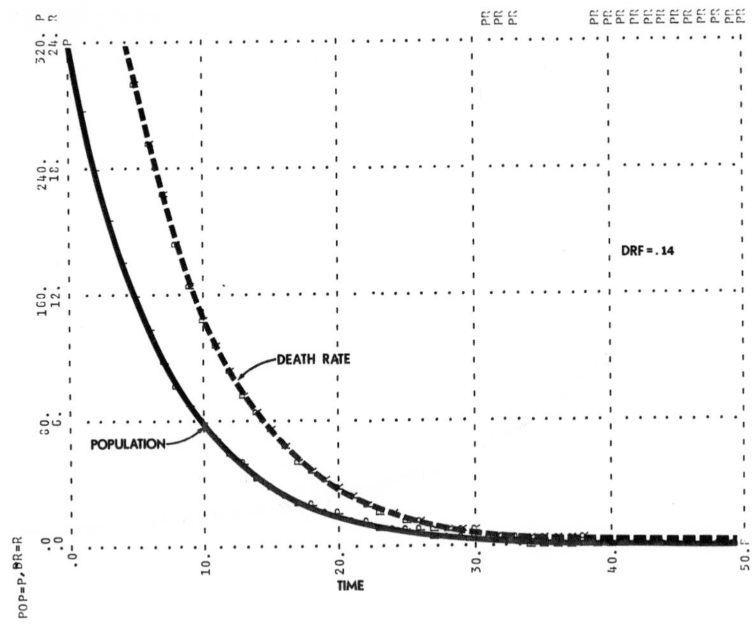

(d)

DRF = 0.14

Exercise 6
Negative Feedback: Application to Inventory Control

The reader should attempt Exercise 6 before reading the case examples in Chapter 3 (sections 3.10, 3.11, and 3.12). This exercise focuses on the basic goal-seeking negative loop using an inventory control scheme.

Time Required: One hour.

INVENTORY MODEL

A dealer likes to maintain a desired level of inventory. When his stock of goods falls below the desired level, he places orders to the distributor to replenish his supply. He stops ordering when his stock reaches the desired level. With too much inventory, he sends the excess back to the distributor (assume this excess never occurs). The causal-loop diagram of the dealer's inventory control system appears in Figure E6-1.

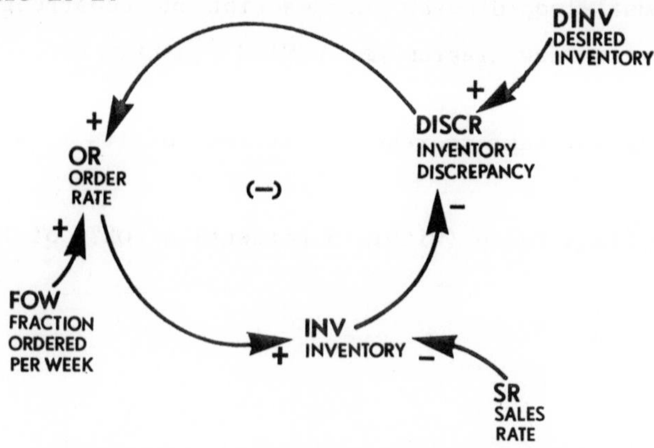

FIGURE E6-1 Inventory control loop

Sales, which deplete the inventory INV, depend on market conditions outside the system boundary. Assume that the dealer has no influence on demand for his goods. The order rate OR replenishes the inventory stock and depends on the dealer's ordering policy. The causal-loop diagram represents one simple ordering scheme. The order rate OR depends on the discrepancy DISCR between INV and desired inventory DINV, as well as on the fraction ordered per week FOW. The fraction ordered per week FOW, a constant with a value of 0.5 for this dealer, measures how quickly the dealer responds to a discrepancy between DINV and INV. If INV suddenly drops, then the dealer places an order equal to 50 percent of the difference between DINV and INV the first week and 50 percent of the difference the next week. Two weeks would pass before the dealer could make up the discrepancy if there were no further changes in inventory. For simplicity, we ignore all delays in shipping, handling goods, and recognizing changes in inventory. An order is immediately filled. The dealer's DINV and his initial INV both equal 200. The sales rate SR initially equals zero and the system rests in equilibrium.

How well does this inventory control policy respond to a sudden rise (step) in sales from zero to 20 items per week? Let the step begin in week 4 and continue for the remaining 16 weeks of the 20-week simulation. Answer the questions below prior to running the model on the computer. Intuitive guesses are sufficient.

E6.1 From the causal-loop diagram and description, construct a corresponding simple flow diagram and DYNAMO equation set.

E6.2 How long will INV take to reach 95 percent of its final value?

E6.3 What is the final value (within 5 percent) of OR? of INV?

Sketch the time shape of OR and INV.

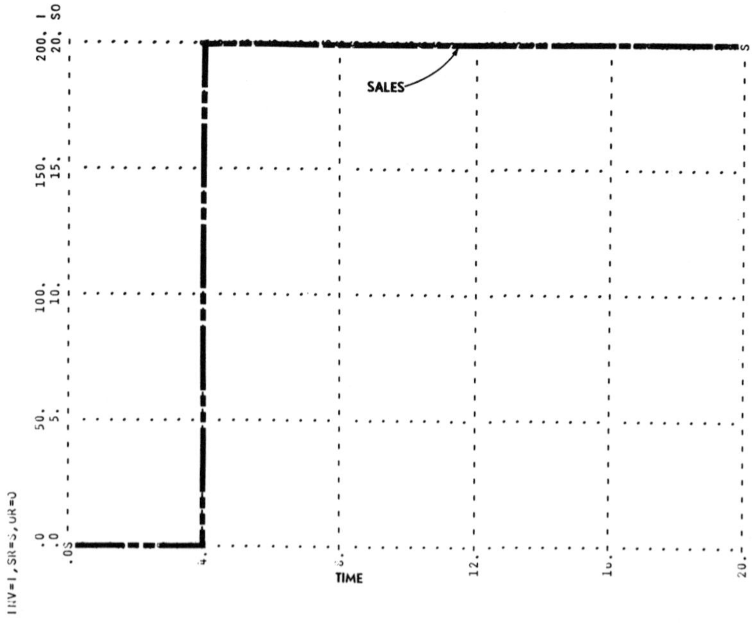

If the fraction ordered per week FOW drops to 0.25, how will system behavior change? Sketch the time shape of OR and INV below. Indicate on your sketch the final values of INV and OR and how much time will pass for them to reach their final values.

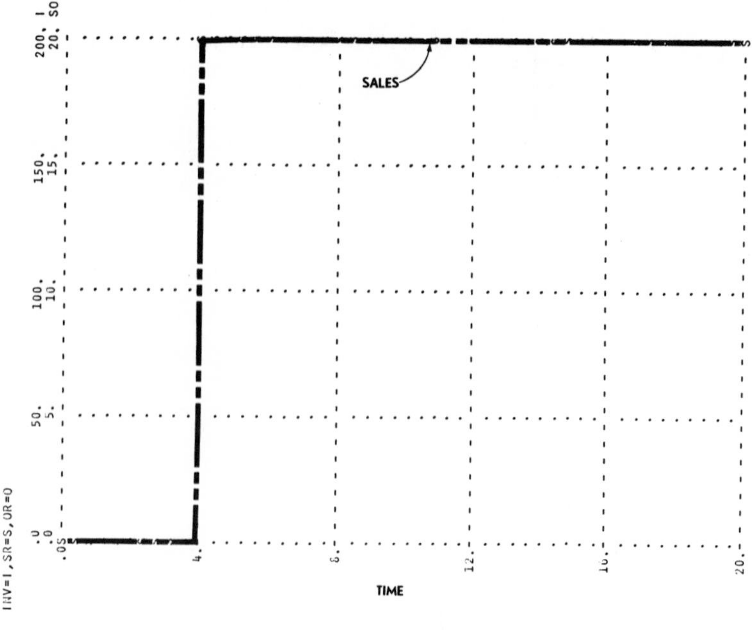

E6.6 If the fraction ordered per week FOW increases to 0.75, how will system behavior change? Sketch the time shape of OR and INV below. Indicate on your sketch the final values of INV and OR and how much time will pass for them to reach their final values.

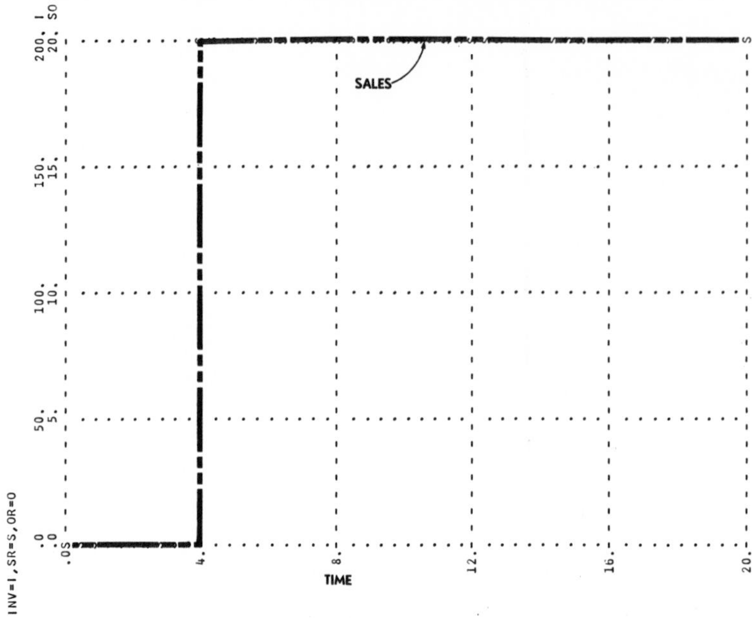

E6.7 If FOW equals 0.5 and the step in sales equals 40 instead of 20, how will system behavior change? Sketch the time shape of OR and INV below.

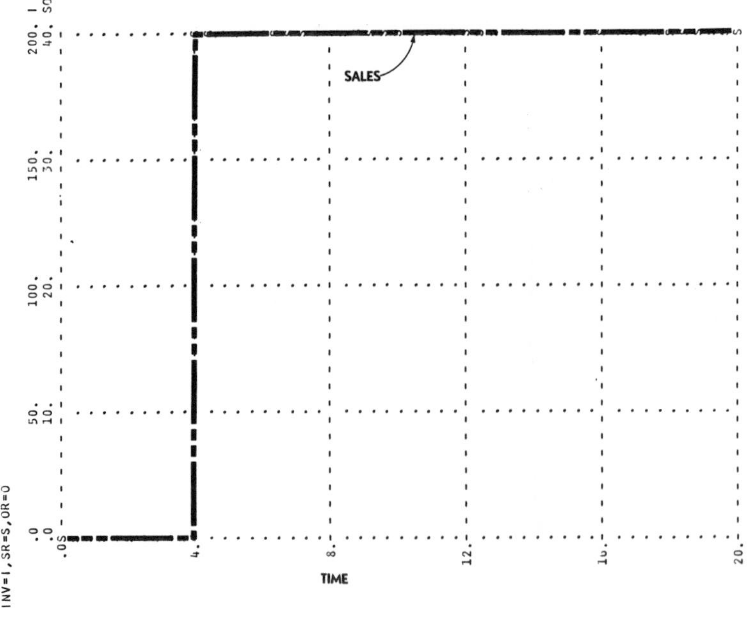

Does this simple ordering scheme fulfill its purpose of maintaining a desired inventory and meeting orders? **E6.8**

Use the computer to check your answers to the preceding questions. Use a DT of 0.1, a LENGTH of 20, and a PLTPER of 0.4. **E6.9**

Solution 6

Chapter 3 (section 3.1) presents a detailed analysis of this exercise.

INVENTORY MODEL

Flow diagram: **S6.1**

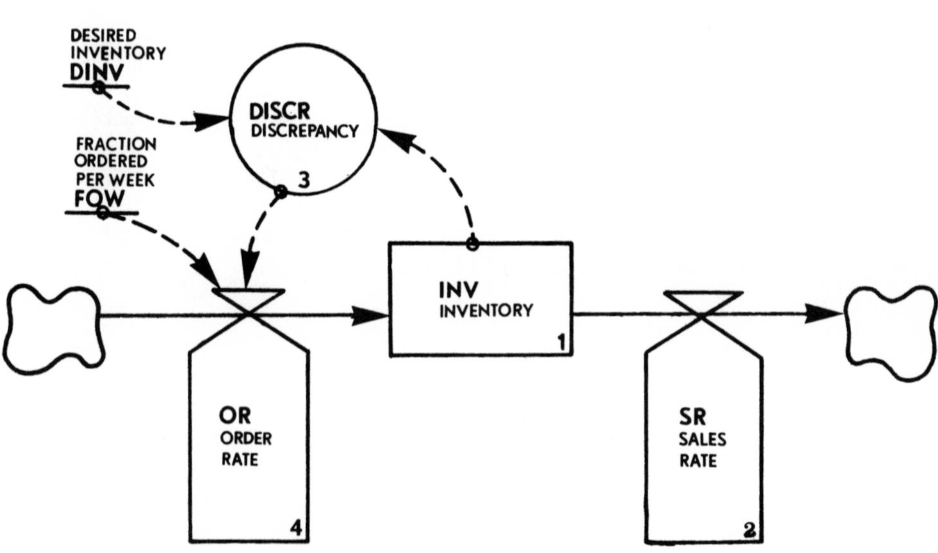

DYNAMO equations:

```
INV.K=INV.J+(DT)(OR.JK-SR.JK)                           1, L
INV=DINV                                                1.1, N
     INV      - INVENTORY (UNITS)
     OR       - ORDER RATE (UNITS/WEEK)
     SR       - SALES RATE (UNITS/WEEK)
     DINV     - DESIRED INVENTORY (UNITS)

SR.KL=STEP(20,4)                                        2, R
     SR       - SALES RATE (UNITS/WEEK)

DISCR.K=DINV-INV.K                                      3, A
DINV=200                                                3.1, C
     DISCR    - DISCREPANCY (UNITS)
     DINV     - DESIRED INVENTORY (UNITS)
     INV      - INVENTORY (UNITS)

OR.KL=FOW*DISCR.K                                       4, R
FOW=.5                                                  4.1, C
     OR       - ORDER RATE (UNITS/WEEK)
     FOW      - FRACTION ORDERED PER WEEK (FRACTION/WEEK)
     DISCR    - DISCREPANCY (UNITS)
DT=.1                                                   4.2, C
PLTPER=.5                                               4.3, C
LENGTH=20                                               4.4, C
PLOT INV=I(0,200)/SR=S,OR=O(0,20)
```

S6.2 Time constant equals 1/FOW or 2 weeks. In approximately 6 weeks INV reaches 95 percent of final value.

S6.3 Final OR value equals 20. Final INV value equals 160.

S6.4 See Figure (a) in S6.9

S6.5 See Figure (b) in S6.9

S6.6 See Figure (c) in S6.9

S6.7 See Figure (d) in S6.9

S6.8 The dealer can never replace the stock depleted during the first 6 weeks while orders are still below sales. Hence, this ordering policy does not allow the dealer to maintain his desired level of inventory. A continuing increase in sales would eventually exhaust his inventory.

S6.9

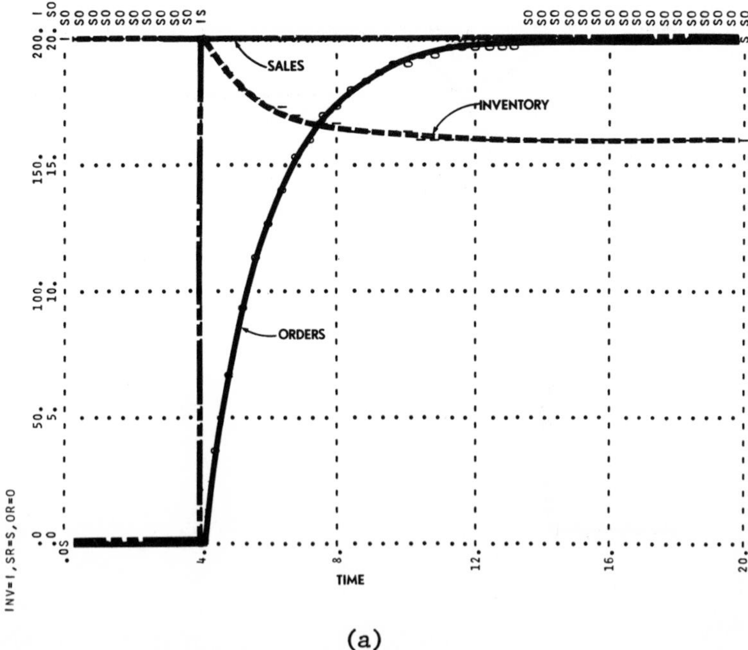

(a)

Base run of inventory model

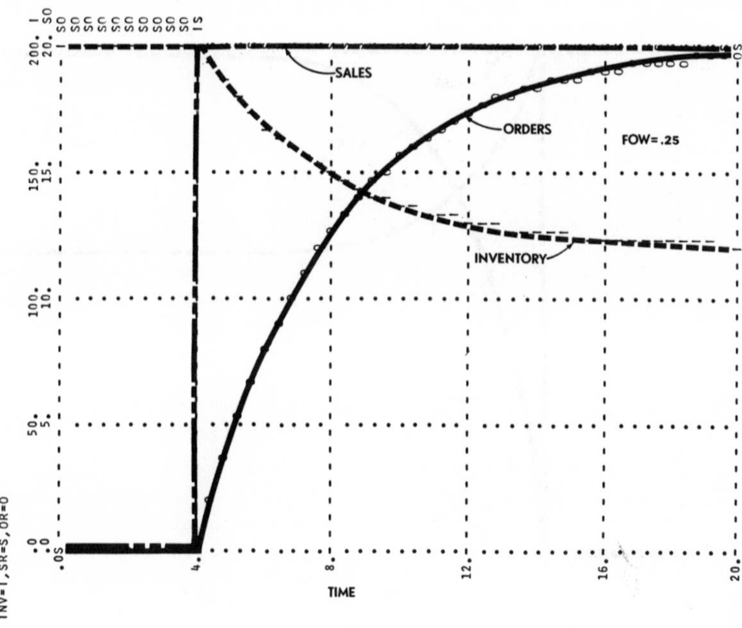

(b)

FOW = 0.25

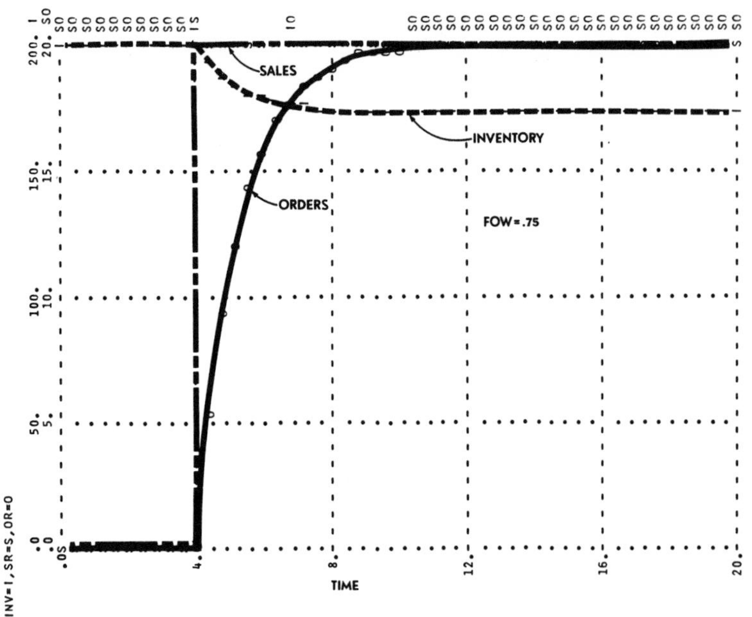

(c)

FOW = 0.75

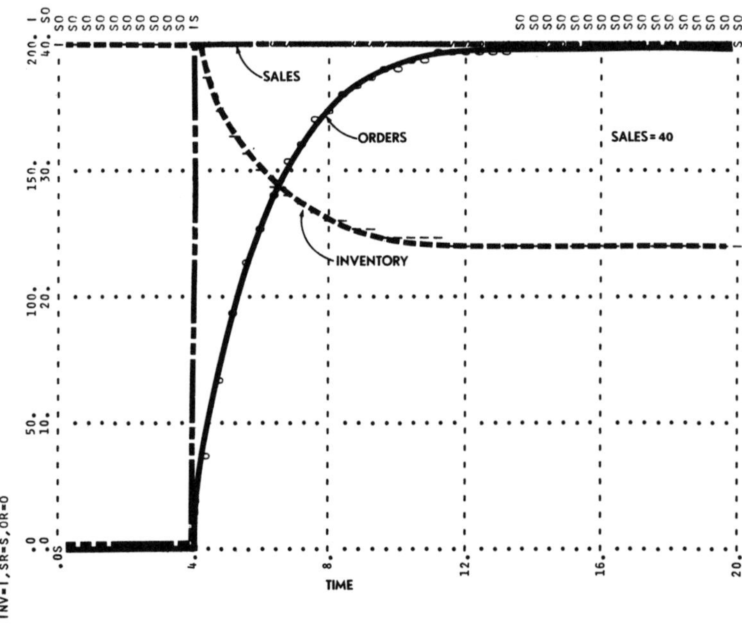

(d)

Sales = 40

Exercise 7
First-Order Linear Systems

(with Dennis L. Meadows)

Exercise 7 provides an in-depth presentation of elementary first-order linear structures found in Chapters 2 and 3. This optional exercise uses an analytical approach to investigate the behavior of both positive and negative feedback structures and requires previous knowledge of elementary calculus.

Time Required: Two hours.

FIRST-ORDER NEGATIVE STRUCTURE

Consider the first-order negative system shown below:

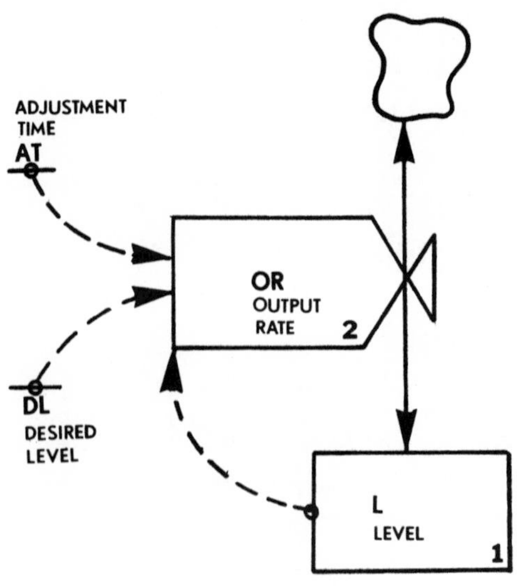

FIGURE E7-1

```
L.K=L.J+(DT)(OR.JK)                                      1, L
L=L(0)                                                   1.1, N
    L      - LEVEL (UNITS)
    OR     - OUTPUT RATE (UNITS/TIME)

OR.KL=(1/AT)(DL-L.K)                                     2, R
DL=CONSTANT                                              2.1, C
AT=CONSTANT                                              2.2, C
    OR     - OUTPUT RATE (UNITS/TIME)
    AT     - ADJUSTMENT TIME (TIME)
    DL     - DESIRED LEVEL (UNITS)
    L      - LEVEL (UNITS)
```

E7.1 Assume the system has been in equilibrium prior to t = 0.

a) What values had OR at t = -10 and t = -5?

b) What values had L at t = -10 and t = -5?

E7.2 Let L(0) = 2, DL = 10, and AT = 3.

a) What equilibrium values will L and OR have?

b) In general, what equilibrium conditions will the simple first-order structure shown above have in terms of L, OR, DL, and AT?

We now wish to investigate the transient behavior of this first-order structure, that is, to study the system behavior over time.

We know that the level represents the process of integration, and the first-order differential equation provides an equivalent representation of this system:

$$dL(t)/dt = (DL-L(t))/AT \qquad (7.1)$$

Given the initial condition L = L(0) at time t = 0, this equation has the general solution (for t > 0)

$$L(t) = DL+(L(0)-DL)e^{(-t/AT)} \qquad (7.2)$$

where e is the base of the natural logarithm. This expression gives a complete description of the time behavior of the system given DL, AT, and L(0). (Equation (7.2) is equivalent to equation (3.1) derived in Chapter 3.)

E7.3 Verify that the solution L(t) in equation (7.2) actually solves the differential equation (7.1) and also satisfies the initial condition.

Plot the time shape of L (using the graph below) for DL = 1, AT = 3 and L(0) = 5. **E7.4**

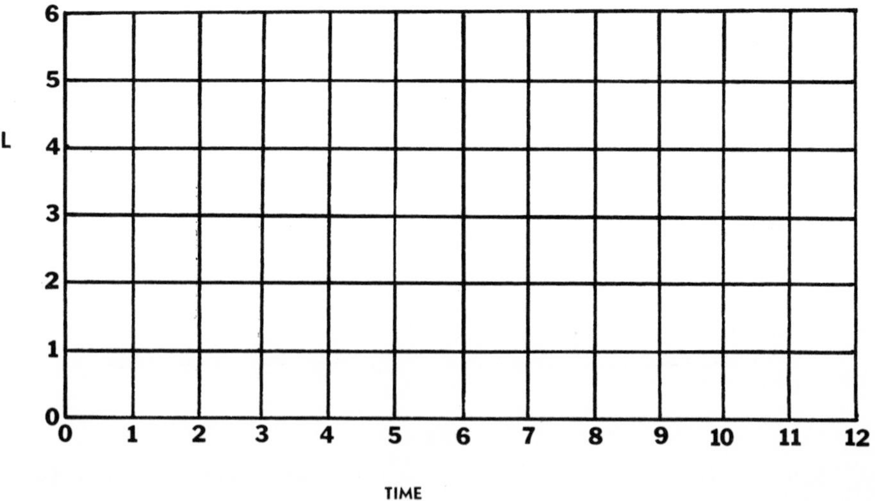

Looking at the special case in question E7.4, answer the following questions: **E7.5**

a) What is the initial slope of L?

b) How far would L differ from its equilibrium value if its initial rate of change continued for a length of time equal to AT (i.e., equal to 3)?

c) How closely does L actually approach its new equilibrium value in the first AT time period?

d) How do you explain the difference?

e) What is the slope of L at t = AT? t = 2*AT? How closely does L approach its equilibrium value if these rates of change continue for one time period equal to AT?

Using the general solution for L(t) in equation (7.2), give the general answers to questions E7.5a) through E7.5c). **E7.6**

E7.7 Sketch below the three transient responses of the first-order negative structure. The three modes arise when

i) L(0) = DL ii) L(0) > DL iii) L(0) < DL

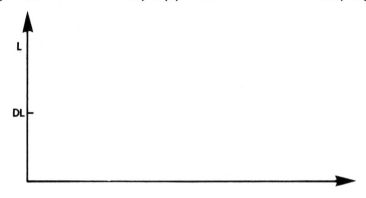

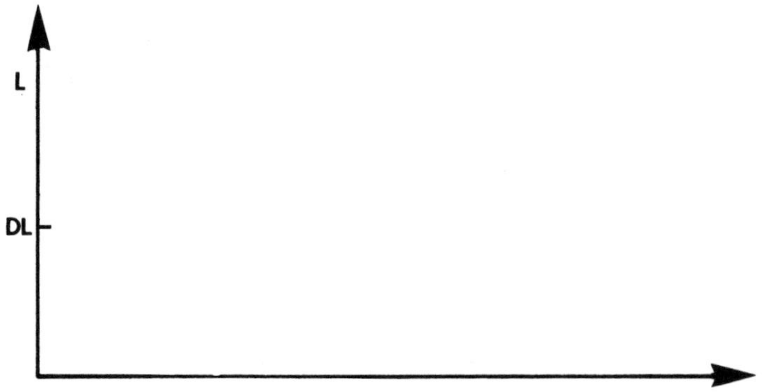

E7.8 The figure below shows the behavior of a system structurally identical to the structure shown in Figure E7-1. The system below has L(0) = 0 and DL = 5. Using the answer to question E7.6, can you graphically determine the time constant AT of this system?

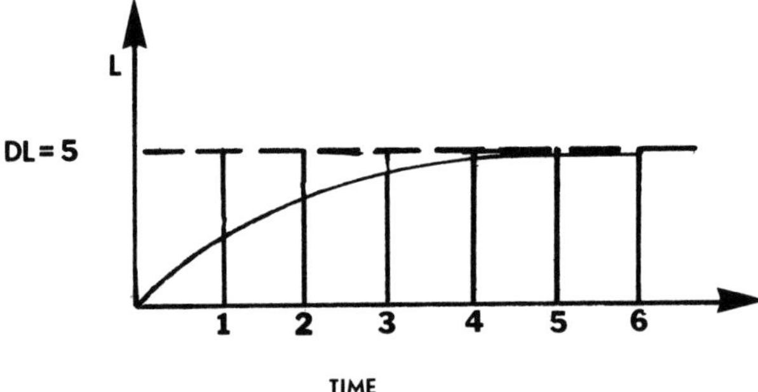

FIRST-ORDER NEGATIVE STRUCTURE WITH DL = 0

Consider a first-order negative system, where DL = 0, similar to that shown below with equations:

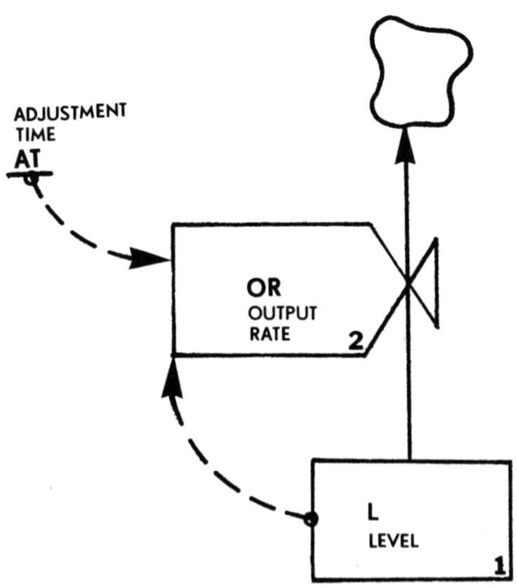

FIGURE E7-2

```
L.K=L.J+(DT)(OR.JK)                                  1, L
L=L(0)                                               1.1, N
    L      - LEVEL (UNITS)
    OR     - OUTPUT RATE (UNITS/TIME)

OR.KL=(1/AT)(-L.K)                                   2, R
AT=CONSTANT                                          2.1, C
    OR     - OUTPUT RATE (UNITS/TIME)
    AT     - ADJUSTMENT TIME (TIME)
    L      - LEVEL (UNITS)
```

E7.9

a) Give the general solution for this system when L = L(0) at t = 0. Use equation (7.2).

b) Plot the time-shape for L for L(0) > 0.

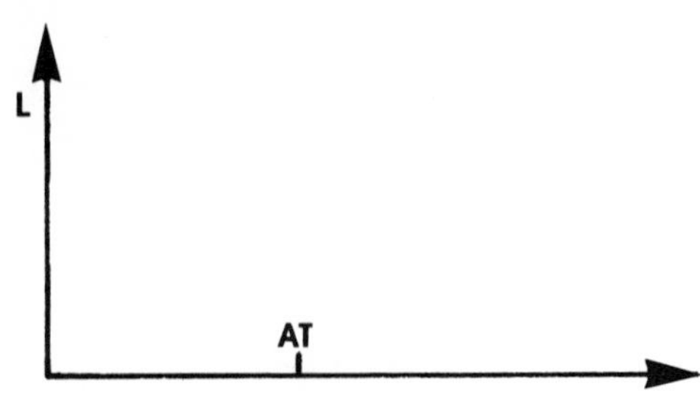

E7.10 We can convert the preceding negative first-order system to a positive first-order system by changing the equation for OR to:

R OR.KL = (1/AT)(L.K) Rate (Units/time)

a) Give the general solution for the resulting system when L = L(0) at t = 0. Use equation (7.2).

b) Plot the time-shape for L for L(0) > 0.

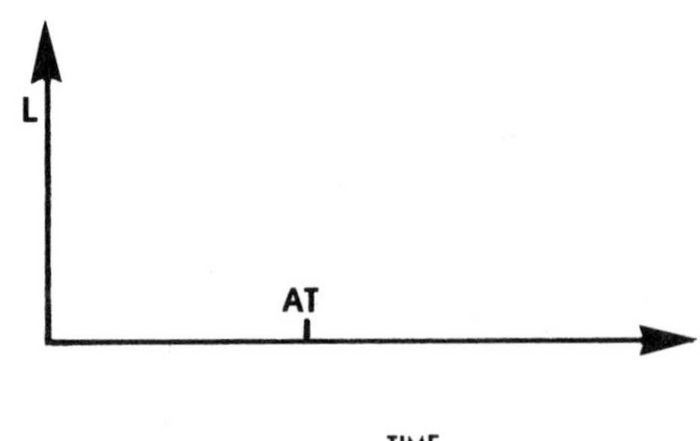

c) Is this system goal-seeking?

RESPONSE TO STEP INPUT

We now wish to consider the first-order negative loop's response to an exogenous input. Consider the system shown below:

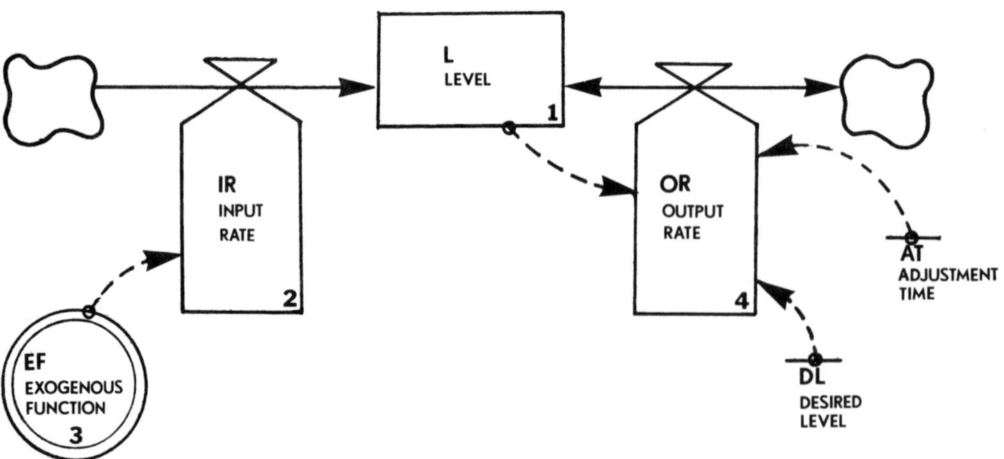

FIGURE E7-3

```
L.K=L.K+(DT)(IR.JK+OR.JK)                                1, L
L=L(0)                                                   1.1, N
     L       - LEVEL (UNITS)
     IR      - INPUT RATE (UNITS/TIME)
     OR      - OUTPUT RATE (UNITS/TIME)

IR.KL=EF.K                                               2, R
     IR      - INPUT RATE (UNITS/TIME)
     EF      - EXOGENOUS FUNCTION (UNITS/TIME)

EF.K=EXOGENOUS                                           3, A
     EF      - EXOGENOUS FUNCTION (UNITS/TIME)

OR.KL=(1/AT)(DL-L.K)                                     4, R
DL=CONSTANT                                              4.1, C
AT=CONSTANT                                              4.2, C
     OR      - OUTPUT RATE (UNITS/TIME)
     AT      - ADJUSTMENT TIME (TIME)
     DL      - DESIRED LEVEL (UNITS)
     L       - LEVEL (UNITS)
```

Assume EF is a step function such that IR = 0 prior to time equals zero and IR = A after that time:

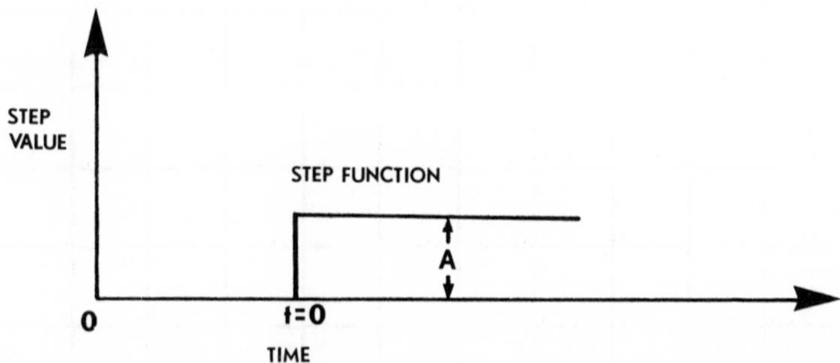

E7.11 Assume the system above remains in equilibrium prior to t = 0.

a) What value has OR prior to t = 0?

b) Let DL = 10 and AT = 3. What new equilibrium value has OR after t = 0? What value must L have in this condition? (Assume A = 1.)

c) Let DL and AT be constants with unspecified values. Express the equilibrium level of L in terms of DL, A, and AT after applying the step input.

The first-order differential equation provides an equivalent representation of the preceding system:

$$dL(t)/dt = ((DL-L(t))/AT)+IR(t) \tag{7.3}$$

where

$$IR(t) = \begin{cases} 0 & t < 0 \\ A & t \geq 0 \end{cases}$$

Given that L = L(0) for t < 0, this equation has the general solution (for $t \geq 0$)

$$L(t) = (DL+A*AT)+(L(0)-(DL+A*AT))*e^{(-t/AT)} \tag{7.4}$$

This expression gives a complete description of system behavior after encountering the step function.

E7.12 Verify that the solution L(t) in equation (7.4) actually solves the differential equation (7.3) and also satisfies the initial condition.

E7.13 Using equation (7.4) plot the time-shape of L for DL = 2, AT = 4, A = 2, and L(0) = 1:

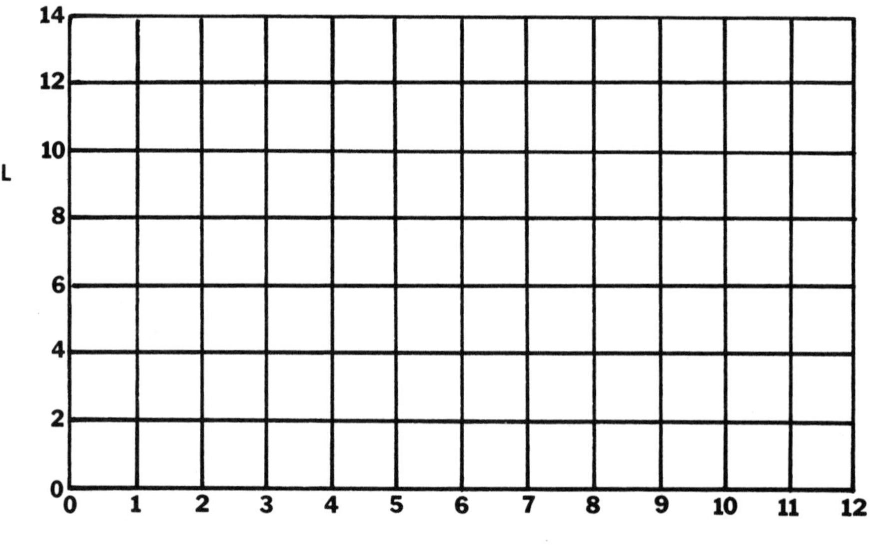

E7.14 Explain why the goal-seeking negative loop allows L to remain larger than DL in the new equilibrium.

Write the expression describing OR as a function of time. Use equation (7.4). **E7.15**

a) Sketch IR and OR from question E7.15 in general as functions of time on the graph below. Assume A > 0 and that the system remained in equilibrium at t < 0 with L(0) = DL. **E7.16**

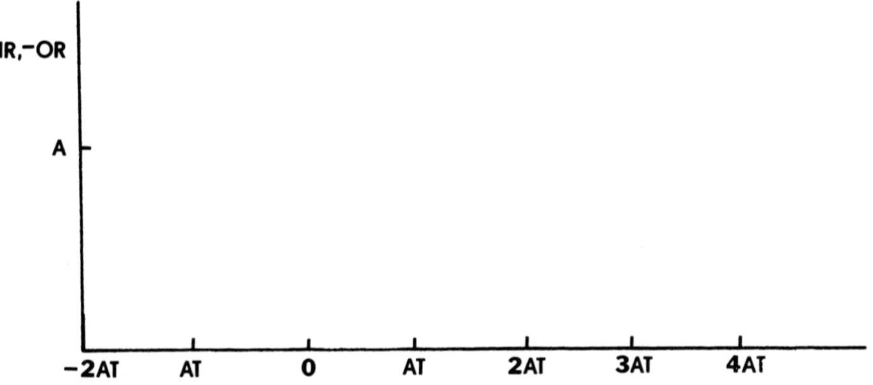

b) Sketch the time-shape of L.

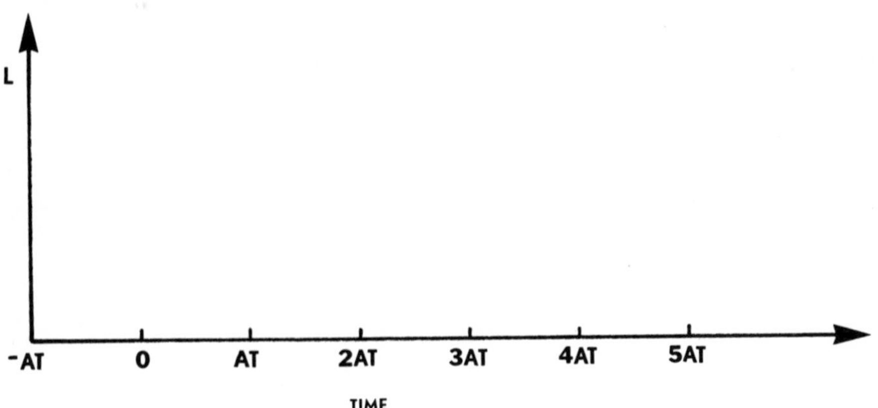

Solution 7

FIRST-ORDER NEGATIVE STRUCTURE

a) OR = 0 at all $t < 0$ **S7.1**

b) OR = 0 implies L = DL at all $t < 0$

a) L = DL = 10, OR = 0 **S7.2**

b) L = DL, OR = 0 for AT not zero.

Differentiating equation (7.2) yields: **S7.3**

$$dL(t)/dt = (-1/AT)(L(0)-DL)e^{-t/AT}.$$

From equation (7.2):

$$(L(0)-DL)e^{-t/AT} = L(t)-DL.$$

Thus,

$$dL(t)/dt = (-1/AT)(L(t)-DL),$$

or,

$$dL(t)/dt = (DL-L(t))/AT.$$

Solving equation (7.2) at t = 0

$$L(t = 0) = DL+(L(0)-DL)e^{\circ} = L(0).$$

S7.4

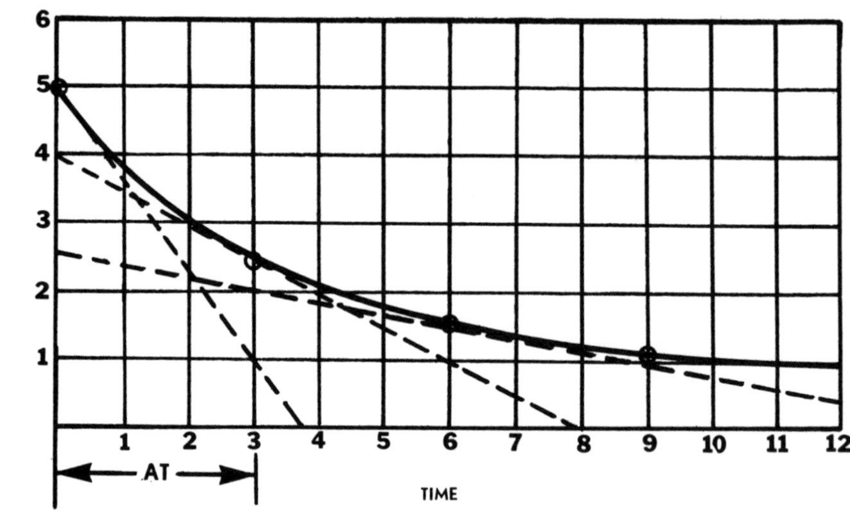

S7.5 a) -4/3

b) It would be at DL.

c) Approximately 63 percent of (L(0)-DL)

d) The rate changes value with the level.

e) Slope at t = AT equals 1.48/3 or 0.49. Slope at t = 2*AT equals 0.54/3 or 0.18. If these rates of change continued for a length of time equal to AT, L would reach DL.

S7.6 a) Initial slope at t = 0 equals $dL(0)/dt = (DL-L(0))/AT$.

b) At an arbitrary time t_1,

$$dL(t_1)/dt = (1/AT)(DL-L(0))e^{-t_1/AT}$$

if $dL(t_1)/dt$ continues for another interval equal to AT, L(t) will approximate

$$\begin{aligned} L(t_1+AT) &= L(t_1)+(dL(t_1)/dt)(AT) \\ &= DL+(L(0)-DL)e^{-t_1/AT}+(DL-L(0))e^{-t_1/AT} \\ &= DL \end{aligned}$$

c)
$$\begin{aligned} L(t = AT) &= DL+(L(0)-DL)e^{-1} \\ &= DL+0.368(L(0)-DL) \\ &= L(0)+0.632(DL-L(0)) \end{aligned}$$

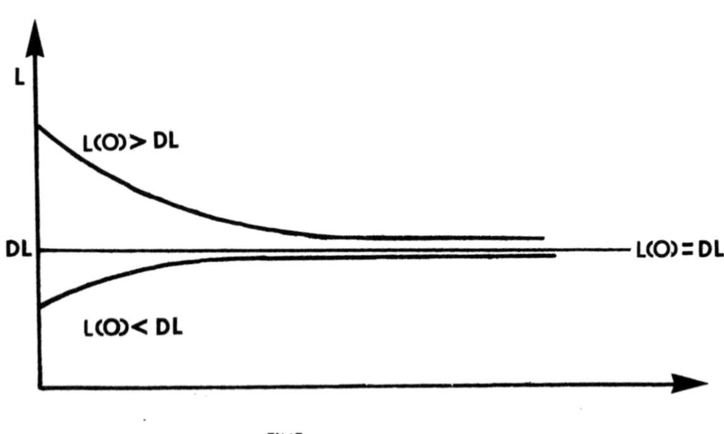

Since L reaches DL after an AT interval when following a tangent, then from the figure below, AT = 2. **S7.8**

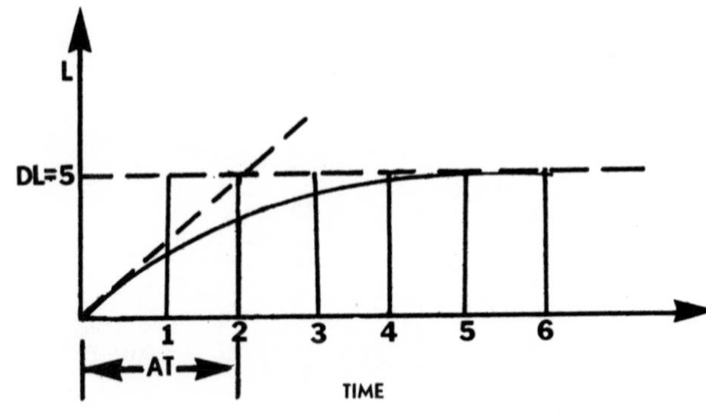

FIRST-ORDER NEGATIVE STRUCTURE WITH DL = 0

a) $L(t) = L(0)e^{(-t/AT)}$ **S7.9**

b)

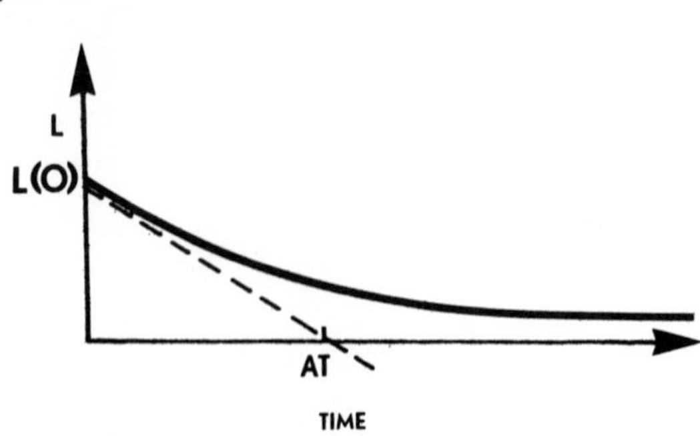

S7.10 a) $L(t) = L(0)e^{(t/AT)}$

b)

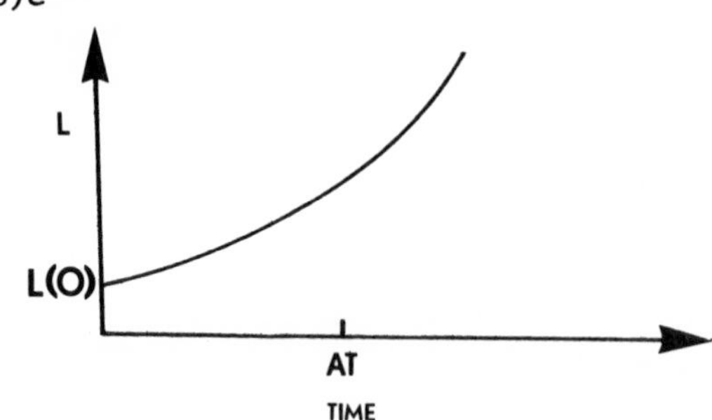

c) No

RESPONSE TO STEP INPUT

S7.11 a) For equilibrium, IR+OR = 0 implies OR = 0 at all t < 0.

b) IR = 1 implies OR = -1 at equilibrium. When OR = -1, then

(DL-L(t))/AT = (10-L(t))/3 = -1.

L then equals 13.

c) OR = -IR = -A implies -A = (DL-L(t))/AT

Hence,

L = DL+A*AT

S7.12 Differentiating equation (7.4) yields

$dL(t)/dt = (L(0)-(DL+A*AT))(-1/AT)e^{(-t/AT)}$

Substituting from equation (7.4)

((DL+A*AT)-L(t))/AT = ((DL-L(t))/AT)+A = ((DL-L(t))/AT)+IR(t)

Solving equation (7.4) at t = 0

L(0) = DL+A*AT+(L(0)-(DL+A*AT)) = L(0)

S7.13

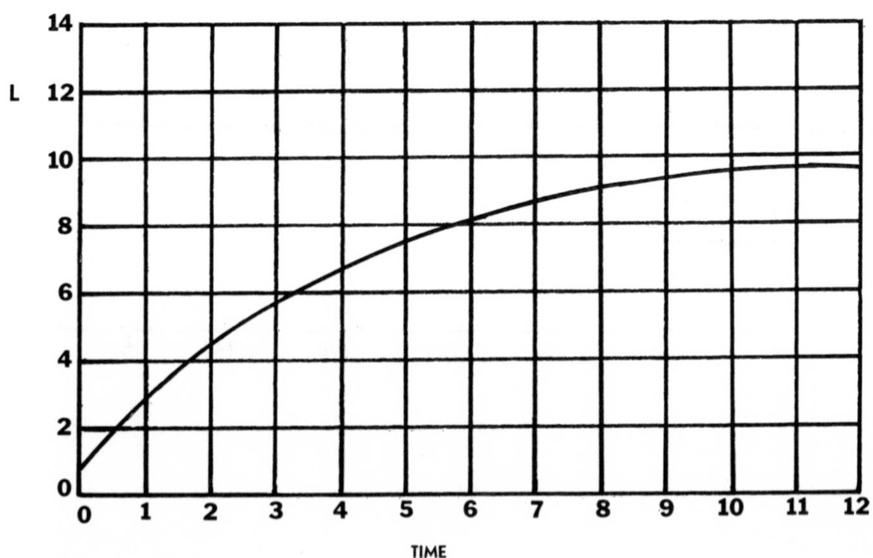

S7.14

Equilibrium results when OR attains a value equal to -IR. In the process of reaching this equilibrium, the level accumulates the difference between OR and IR over time.

S7.15

$$OR = (DL-L(t))/AT = DL/AT-(1/AT)[(DL+A*AT)+(L(0)-(DL+A*AT))e^{-t/AT}]$$
$$= -A(1-e^{-t/AT})+(1/AT)(DL-L(0))e^{-t/AT}$$

S7.16

a) For L(0) = DL,
$OR = -A(1-e^{-t/AT})$

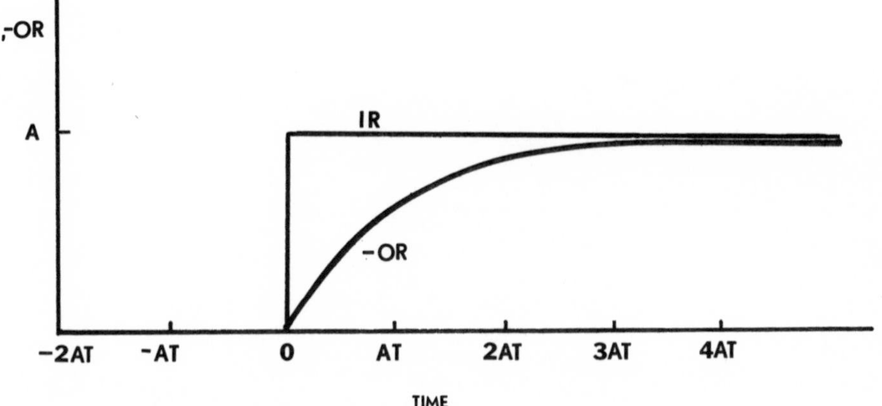

b)

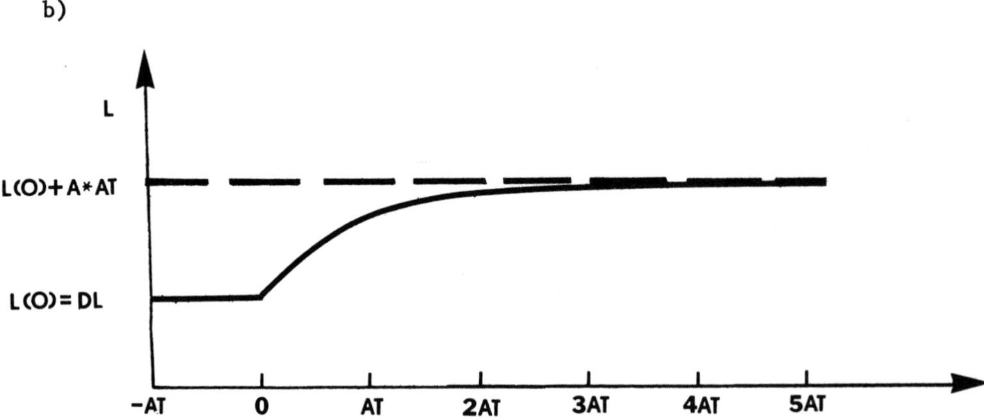

Exercise 8
Simple Structures

Exercise 8 reviews the elementary structure concepts developed in Chapters 2 through 5. The reader encountering difficulty with Exercise 8 should review the appropriate background material.

Time Required: One hour.

INDUSTRIAL CAPITAL MODEL

Write DYNAMO equations for the industrial capital model below. **E8.1**

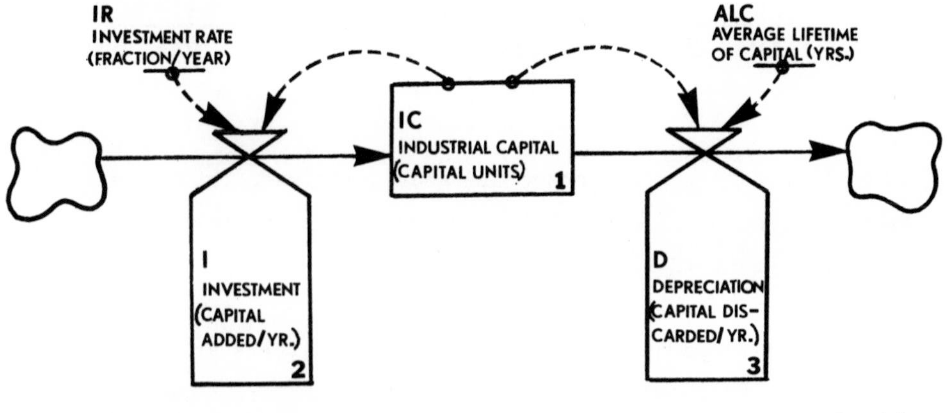

(Leave the values of constants and initial conditions unspecified.)

What possible modes of behavior can the system exhibit? **E8.2**

____ 1. Oscillations - undamped

____ 2. No growth

____ 3. Linear growth

____ 4. Exponential growth

____ 5. Exponential decay

____ 6. Sigmoidal (S-shaped) growth

____ 7. Overshoot (Growth and Decay)

____ 8. Other (Please specify)

E8.3 Hypothetical data for the growth of capital appear in the following plot.

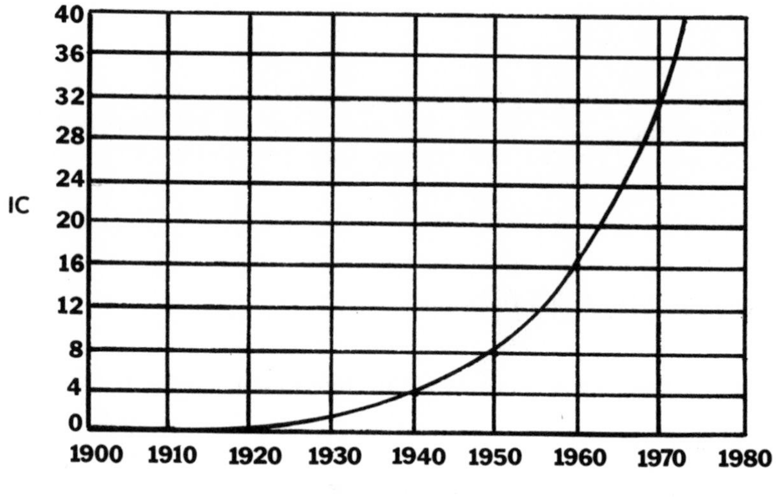

Using this data and the structure above, estimate the necessary average lifetime of capital ALC to produce the historic trend. The investment rate IR equals 12 percent per year.

E8.4 For IR at 11.5 percent per year and ALC at 5 years in the structure in question E8.1, sketch capital growth on the graph below, given an initial value of 36 capital units. Indicate the final equilibrium value and approximate time required to reach that value.

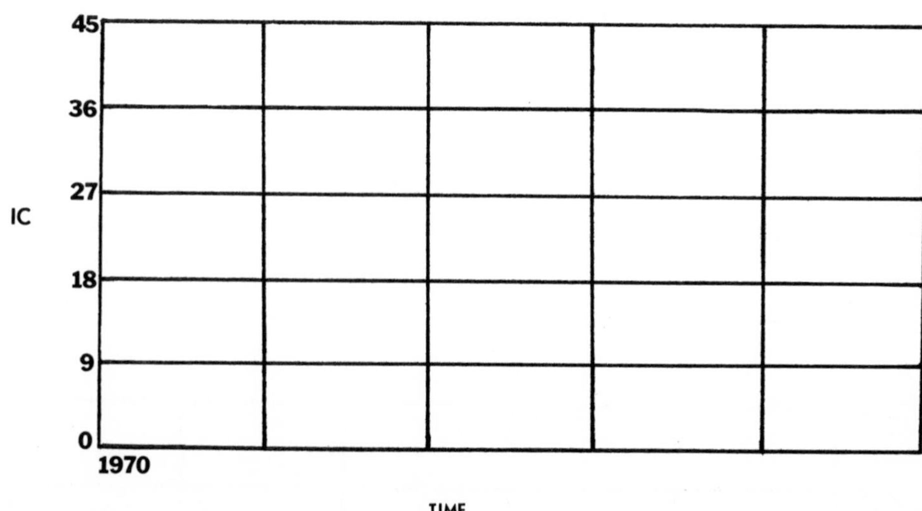

Suppose that IR from the structure in question E8.1 is a function of IC, as shown in the table below. **E8.5**

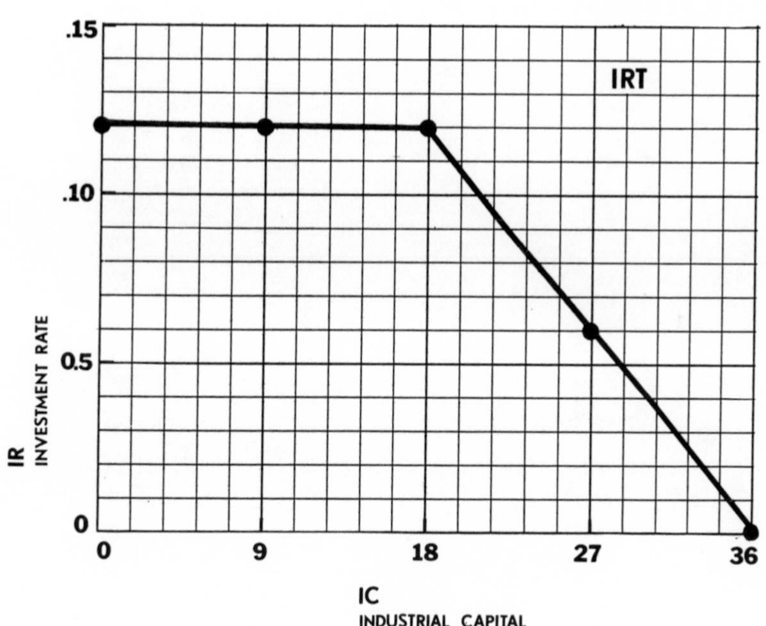

For ALC equal to 20 years, sketch model behavior with the hypothesized IR relationship to IC. Use the grid below and start the system at 9 capital units. Indicate if and at what value IC goes to equilibrium.

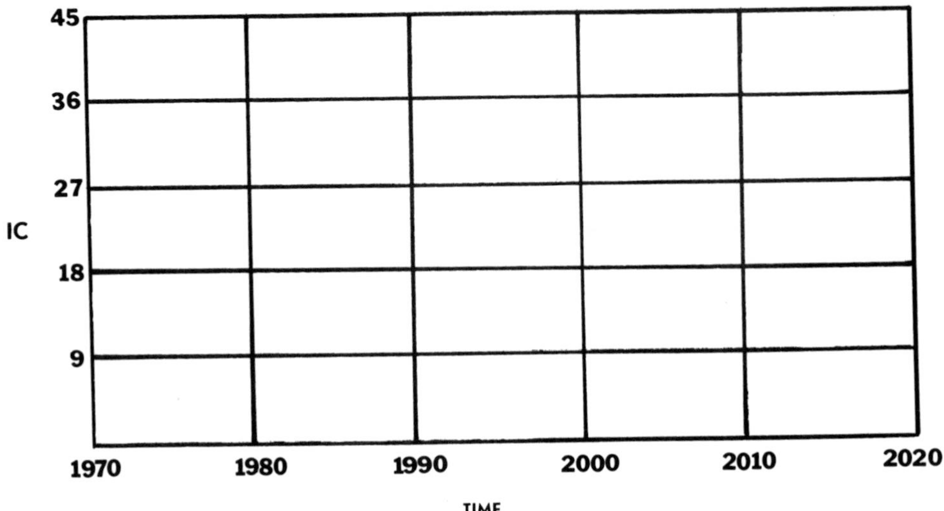

Solution 8

INDUSTRIAL CAPITAL MODEL

S8.1

```
IC.K=IC.J+(DT)(I.JK-D.JK)                                    1, L
IC=UNSPECIFIED                                               1.1, N
    IC      - INDUSTRIAL CAPITAL (CAPITAL UNITS)
    I       - INVESTMENT (UNITS/YEAR)
    D       - DEPRECIATION (UNITS/YEAR)
I.KL=IR*IC.K                                                 2, R
IR=UNSPECIFIED                                               2.1, C
    I       - INVESTMENT (UNITS/YEAR)
    IR      - INVESTMENT RATE (FRACTION/YEAR)
    IC      - INDUSTRIAL CAPITAL (CAPITAL UNITS)
D.KL=IC.K/ALC                                                3, R
ALC=UNSPECIFIED                                              3.1, C
    D       - DEPRECIATION (UNITS/YEAR)
    IC      - INDUSTRIAL CAPITAL (CAPITAL UNITS)
    ALC     - AVERAGE LIFETIME OF CAPITAL (YEARS)
```

S8.2

____ 1. Oscillations - undamped

X 2. No growth

____ 3. Linear growth

X 4. Exponential growth

X 5. Exponential decay

____ 6. Sigmoidal (S-shaped) growth

____ 7. Overshoot (Growth and Decay)

____ 8. Other (Please specify)

S8.3 The doubling time roughly equals 10 years. The level behaves according to an equation of the form:

$$IC(t) = IC(0)e^{rt}$$

where

r = IR-1/ALC

t = time

IC(0) = initial level value

To find the doubling time T_d,

$$T_d = 0.7(1/r)$$

since 1/r equals the time constant.

Hence,

$$r = 0.7/T_d$$

but,

$$r = IR-1/ALC = 0.7/T_d$$

Therefore,

$$ALC = 1/(IR-0.7/T_d)$$

$$= 1/(0.12-0.7/10)$$

$$= 1/0.05 = 20 \text{ years}$$

S8.4 $r = IR-1/ALC = 0.115-1/5 = -0.085$

Since r is negative, we can expect exponential decay. The time constant is 12 years. The approximate settling time equals three time constants or 36 years. The equilibrium value must approach zero since e^{rt} approaches zero as t approaches infinity for negative r.

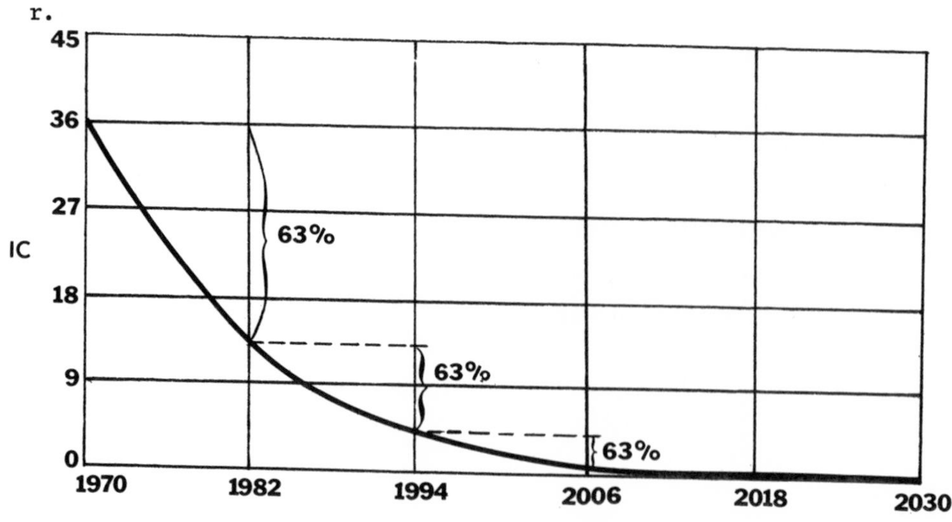

S8.5

r = IR-1/ALC

For IC between 0 and 18,

$$r = 0.12-1/20 = 0.07$$

and doubling time

$$T_d = (1/0.07)(0.7)$$
$$= 10 \text{ years}$$

For IC between 18 and 36

$$IR = 0.24-(0.12/18)IC$$
$$r = 0.24-(0.12/18)IC-1/ALC$$
$$r = 0.19-(0.12/18)IC$$

A plot of r versus IC yields:

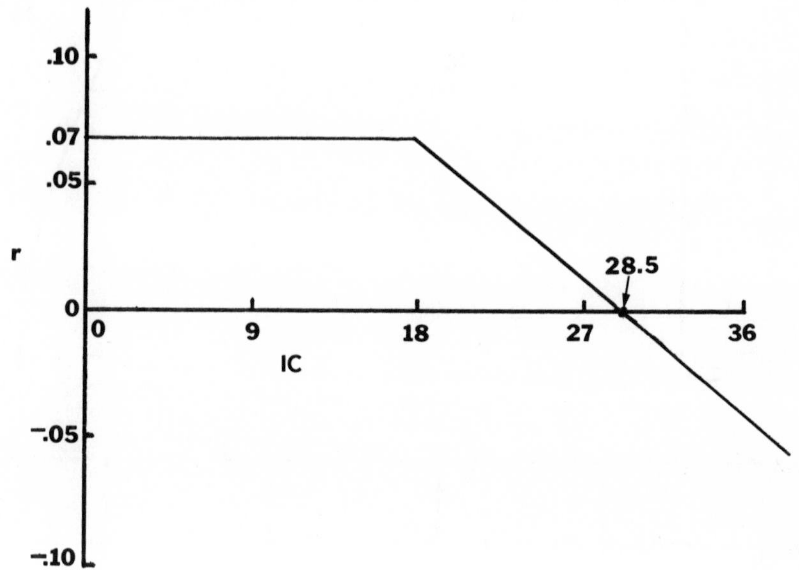

A plot of the net rate NR (I–D) as a function of IC appears below. The rate-level structure produces S-shaped growth.

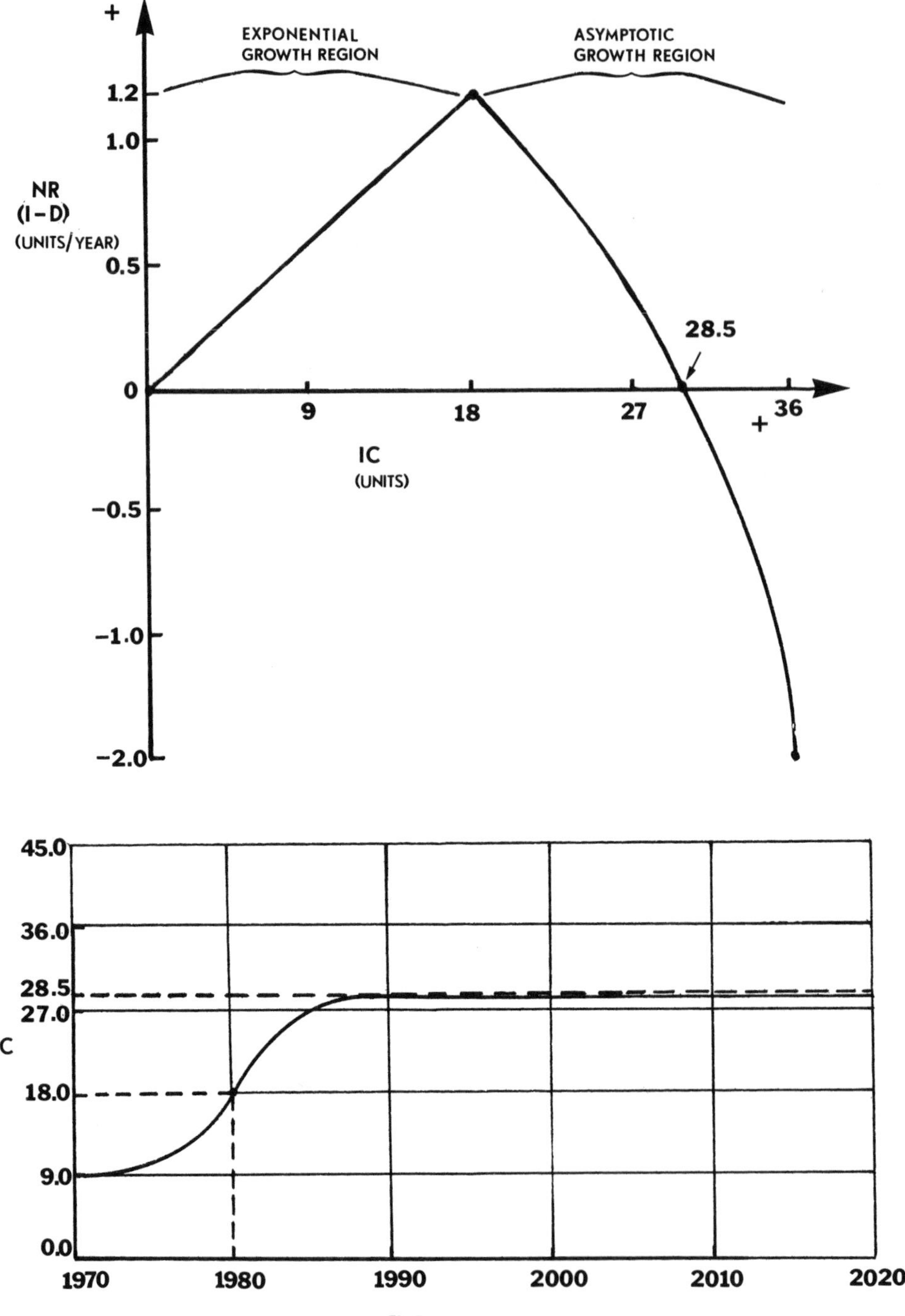

Part Three
Exercises in Analysis and Conceptualization

Exercise 9
Delays:
Exercise and Supplementary Notes

by Dennis L. Meadows

Exercise 9 explores some response properties of exponential delays. The reader should review Chapter 8 of Principles of Systems and Chapter 9 and Appendices E, G, and H in Industrial Dynamics before attempting this exercise. Additional notes on delays appear below. The first part develops the output response of first-, second-, and third-order delays for a pulse input. It requires knowledge of elementary calculus to understand the derivations included in the appendices to this exercise. However, the conclusions can be studied without necessarily following the derivations. The second part develops response characteristics of a delay to a sinusoidal input and introduces open-loop analysis using gain and phase shift.

PULSE-INPUT RESPONSE

An exponential delay transforms an input time series into an output series. The operation depends only upon the order and time constant of the delay. The order denoted by n, is the number of integrations or levels in the delay. The time constant (or delay time) is denoted by T. Given T and n, we can quantitatively express the relation between any specified input and the delay's output. Important characteristics of exponential delays become clear when we study their output response to a pulse input. The following calculations yield the time form of a delay output as a function of the order (n = 1,2,3), time constant, and magnitude of the input pulse.

Nomenclature. When the total delay T is distributed evenly among n levels of the delay, the instantaneous outflow from the i-th level in an n-th order exponential delay equals:

$$OUTi_n(t) = (Li_n(t))/(T/n)$$

where

$OUTi_n(t)$ - Outflow rate from level i in an n-th order delay at time t.

$Li_n(t)$ - Quantity in level i of an n-th order delay at time t.

T - Time constant.

n - Order of the delay.

The instantaneous net change in the i-th level in an n-th order delay equals:

$$\frac{dLi(t)}{dt} = INi_n(t)-OUTi_n(t) \qquad (9.1)$$

However, the input to level i, of course, equals only the output from level i-1. Thus:

$$\frac{dLi(t)}{dt} = (L(i-1)_n(t))/(T/n)-Li_n(t)/(T/n) \qquad (9.2)$$

For i = 1, the input equals the exogenous input to the delay.

<u>First-order Delay Response</u>. Consider a first-order exponential delay with time constant T as shown in Figure E9-1.[1] Assume that the delay has initial quantity of $L1_1(0)$ (as a result of a pulse input), and that no subsequent input occurs.

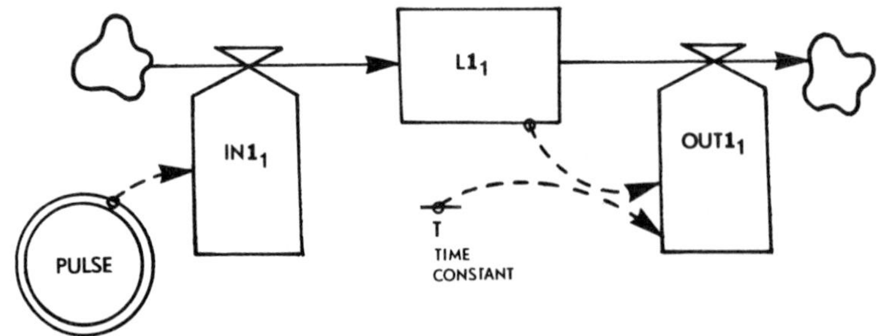

FIGURE E9-1 First-order delay

[1]The first-order delay exactly replicates the zero-goal negative feedback structure found in Chapter 3, section 3.7.

Thus, $\frac{dL1_1(t)}{dt} = -OUT1_1$

$= -L1_1(t)/T$

Separating variables and integrating both sides yields:

$$L1_1(t) = L1_1(0)e^{-t/T} \quad (9.3)$$

Hence, the level quantity in a first-order delay with no exogenous input varies through time as in equation (9.3). The output rate $OUT1_1$ is simply proportional to $L1_1$ over time.

From equation (9.3), the maximum value of $L1_1$, occurs at t = 0. The maximum value is $L1_1(0)$. For each time period equal in length to T, the level decreases by about 63 percent. After three T time periods, over 95 percent of the initial quantity has dissipated from a first-order delay. Figure E9-4 shows output response of the delay to the pulse input. Appendix A proves that T equals the average time spent by a unit in the delay.

<u>Second-order Delay Response</u>. Consider a second-order exponential delay with total time constant T divided evenly, as shown in Figure E9-2, among the two levels. The pulse input at t = 0 is $L1_2(0)$.

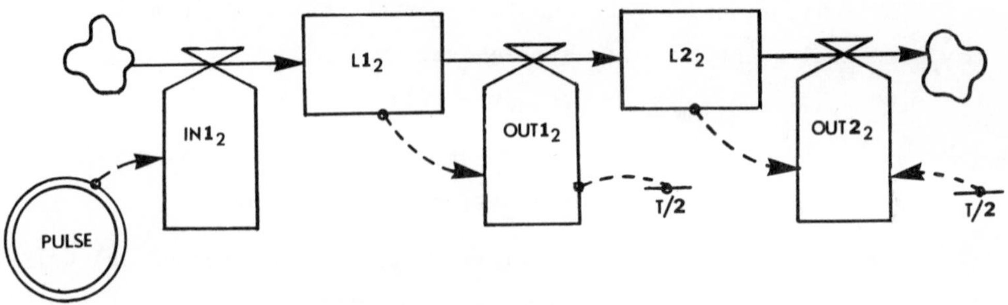

FIGURE E9-2 Second-order delay

We now have:

$$OUT1_2(t) = IN1_2 = L1_2(t)/(T/2)$$

$$OUT2_2(t) = L2_2(t)/(T/2)$$

$$\frac{dL2_2(t)}{dt} = L1_2(t)/(T/2)-L2_2(t)/(T/2) \qquad (9.4)$$

Equation (9.5) represents the solution to equation (9.4). (See Appendix B for the derivation.)

$$L2_2(t) = (1/T)(2)L1_2(0)(t)e^{-2t/T} \qquad (9.5)$$

We can find the maximum $L2_2$ value by setting the derivative of equation (9.5) equal to zero.

$$\frac{dL2_2(t)}{dt} = (1/T)(2)L1_2(0)(e^{-2t/T}-(2/T)(t)e^{-2t/T}) = 0$$

This implies,

$$t = T/2$$

Thus,

$$L2_2(t = T/2) = (2/T)(T/2)L1_2(0)e^{-1}$$

$$= L1_2(0)/e^1$$

$$= (0.368)L1_2(0)$$

The maximum quantity in $L2_2$ thus occurs at t = T/2 and has a value independent of T.

The maximum output rate equals the rate resulting from the maximum level quantity calculated above.

$$[OUT2_2(t)]_{maximum} = (2/T)L2_2(t = T/2) = (0.74/T)L1_2(0)$$

This value decreases as the time constant T increases. Figure E9-4 illustrates output response of the second-order delay.

<u>Third-order Delay Response</u>. Consider a third-order exponential delay, with time constant T, encountering an input pulse of $L1_3(0)$ at t = 0, as in Figure E9-3.

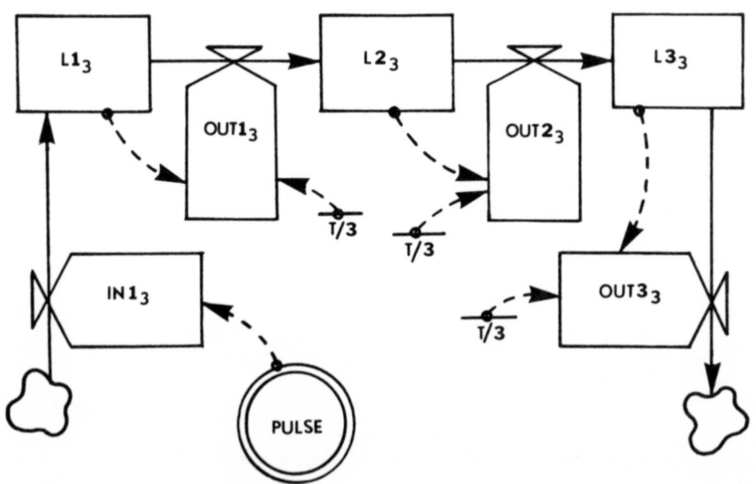

FIGURE E9-3 Third-order delay

Adapting from equation (9.3): $L1_3(t) = L1_3(0)e^{-3t/T}$

Adapting from equation (9.5): $L2_3(t) = (1/T)(3)L1_3(0)(t)e^{-3t/T}$ (9.6)

From equation (9.2):

$$\frac{dL3_3(t)}{dt} = (T/3)L2_3(t)-(T/3)L3_3(t) \quad (9.7)$$

Equation (9.8) represents the solution to equation (9.7). (See Appendix C for derivation.)

$$L3_3(t) = (1/T^2)(4.5)L1_3(0)(t^2)e^{-3t/T} \quad (9.8)$$

Maximum $L3_3(t)$ occurs at:

$$\frac{dL3_3(t)}{dt} = 0$$

which yields

$$t = 2T/3$$

The quantity in $L3_3$ at $t = 2T/3$ equals:

$$(1/T^2)(4.5)L1_3(0)(4T^2/9)e^{-2} = (2/e^2)L1_3(0).$$

Since the delay output is directly proportional to the quantity in its last level, the output of a third-order delay also peaks at $t = 2T/3$.

Decreasing the time constant T lessens the time for the peak output to occur. The maximum output rate equals $(0.8/T)L1_3(0)$. This maximum is a decreasing function of T. Figure E9-4 illustrates output response of the third-order delay.

Summary.

1. We can compare the maximum output of the second- and third-order delay since $L1_3(0) = L1_2(0)$.

Thus,

$$\frac{[OUT3_3(t)]_{maximum}}{[OUT2_2(t)]_{maximum}} = (6L1_3(0)/(T)e^2)/(2L1_2(0)/(T)e^1) = 3/e \approx 1.1$$

Increasing n from 2 to 3, for constant T, increases the maximum output rate by about 10 percent as shown in Figure E9-4.

2. Comparing times required to attain maximum output for the second- and thiid-order delays, we get:

$$\frac{t(\text{maximum output of third-order delay})}{t(\text{maximum output of second-order delay})} = (2/3)T/(1/2)T$$

$$= 1.33.$$

Increasing the delay order from n = 2 to n = 3 displaces the peak output about 33 percent as shown in Figure E9-4.

3. The second- and third-order delays simply illustrate specific cases of the general nth-order delay. As shown in Figure E9-4, as n approaches infinity the height of the output becomes equal to the height of the input. This response occurs T time units after the input.

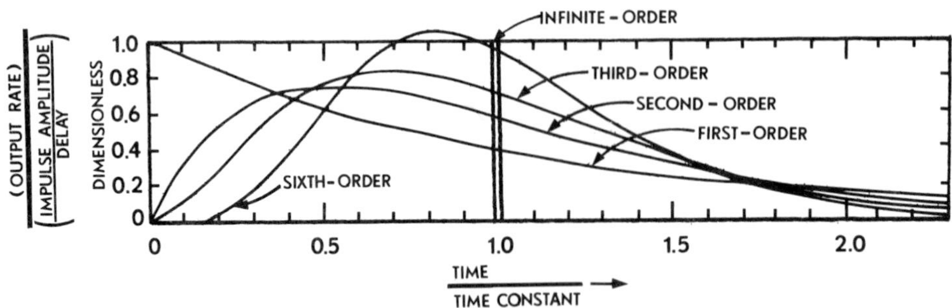

FIGURE E9-4 Exponential-delay responses to a unit impulse

Source: Forrester, Industrial Dynamics, p. 92.

SINUSOIDAL INPUT RESPONSE.

Before exploring the response of delays to sinusoidal inputs, we first need to examine the response of a level to a sinusoidal input rate, understand definitions of gain and phase shift, and become familiar with open-loop analysis.

Sinusoidal Input Response of a Level. The response of the level L in Figure E9-5 to a sinusoidally varying rate R can be determined by integrating the input rate equation:

$$R(t) = (A)\sin(2\pi t/P)$$

where,

P = period of input sinusoid

A = amplitude of input sinusoid.

Thus,

$$L(t) = \int_0^t R(\tau)d\tau \qquad \text{(where } \tau \text{ is a dummy variable)}$$

$$L(t) = \int_0^t (A)\sin(2\pi\tau/P)d\tau$$

$$= (A)(P/2\pi)\cos(2\pi t/P)$$

$$= (A)(P/2\pi)\sin(2\pi t/P+\pi/2)$$

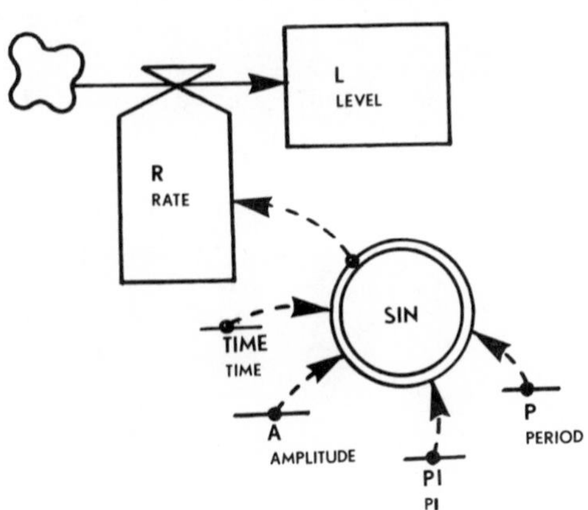

FIGURE E9-5 Sinusoidal input to level

```
L  L.K = L.J+(DT)(R.JK)                        (Units)
N  L = 0                                       (Units)
R  R.KL = A*SIN(2*PI*TIME.K/P)                 (Units/time)
C  PI = 3.14                                   (Dimensionless)
C  P = period of input sinusoid                (Time)
C  A = amplitude of input sinusoid             (Units/time)
```

The level lags the input rate by $\pi/2$ or 90 degrees and the amplitude is $(A)(P/2\pi)$.

Gain and Phase Shift.

Gain is defined as the ratio of the amplitude of the output to the amplitude of the input after the equilibrium or steady-state has been reached.

The ratio of the amplitude of level L (the output) and the amplitude of the rate R (the input) gives the gain for the system described in Figure E9-5:

$$\text{Gain} = \frac{\text{Output amplitude}}{\text{Input amplitude}}$$

$$= ((A)(P/2\pi))/A$$

$$= P/2\pi$$

$$= P/6.28$$

The phase shift is the amount by which the output shifts or displaces in time from the input once the steady-state has been reached.

The phase shift in the example above equals $\pi/2$ or 90 degrees (one-fourth of the input period). That is, the output sinusoid shifts one-fourth of a cycle (period) from the input. The output and input have the same period P. Figure E9-6 illustrates the gain and phase shift relationships.

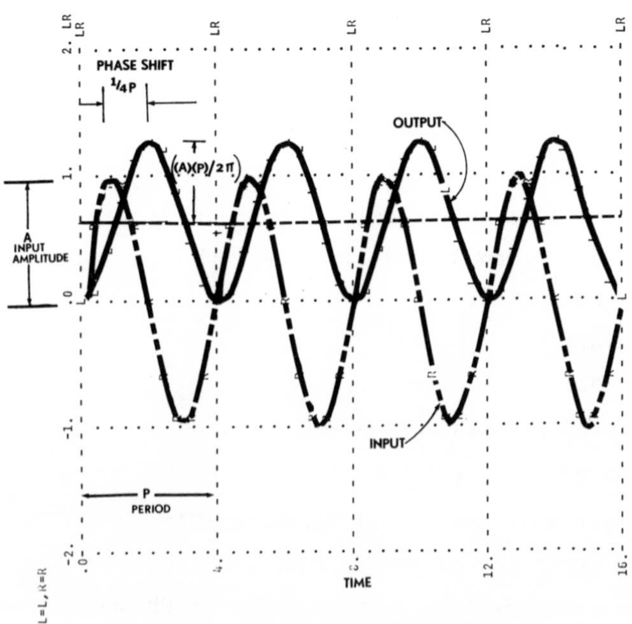

FIGURE E9-6 Phase and gain relationship (steady-state)

Open-loop Analysis. Open-loop analysis investigates response of cascaded rates, levels and auxiliaries to exogenous inputs. It is based on the multiplicative and additive properties of gain and phase shift. These properties can be summarized in the rule given below:

When two or more structures, such as the one in Figure E9-5, cascade, the gain of the total system equals the product of the gains across the individual structures. The phase shift of the system equals the sum of the phase shifts across the elements.[2]

We illustrate the open-loop analysis with the following example.

---

[2]For proof of the multiplicative and additive properties of gains and phase shift, see Amar G. Bose and Kenneth N. Stevens, *Introductory Network Theory* (New York: Harper & Row, 1965), Chapter 4.

Consider a system such as shown in Figure E9-7, containing three levels and three rates cascaded together. By employing the above rule, we can generate the phase and gain relationships between each rate and level variable and across the entire network. The table in Figure E9-8 summarizes the results. The gain and phase shift values in the second and third columns are relative to the previous variable. Thus, for P = 12 and A = 1 the gain and phase shift between SR and I equals P/6.28 or 1.91 and 90 degrees respectively. To determine gain and phase shift between HR and I, for example, we separate the HR equation into two parts:

```
            Part 1      Part 2
HR.K = UR*DI/AT   -(UR/AT)I.K
```

Oscillation in HR results exclusively from oscillation of I in Part 2. No integration occurs. The components in Part 1 are all constants. The steady-state gain must equal UR/AT or 10/6 as in the table. The negative sign in Part 2 of the HR equation indicates that the maximum value of I corresponds to a minimum value of HR and vice-versa. Oscillations of HR are 180 degrees, or half a period, out of phase with I. Thus, the phase shift equals 180 degrees.

Columns 4 and 5 in the table contain values of the gain and phase shift for the element relative to SR, the input. Column 4 gains are obtained by multiplying the column 2 gains together. Column 5 shifts are obtained by adding the column 3 values. For example, the total gain G between SR and WF equals the product of the gains between I and SR, HR and I, and WF and HR, or:

G = 1.91*1.67*1.91

= 6.09.

The total phase shift Ø between SR and WF equals the sum of the phase shifts between I and SR, HR and I, and WF and HR, or:

Ø = 90°+180°+90°

= 360°

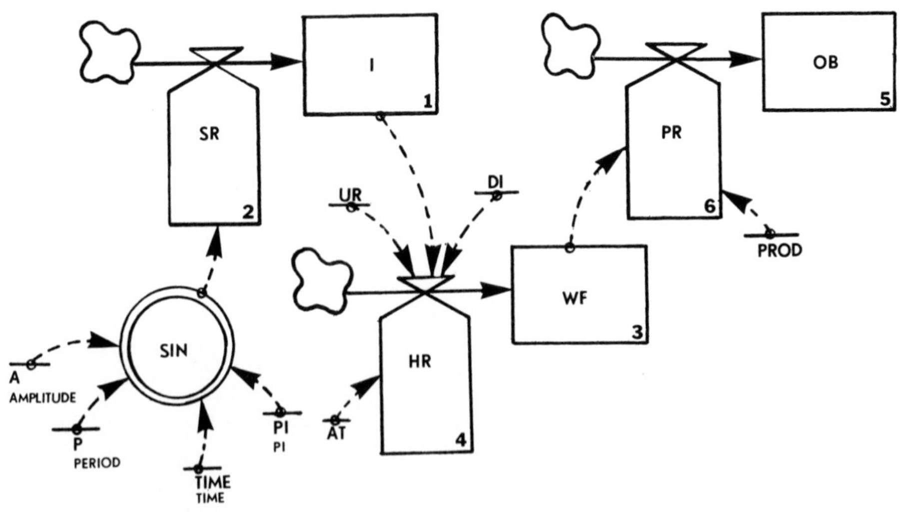


```
L  I.K=I.J+(DT)(SR.JK)                 1
N  I=0                                 1.1
R  SR.KL=A*SIN(2*PI*TIME.K/P)          2
C  A=1                                 2.1
C  PI=3.14                             2.2
C  P=12                                2.3
L  WF.K=WF.J+(DT)(HR.JK)               3
N  WF=0                                3.1
R  HR.KL=UR*(DI-I.K)/AT                4
C  DI=1.91                             4.1
C  UR=10                               4.2
C  AT=6                                4.3
L  OB.K=OB.J+(DT)(PR.JK)               5
N  OB=0                                5.1
R  PR.KL=WF.K*PROD                     6
C  PROD=.2                             6.1
```

FIGURE E9-7 Cascaded system

① Element	② Gain	③ Phase Shift	④ Gain Relative to SR	⑤ Phase Shift Relative to SR	Notes
SR			1.00	0°	start
I	(P/2π)1.91	90°	1.91	90°	integration
HR	(UR/AT)1.67	180°	3.19	270°	unit change sign change
WF	(P/2π)1.91	90°	6.09	360° (equivalent to 0° phase shift)	integration
PR	(PROD)0.20	0°	1.22	0°	unit change
OB	(P/2π)1.91	90°	2.32	10°	integration

FIGURE E9-8 Table of gain and phase values

Figure E9-9, the system simulation, verifies values found in Figure E9-8.

To apply open-loop analysis, we must meet three important restrictions.

1. The rate, level, and auxiliary network cannot form a closed path, but instead must be open ended. We can arbitrarily break a closed loop between any two elements to form an open loop. The open path, however, can include such feedback structures as delays (or structures convertible into delays) as we will see in the next section.

2. The open-loop analysis is applicable to equilibrium or steady-state only. Gains and phase shifts are based on the equilibrium values of variables.[3]

3. Nonlinear table functions are linearized in their operating range. Excursions into nonlinear ranges change phase-gain relationships.

---

[3]All the computer runs pertaining to open-loop analysis in this exercise show only the steady-state behavior of the system. The beginning portion has been ommitted as the transients may predominate in this region. The term steady-state refers to oscillations in the system which are neither convergent nor divergent.

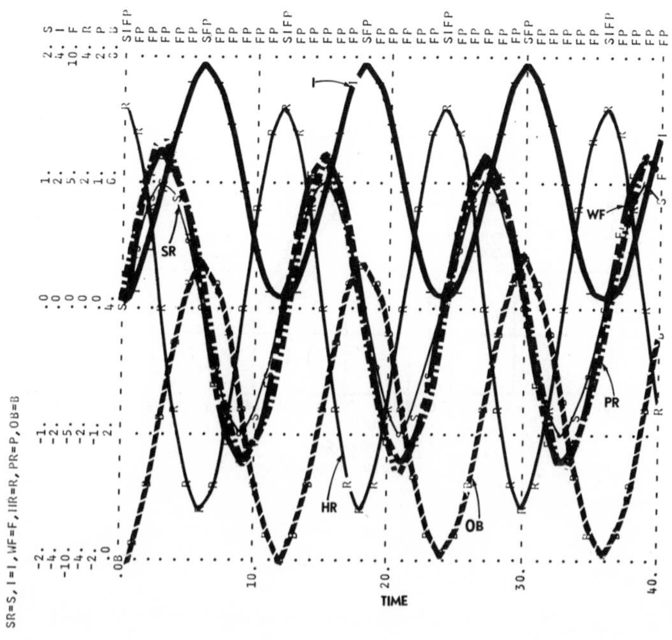

FIGURE E9-9 Simulation of cascaded systems without delays (steady-state)

We now turn to the open-loop analysis of networks containing delays.

<u>Sinusoidal Input Response of a Delay</u>. We can determine the response of a delay shown in Figure E9-10 to a sinusoidally varying input by using the gain and phase charts of Figure E9-11.

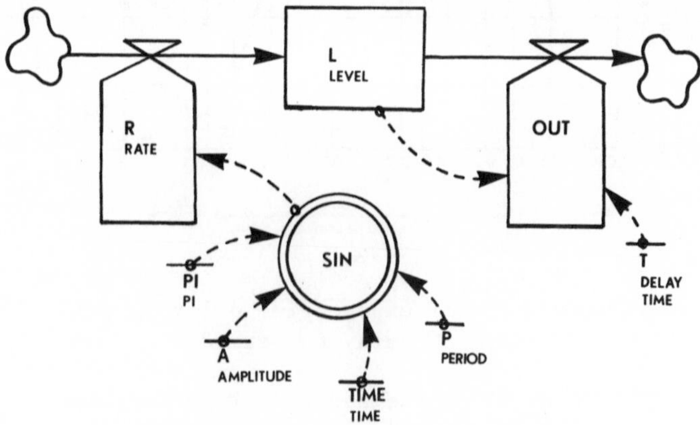

FIGURE E9-10 Sinusoid input to delay

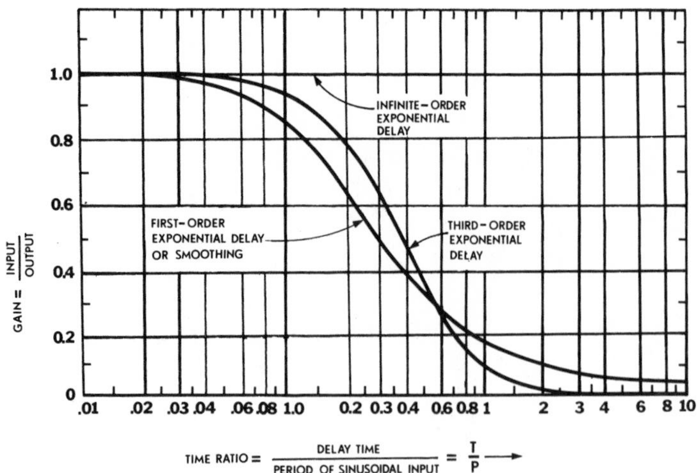

(a)

Ratio of output amplitude to input amplitude versus ratio of delay time to input period.

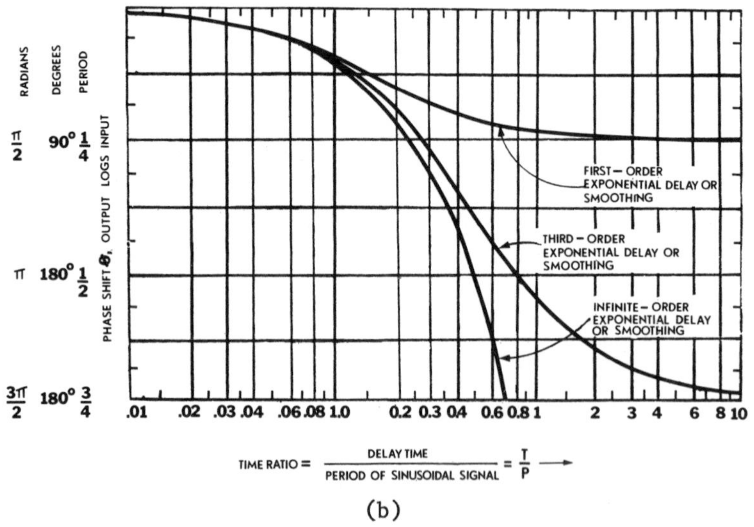

(b)

Phase shift of exponential delay versus ratio of delay to period.

FIGURE E9-11 Phase and gain relationships

Source: Forrester, Industrial Dynamics, p. 417.

System equations are:

```
L L.K=L.J+(DT)(R.JK-OUT.JK)            (Units)
N L=0                                  (Units)
R R.KL=A*SIN(2*PI*TIME.K/P)            (Units/time)
C PI=3.14                              (Dimensionless)
C P=period of input                    (Time)
C A=amplitude of input signal          (Units/time)
R OUT.KL=L.K/T                         (Units/time)
C T=delay time                         (Time)
```

Find the gain and phase shift between the input rate and output rate by:

1. calculating the ratio of the delay time (or time constant T) to the input period P;

2. locating the ratio on the abscissa in Figure E9-11; and

3. reading the appropriate gain (Figure E9-11 (a)) and phase value (Figure E9-11 (b)) according to the order of the delay from the curve.

Open-loop analysis can be used for calculating gain and phase shift of networks containing multiple levels and delays. However, all feedback structure along the network must be represented as a delay. Otherwise, open-loop analysis can not apply. The following example illustrates use of open-loop analysis on a network of rates and levels.

Sinusoidal-input Response of Network Containing a Delay. We modify the Figure E9-7 model so that HR depends on WF as well as I. Figure E9-12 shows the new structure. The new rate equation for HR is:

```
R HR.KL=(UR*(DI-I.K)/AT)-(WF.K/TT)                   4
C TT=2                                               4.4
```

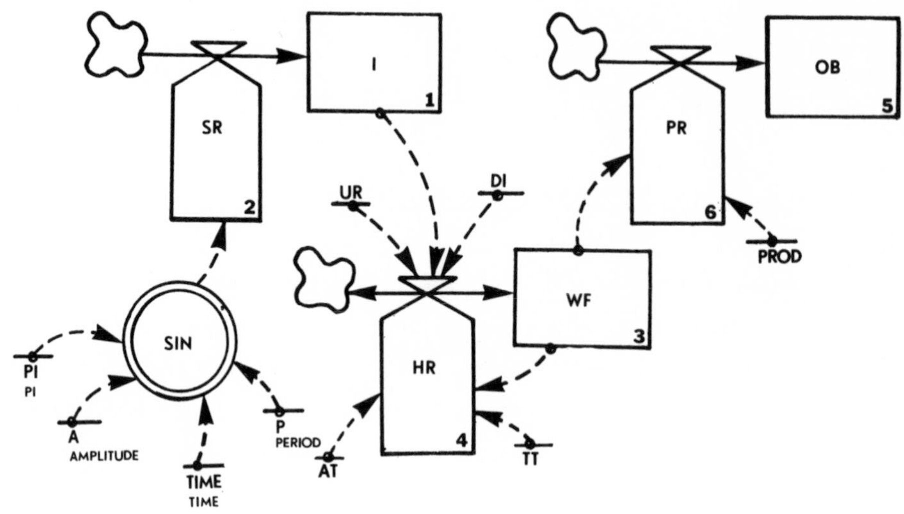

FIGURE E9-12 Cascaded system with feedback loop

HR can be expressed as the sum of two parts:

HR = HR1+HR2

where

HR1 = UR*(DI-I)/AT (as in Figure E9-7)

HR2 = -WF/TT

We can then redraw the DYNAMO flow diagram as in Figure E9-13. This diagram and the revised equation set which it represents are dynamically equivalent to the structure in Figure E9-12.

We can separately evaluate the phase shift and gains relative to SR in both parts of HR. HR1 exactly matches HR in the model of Figure E9-7; the results in the table of Figure E9-8 still apply.

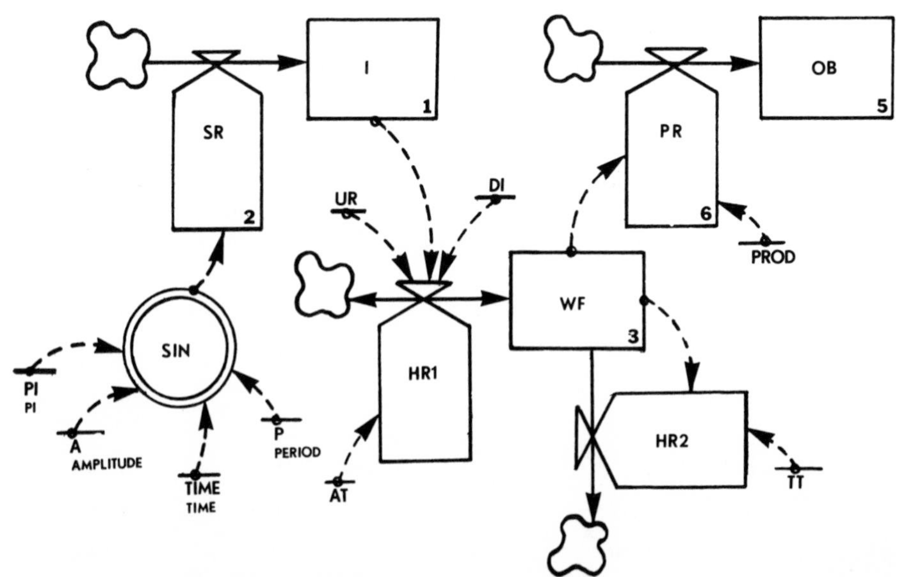

FIGURE E9-13 Redrawn cascaded system with feedback loop

The first-order delay composed of WF and HR2 also introduces a gain and phase shift. We want to know the gain and phase shift of WF versus SR. However, the phase and gain charts relate to the phase shift and gain of a delay's output rate (HR2) relative to its input rate (HR1). Hence, we must first calculate the phase and gain across the delay. The time constant TT of the delay equals 2, and the period P of the input equals 12. Across a first-order delay where the ratio of delay time to period equals 0.16, the gain equals 0.7 and the phase shift equals 45 degrees as seen in the charts in Figure E9-11. No phase shift occurs between HR2 and WF so WF must also exhibit a 45 degree phase shift relative to its input (HR1). The gain between WF and HR2 equals 1/TT since HR2/WF = 1/TT. Since the gain from HR1 to HR2 equals 0.7, the gain from HR1 to WF equals 0.7*TT or 1.4. The gain between HR1 and WF in this case is smaller than in the case of no feedback.[4] The table in Figure E9-14 summarizes the gain and phase

---

[4]As seen in Figure E9-11 (a), as the delay time T increases relative to the period P of the input, the output of the first-order delay is attenuated. The delay acts like a filter. Or similarly, for a given T, low frequency inputs (T/P < 0.1) pass through the delay undisturbed while high frequency inputs (T/P > 1) do not.

shift values. The amplitude and phase shift values of I, HR, WF, PR and OB seen in the simulation run of Figure E9-15 agree with the preceding analysis.

① Element	② Gain	③ Phase Shift	④ Gain Relative to SR	⑤ Phase Shift Relative to SR	Notes
SR	-	-	1.00	0°	start
I	1.91	90°	1.91	90°	integration
HR1	1.67	180°	3.19	270°	unit change sign change
WF	1.40	45°	4.46	315°	delay
PR	0.20	0°	0.89	315°	unit change
OB	1.91	90°	1.70	405° (equivalent to 45° phase shift)	integration

FIGURE E9-14 Table of gain and phase values

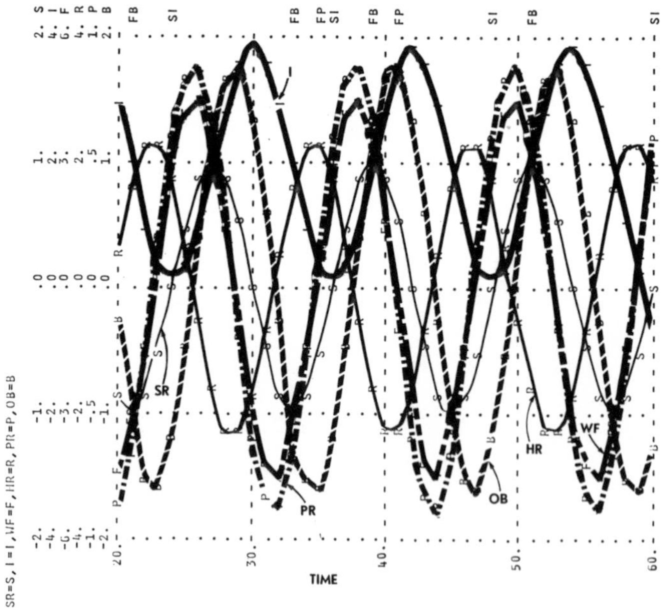

FIGURE E9-15 Simulation of cascaded system with feedback (steady-state)

APPENDIX A: DERIVATION OF AVERAGE TIME A UNIT SPENDS IN FIRST-ORDER DELAY.

Suppose we insert a number of units equal to $L1_1(0)$ into a first-order delay. How long, on the average, does each unit spend in the delay?

The level quantity in the first-order delay varies through time as follows:

$$L1_1(t) = L1_1(0)e^{-t/T} \tag{9.3}$$

The absolute value of the outflow rate, therefore, equals:

$$OUT_1(t) = (1/T)L1_1(0)e^{-t/T}$$

The average time spent in the delay is determined by weighting the outflow rate by time t, integrating over all t, and dividing by $L1_1(0)$ (the initial contents of the delay). This procedure can be interpreted simply for the case of a discrete delay. Suppose that we mail 10 items. Four of the items arrive at the destination after 5 days; 4 arrive after 8 days; and 2 arrive on the ninth day. The total time spent in transit by the items is then 4(5)+4(8)+2(9), or 70 days. The _average_ time spent in transit is then (70/10), or 7 days. Extending this idea to the case of a continuous delay, we see that the average delay time is

$$\int_0^\infty (1/L1_1(0))(t)OUT_1(t)dt = (1/T)\int_0^\infty (t)e^{-t/T}dt \tag{A1}$$

Integrating equation (A1) by parts yields:

$$-(t)(e^{-t/T}) \Big|_0^\infty + \int_0^\infty e^{-t/T}dt = -(T)e^{-t/T} \Big|_0^\infty = T$$

Hence, T equals the average residence time of a unit in the delay.[5]

---

[5]This proof has been contributed by Nathaniel J. Mass, System Dynamics Laboratory, at the Sloan School of Management, Massachusetts Institute of Technology, Cambridge, Massachusetts.

APPENDIX B: DERIVATION OF SECOND-ORDER DELAY RESPONSE.

Assume for a moment a solution to equation (9.4) in the form:

$$L2_2(t) = Cte^{-2t/T} \tag{B1}$$

If the solution works, then the existence and uniqueness theorem guarantees the solution will be the only possible one. This equation satisfies the requirement that $L2_2$ approach zero as t approaches either zero or infinity. We derive C by taking the time derivative of equation (B1):

$$\frac{dL2_2(t)}{dt} = Ce^{-2t/T}-(2C/T)(t)e^{-2t/T}$$

Substituting equation (B1) into equation (9.4) yields:

$$L1_2(t)/(T/2)-L2_2(t)/(T/2) = Ce^{-2t/T}-(2C/T)(t)e^{-2t/T} \tag{B2}$$

Adapting equation (9.3) for the case where the time constant equals T/2:

$$L1_2(t) = L1_2(0)e^{-2t/T},$$

and employing the assumed form for $L2_2$ permits us to write equation (B2) as:

$$(T/2)L1_2(0)e^{-2t/T}-(T/2)(C)(t)e^{-2t/T} = Ce^{-2t/T}-(1/T)(2)(C)(t)e^{-2t/T},$$

or,

$$C = (2)L1_2(0)/T.$$

Then (B1) can be written as:

$$L2_2(t) = (1/T)(2)L1_2(0)(t)e^{-2t/T}$$

APPENDIX C: DERIVATION OF THIRD-ORDER DELAY RESPONSE.

Assume a solution to equation (9.7) in the form:

$$L3_3(t) = Ct^2e^{-3t/T} \tag{C1}$$

This solution satisfies the condition that $L3_3$ approaches zero as t approaches either zero or infinity. From equations (9.6), (9.7) and (C1):

$$\frac{dL3_3(t)}{dt} = (9/T^2)L1_3(0)(t)e^{-3t/T}-(3C/T)(t^2)e^{-3t/T}$$

The derivative of equation (C1) equals:

$$\frac{dL3_3(t)}{dt} = -(3C/T)(t^2)e^{-3t/T}+(2C)(t)e^{-3t/T}.$$

Thus,

$$(2C)(t)e^{-3t/T}-(3C/T)(t^2)e^{-3t/T} = (9/T^2)L1_3(0)(t)e^{-3t/T} -(3C/T)(t^2)e^{-3t/T},$$

or,

$$C = (9/2T^2)L1_3(0) = (4.5/T^2)L1_3(0),$$

and equation (C1) becomes:

$$L3_3(t) = (1/T^2)(4.5)L1_3(0)(t^2)e^{-3t/T}. \tag{C2}$$

The derivative of equation (C2) with respect to t satisfies equation (9.6), thereby confirming the assumed solution in equation (C1).

E9.1 The apartment finishing rate AFR is a third-order delay of the apartment construction rate ACR as shown below. The apartment construction delay ACD is 24 months. Draw the response of AFR to the cyclical ACR behavior given in (b). Be sure to specify the phase shift Ø and gain G.

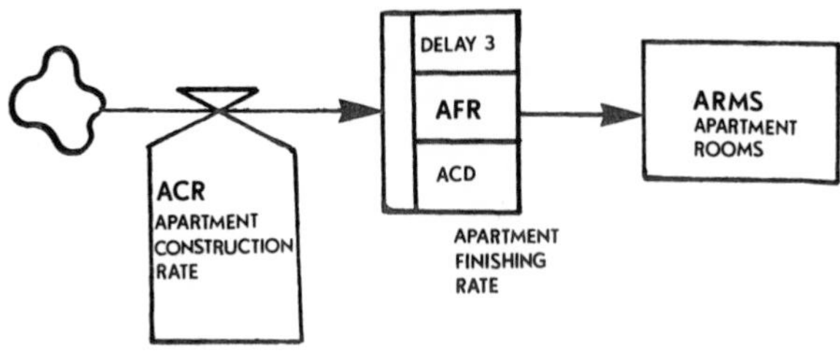

(a)

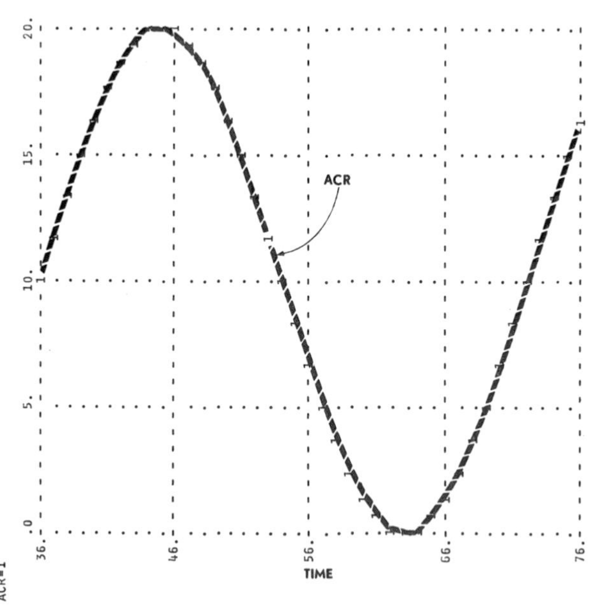

(b)

E9.2

The system below is initially in steady state.

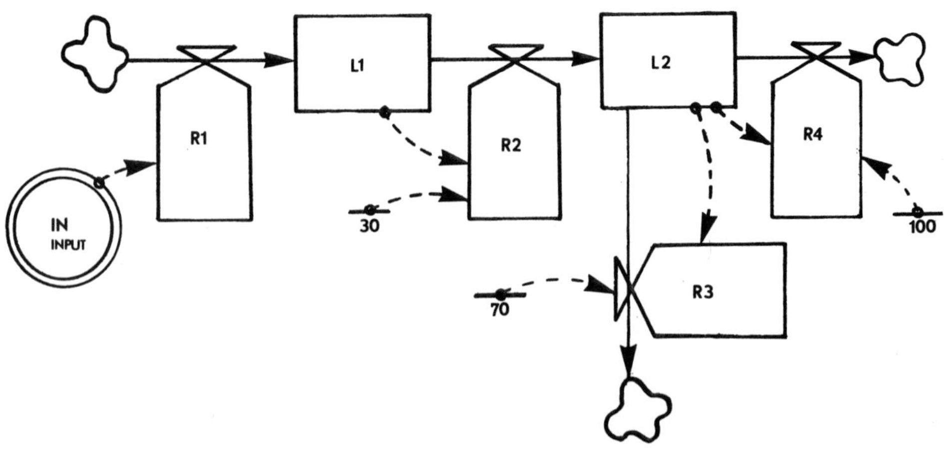


```
L L1.K=L1.J+(DT)(R1.JK-R2.JK)
L L2.K=L2.J+(DT)(R2.JK-R3.JK-R4.JK)
R R1.KL=IN.K
R R2.KL=L1.K/30
R R3.KL=L2.K/70
R R4.KL=L2.K/100
```

(a)

a) What are L1, L2, R1, R2, R3, and R4 equal to if IN equals 5?

b) Given the sinusoidal input IN below, sketch the response of R2 and R3 over time. Assume initially that both R2 and R3 equal zero. Specify the amplitude, period, and phase shift for each.

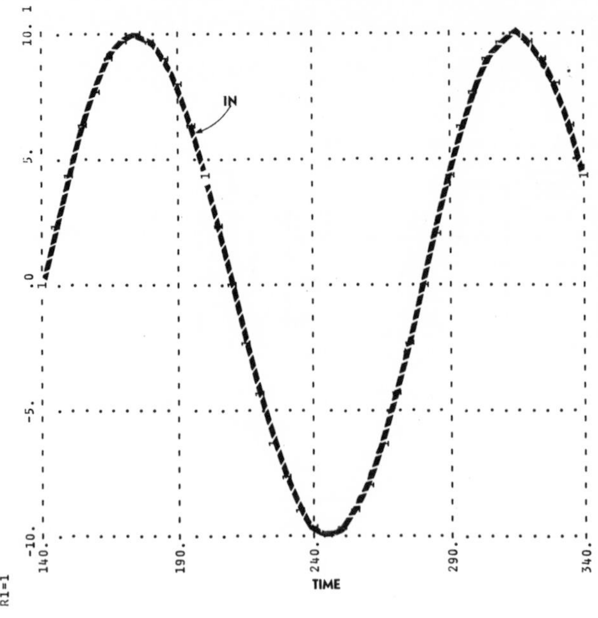

(b)

E9.3 The production process of a plant uses a certain raw material. Exogenous influences determine the usage rate of this material. To insure continuous production, the company keeps an inventory of the raw material. An ordering delay OD of 3 months occurs between the ordering of additional raw material and its arrival in the inventory. In an attempt to guard against inventory shortages, the management has decided to employ the formal policy of ordering exactly the amount of material consumed:

order rate OR = usage rate UR

A representation of the system is shown below.

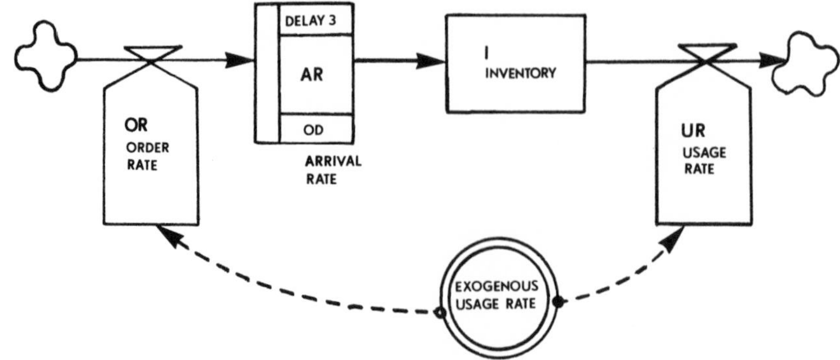

Starting with a non-zero inventory, how will this formal policy work out when faced with growth in usage? Assume the exogenously determined usage rate increases linearly with time. Include a sketch of inventory I over time in your answer.

E9.4 Assume the variables in the system below are in steady state with OR = 1. On Figure (b) sketch the response of rent RENT and average rent AR to a step change in occupancy ratio OR from 1.0 to 1.5.

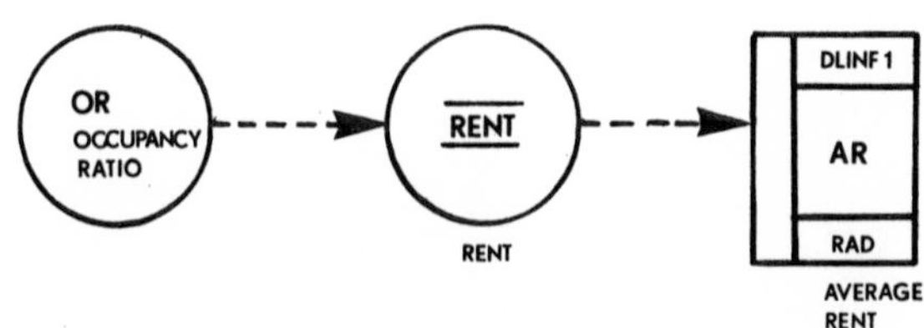

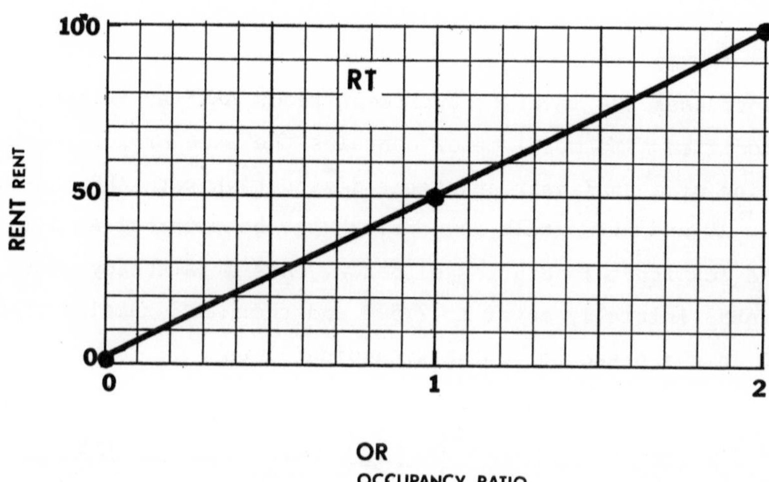


```
RENT.K=TABHL(RT,OR.K,0,2,1)
RT=0/50/100
AR.K=DLINF1(RENT.K,RAD)
RAD=6
```

(a)

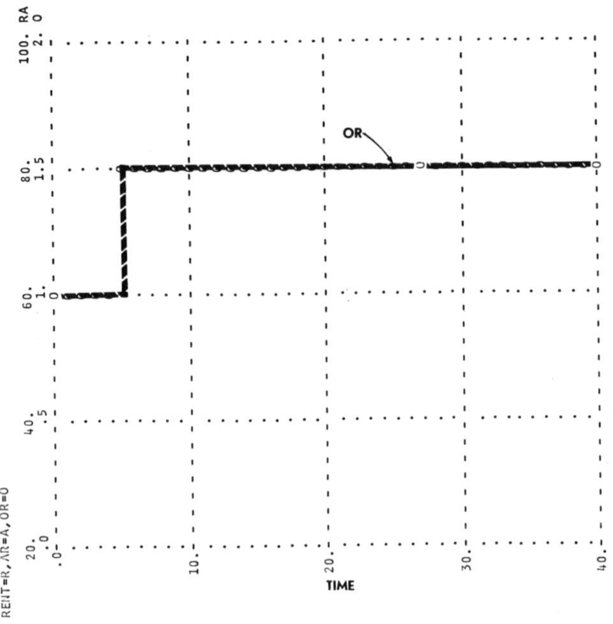

(b)

E9.5 AFR increases the level of apartment rooms ARMS as in question E9.1. The modified ARMS model below contains the same structure with addition of a nonlinear apartment destruction rate ADR. Assume ADR equals AD and ACD equals 24. Sketch the behavior of ARMS in response to the ACR input in Figure (b) below. Assume a steady state system with ARMS initially equal to 2,500 and that ARMS remains close to 2,500 in the presence of the sinusoidal input.

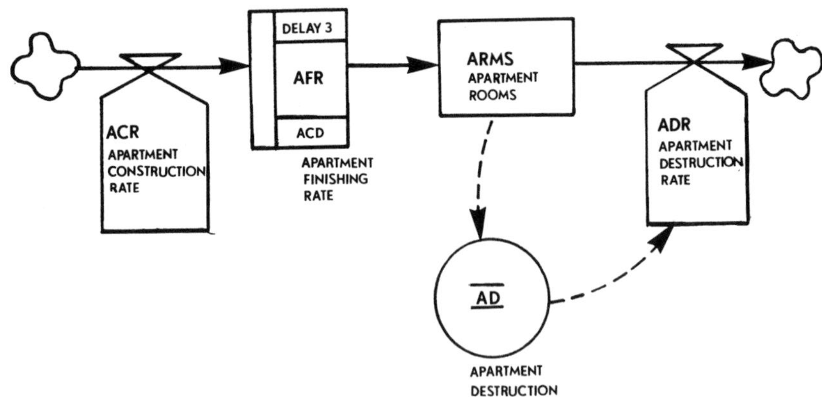

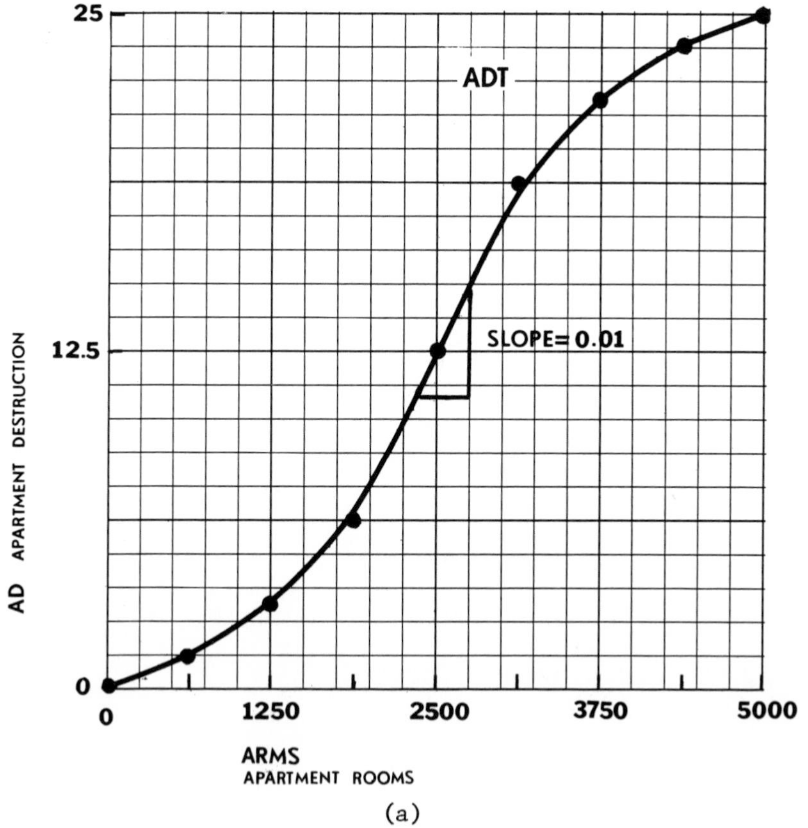
25
12.5
0
AD APARTMENT DESTRUCTION
ADT
SLOPE= 0.01
0
1250
2500
3750
5000
ARMS
APARTMENT ROOMS

(a)

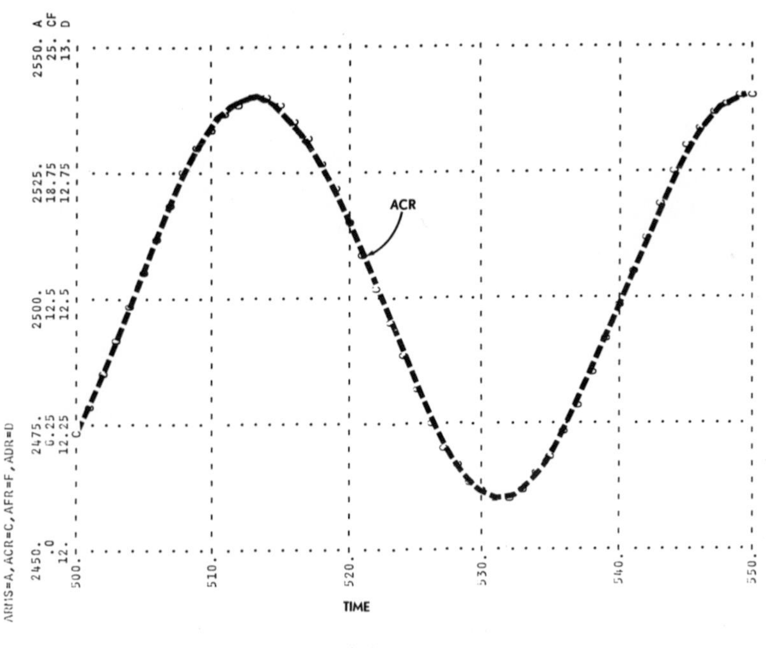
ARMS=A,ACR=C,AFR=F,ADR=D
2450. A
.0 CF
12. D
2475.
6.25
12.25
2500.
12.5
12.5
2525.
18.75
12.75
2550.
25.
13.
ACR
500.
510.
520.
530.
540.
550.
TIME

(b)

E9.6 Given the structure below, sketch the response of DR if the fluctuation in RENT consists of two sinusoidal components, one with a monthly period and one with a yearly period, as given in Figure (b). Assume the system variables initially in equilibrium. Explain why the smooth function acts like a filter.

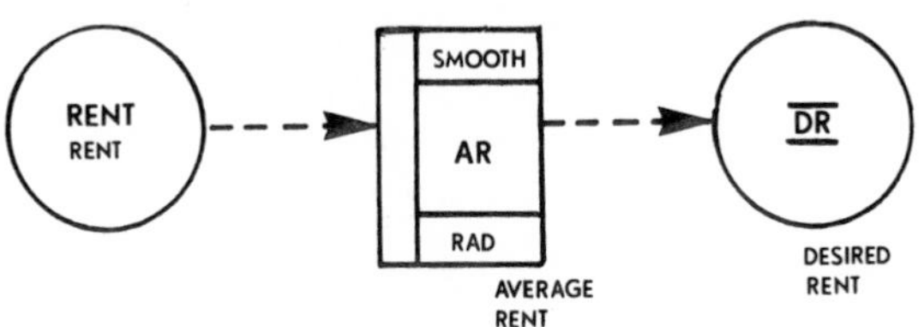


```
A AR.K=SMOOTH(RENT.K,RAD)
C RAD=6
A RENT.K=50+2*SIN(2*PI*TIME.K/1)+10*SIN(2*PI*TIME.K/12)
C PI=3.14
A DR.K=TABLE(DRT,AR.K,0,100,50)
T DRT=0/2500/5000
```

(a)

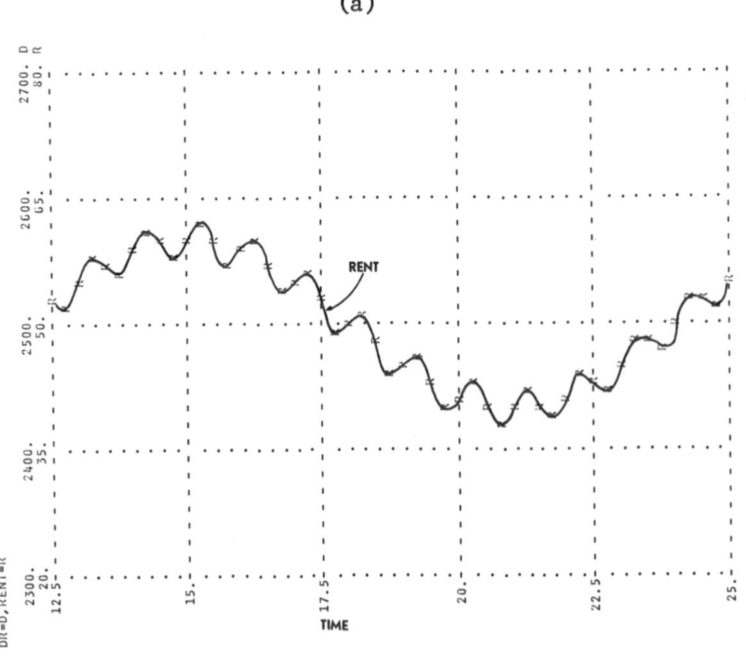

b)

E9.7

Given the input X over time shown below, draw the time-shape of the output Y. Indicate the height, slope, and turning point of Y.

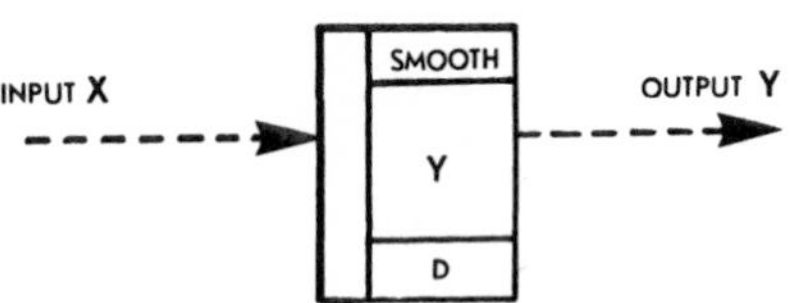


```
A Y.K=SMOOTH(X.K,D)
C D=3
A X.K=TABLE(XT,TIME.K,0,125,25)
T XT=0/10/0/-10/0/10
```

(a)

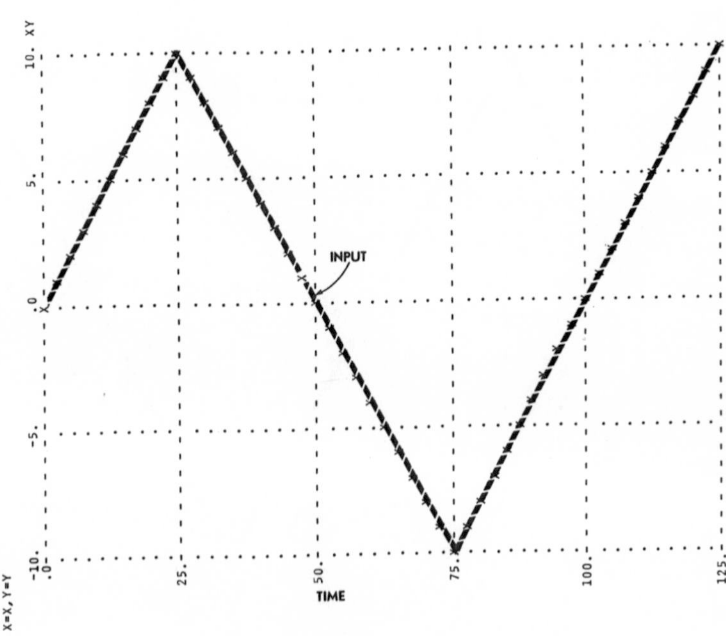

(b)

E9.8 Given the input X over time shown below, draw the time-shape of the output Y. Indicate the height, slope, and turning point of Y.

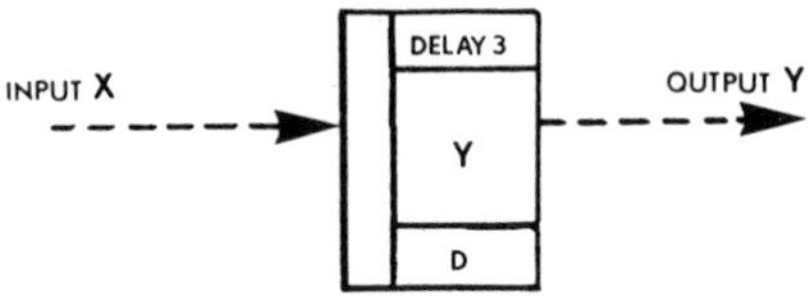


```
A Y.K=DELAY3(X.K,D)
C D=50
A X.K=TABLE(XT,TIME.K,0,125,25)
T XT=0/10/0/-10/0/10
```

(a)

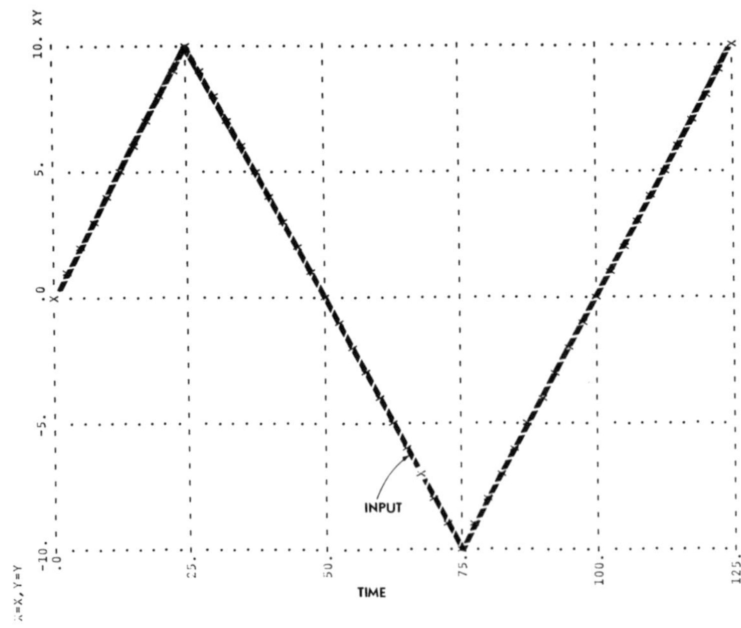

(b)

Solution 9

(Note: In the following solutions, T/P equals the ratio of the time constant T of the delay to the period P of the input sinusoid. Figure E9-11 can be used to determine phase and gain values from the T/P value.)

For T/P = ACD/P = 24/36 **S9.1**

Gain G = 0.18

Phase Shift Ø = 150° = (7/16)P or 16 months.

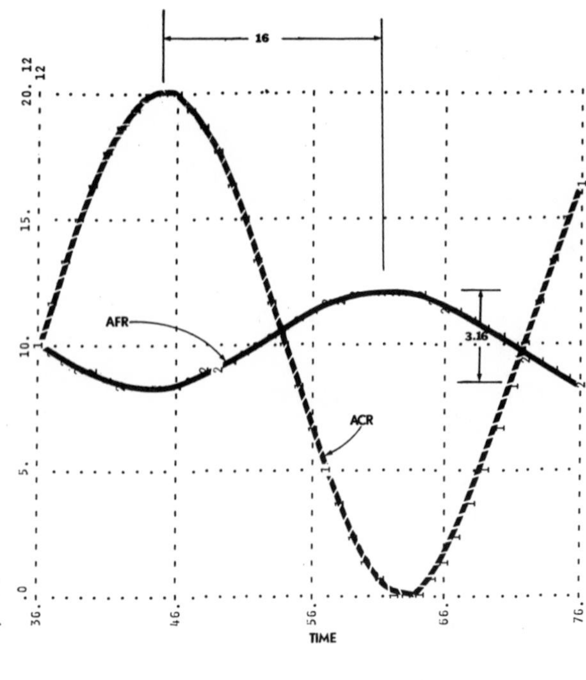

S9.2

a) For IN = 5 and system in steady state:

R1 = R2 = 5

R2 = R3+R4 = L1/30

Hence,

L1 = 30(R2) = 150

R3+R4 = L2(1/70+1/100) = 5

Hence,

L2 = 206

R3 = L2/70 = 2.94

R4 = L2/100 = 2.06

b) Calculation of gain G_1 between R1 and R2:

T/P = 30/140 = 0.21

Gain G_1 = 0.62

Phase shift $\emptyset_1$ = (5/32)P = (5/32)(140) = 22.

Calculation of gain G_2 between R2 and (R3+R4):

R3+R4 = L2/(1/70+1/100)

= L2(170/7000)

= L2(0.0243)

Equivalent time constant = 1/0.0243 = 41

T/P = 41/140 = 0.295

Gain G_2 = 0.48

Phase Shift $\emptyset_2$ = (3/16)P = 26.

Calculation of gain G_3 between (R3+R4) and L2:

R3+R4 = L2/41

L2 = 41(R3+R4)

Gain G_3 = 41

Phase Shift $\emptyset_3$ = 0.

Calculation of gain G_4 between L2 and R3:

R3 = L2/70

Gain G_4 = 1/70

Phase Shift $\emptyset_4$ = 0.

Calculation of total gain G_T between R1 and R3:

$G_T = (G_1)(G_2)(G_3)(G_4)$ = (0.62)(0.48)(41)(1/70) = 0.175

but G_T = R3/R1 = 0.175,

thus,

R3 = 0.175(10) = 1.75

Total Phase Shift $\emptyset_T$:

$$\emptyset_T = \emptyset_1 + \emptyset_2 + \emptyset_3 + \emptyset_4 = 22° + 26° + 0° + 0° = 48°$$

The relationship between R1, R2 and R3 appears below:

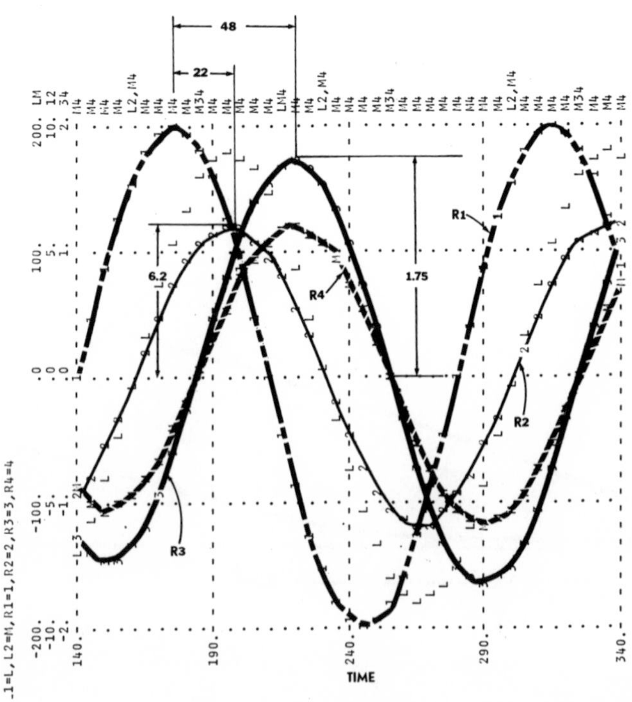

Because of the delay between OR and AR, AR will be less than OR S9.3
and UR at any point in time as long as UR grows linearly over time. A net outflow, the difference between UR and AR, always draws down the inventory I. After roughly 6 months, or two delay times, a constant outflow will produce a continual linear decrease in inventory which falls to zero as shown. Obviously, the policy fails to achieve its goal of maintaining an adequate inventory over the long run.

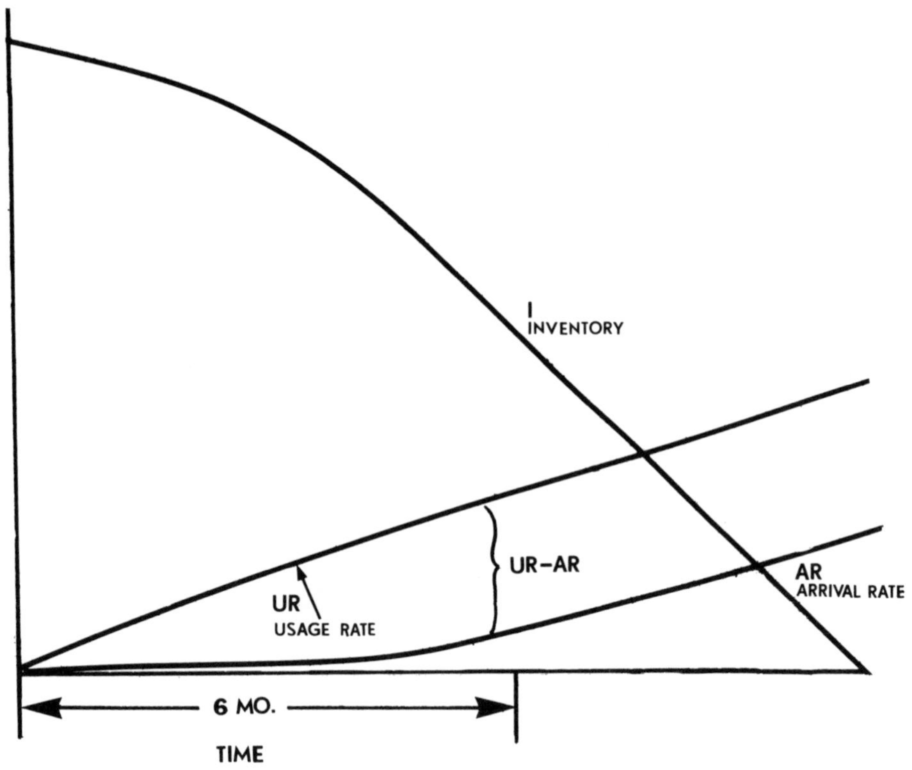
I
INVENTORY
UR
USAGE RATE
UR–AR
AR
ARRIVAL RATE
6 MO.
TIME

S9.4

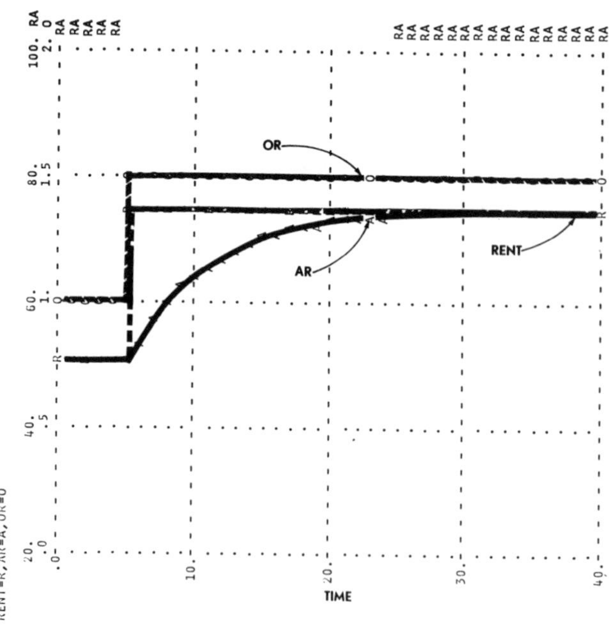
OR
AR
RENT
TIME
RENT=R, AR=A, OR=O

The portion of the AD table in which the system operates can be linearly approximated by: S9.5

AD = Y+0.01*ARMS

where Y equals the intercept value of AD. The time constant from the above relationship equals 1/0.01 or 100. The gain between ADR and AFR from T/P = 100/36 equals 0.06. The phase shift equals 90 degrees or 9 months. The following table summarizes the gain and phase shift values.

	ACR to AFR	AFR to ADR	ADR to ARMS	Total ACR to ARMS
Gain	0.18	0.06	100	1.08
Phase Shift	16	9	0	25

Since ARMS/ACR = Gain = 1.08,

ARMS = (1.08)ACR.

The response over time appears below.

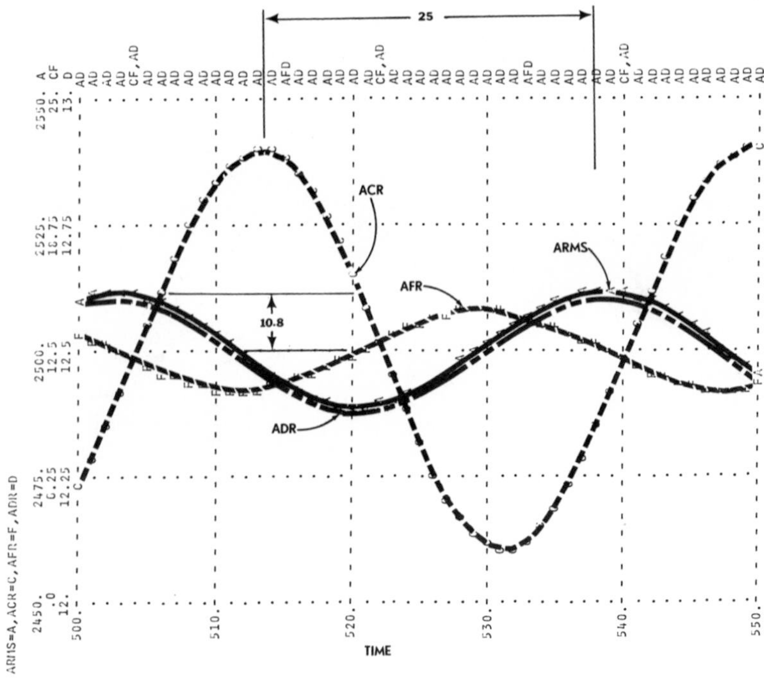

S9.6 Calculation of gain and phase shift of AR relative to RENT:

	One-month Cycle	Twelve-month Cycle
T/P	6	0.5
Gain	0.02	0.3
Phase Shift	90°	70°
Amplitude	(0.02)(2)	(0.3)(10)

Since DR = 50*AR, the 1-month cycle amplitude of DR is 50(0.02)(2) or 2. The amplitude of the 12-month cycle is (50)(0.3)(10) or 150. We can ignore the 1-month cycle, which is strongly attenuated (or filtered out) relative to the 12-month cycle, as shown below. Exponential smoothing filters out high frequency oscillations.

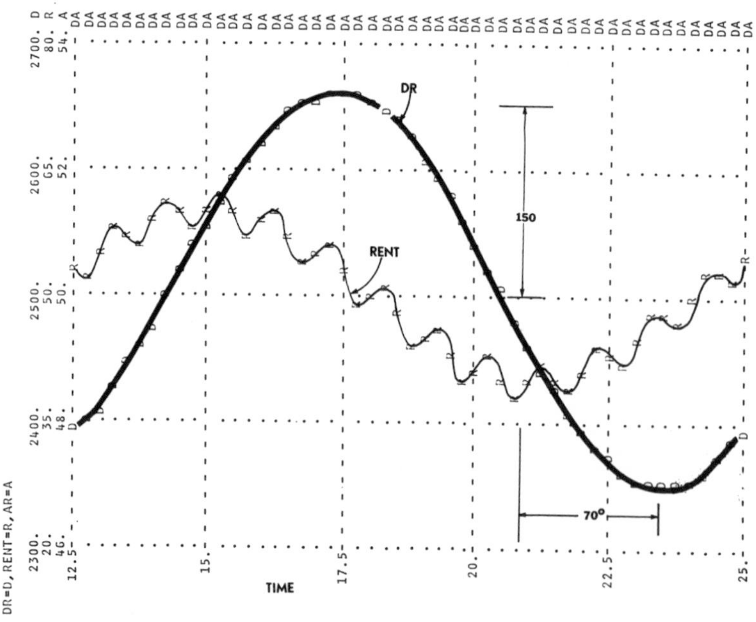

The input X can be approximated by a sinusoid with a period P of 100. Using a first-order delay we can find the gain G and phase shift Ø for a T/P value of 3/100 (or 0.03): **S9.7**

Gain = 0.95

Phase Shift Ø = (1/32)P = 3.14

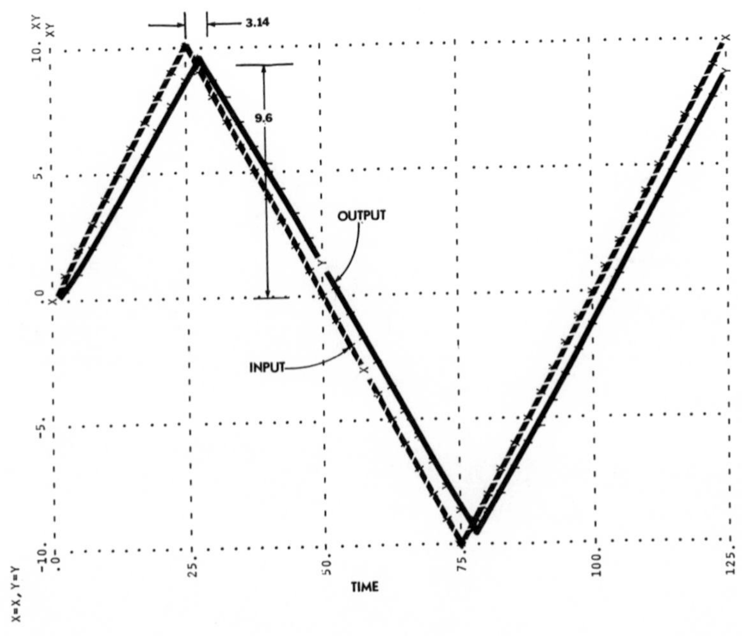

S9.8 Again, assume a sinusoidal input X for T/P equal to 0.5 and a third-order delay:

Gain = 0.3

Phase Shift Ø = (3/8)P = 37.5

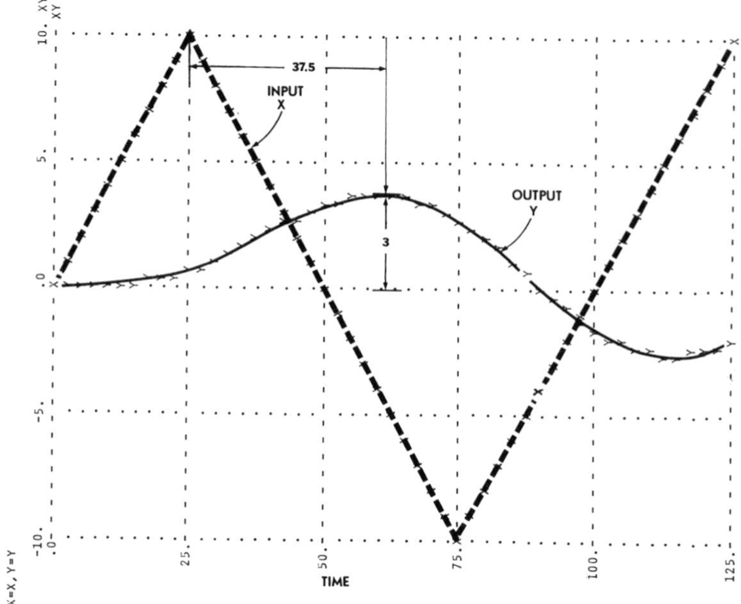

Exercise 10
Commodity Production Cycle Model

by Dennis L. Meadows

This exercise presents a general model of commodity production cycles.[1] *The first part focuses on conceptualization in terms of flow diagramming and equation writing. The student will make structural additions to the basic model outlined in the exercise. The second part investigates response of elements of the commodity model to cyclical inputs. It requires familiarity with the open-loop analysis previously introduced in Exercise 9.*

INTRODUCTION

Any raw material with the following characteristics is a commodity:

1. It is undifferentiable. Thus no producer can obtain higher price or better trading terms through advertising or product modification. He must accept the open market price which prevails at the time of sale.
2. Variable production costs (labor and materials) are small compared to fixed costs. Thus, in the short term, output of the commodity will be relatively insensitive to price changes.
3. For commodity users, the commodity price is only a small fraction of the final product cost. Consumption is relatively price inelastic.

---

[1]The model is based on Dennis L. Meadows, The Dynamics of Commodity Production Cycles (Cambridge: Wright-Allen Press, 1970).

More than twenty important commodities, (e.g. oil, cocoa, sugar, jute and tin) change hands in the commodity trade. The commodity trade accounts for 90 percent of the foreign exchange earnings of underdeveloped countries. The unstable price and production rate of commodities generally varies from one year to the next by 5 to 25 percent. Beef prices, as shown below, illustrate this tendency.

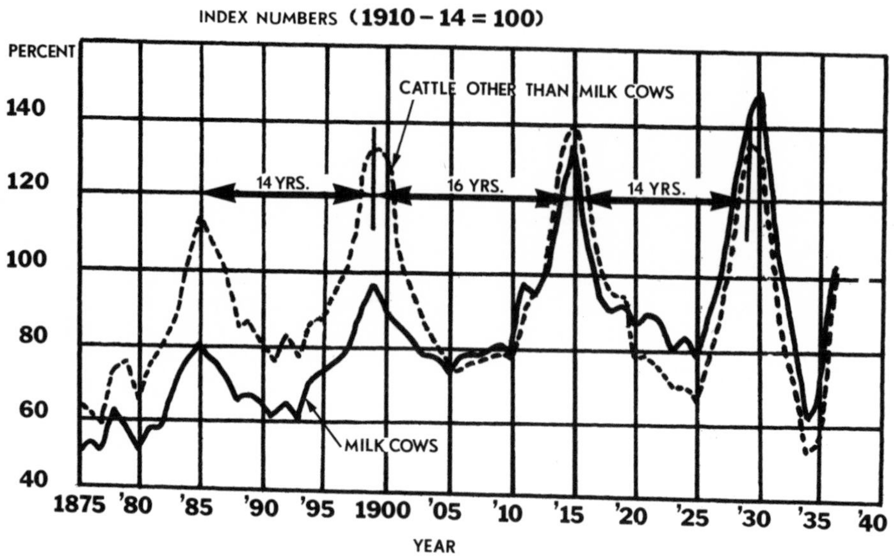

FIGURE E10-1
Beef cycles

Production cycles cause immense difficulties for the producers, distributors, manufacturers and consumers. As a result, individual countries and international agencies have invested many millions of dollars in various stabilization schemes. None have been totally successful and most have been absolute failures. The following pages describe in brief a possible structure underlying the long-term commodity fluctuations.

MODEL DESCRIPTION

We base this model on the interactions among three market sectors: production, distribution, and consumption. Price (per unit) links the three sectors. Producers try to adjust their production capacity to the profit maximizing level for a given market price. Distributors try to adjust the market price to maintain an optimal inventory. Consumers respond to the market price as they attempt to maximize their own utility. Figure E10-2 shows the model structure.

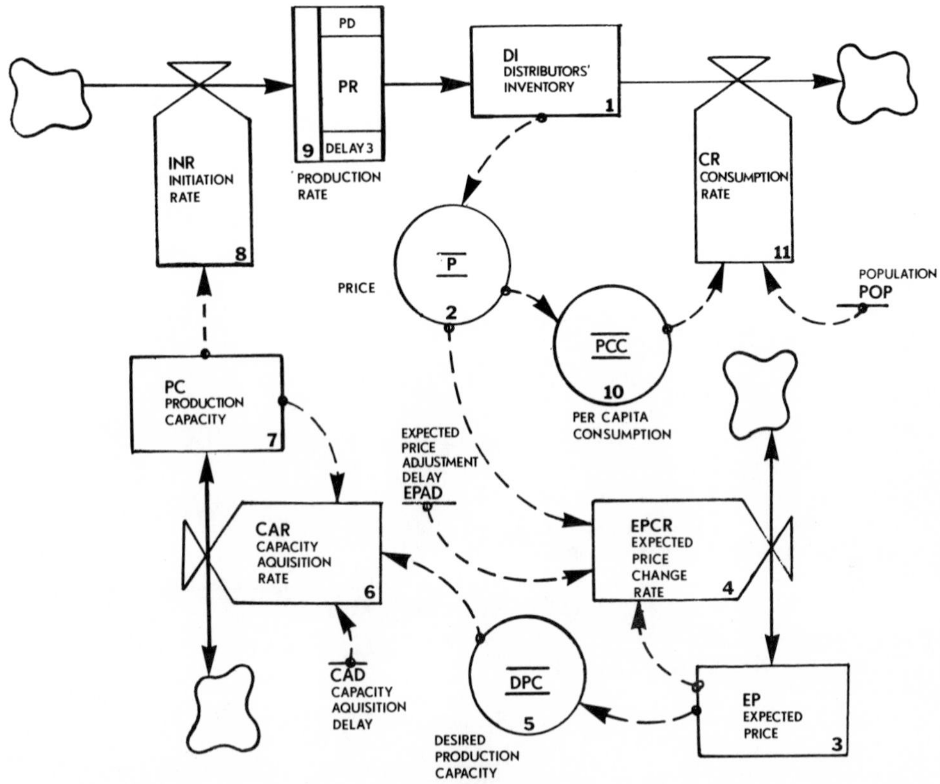

FIGURE E10-2

Commodity model flow diagram

The DYNAMO equations of the model are:

```
 INVENTORY-PRICE SECTOR

DI.K=DI.J+(DT)(PR.JK-CR.JK)                                  1, L
DI=DII                                                       1.1, N
DII=6000                                                     1.2, C
     DI      - DISTRIBUTORS' INVENTORY (UNITS)
     PR      - PRODUCTION RATE (UNITS/MONTH)
     CR      - CONSUMPTION RATE (UNITS/MONTH)
     DII     - INITIAL INVENTORY (UNITS)
```

DYNAMO equations for commodity model (cont'd)

```
P.K=TABLE(PT,DI.K,0,10000,2000)                              2, A
PT=100/94/80/50/20/10                                        2.1, T
    P       - MARKET PRICE (DOLLARS/UNIT)
    PT      - PRICE TABLE
    DI      - DISTRIBUTORS' INVENTORY (UNITS)

 PRODUCTION SECTOR

EP.K=EP.J+(DT)(EPCR.JK)                                      3, L
EP=EPI                                                       3.1, N
EPI=50                                                       3.2, C
    EP      - EXPECTED PRICE (DOLLARS/UNIT)
    EPCR    - EXPECTED PRICE CHANGE RATE (DOLLARS/UNIT/
                MONTH)
    EPI     - INITIAL EXPECTED PRICE (DOLLARS/UNIT)

EPCR.KL=(P.K-EP.K)/EPAD                                      4, R
EPAD=5                                                       4.1, C
    EPCR    - EXPECTED PRICE CHANGE RATE (DOLLARS/UNIT/
                MONTH)
    P       - MARKET PRICE (DOLLARS/UNIT)
    EP      - EXPECTED PRICE (DOLLARS/UNIT)
    EPAD    - EXPECTED PRICE ADJUSTMENT DELAY (MONTHS)

DPC.K=TABLE(DPCT,EP.K,0,100,20)                              5, A
DPCT=0/40/200/1000/1200/1280                                 5.1, T
    DPC     - DESIRED PRODUCTION CAPACITY (UNITS/MONTH)
    DPCT    - DESIRED PRODUCTION CAPACITY TABLE
    EP      - EXPECTED PRICE (DOLLARS/UNIT)

CAR.KL=(DPC.K-PC.K)/CAD                                      6, R
CAD=4                                                        6.1, C
    CAR     - CAPACITY ACQUISITION RATE (UNITS/MONTH/
                MONTH)
    DPC     - DESIRED PRODUCTION CAPACITY (UNITS/MONTH)
    PC      - PRODUCTION CAPACITY (UNITS/MONTH)
    CAD     - CAPACITY ACQUISITION DELAY (MONTHS)

PC.K=PC.J+(DT)(CAR.JK)                                       7, L
PC=PCI                                                       7.1, N
PCI=600                                                      7.2, C
    PC      - PRODUCTION CAPACITY (UNITS/MONTH)
    CAR     - CAPACITY ACQUISITION RATE (UNITS/MONTH/
                MONTH)
    PCI     - INITIAL PRODUCTION CAPACITY (UNITS/MONTH)
INR.KL=PC.K                                                  8, R
    INR     - INITIATION RATE (UNITS/MONTH)
    PC      - PRODUCTION CAPACITY (UNITS/MONTH)

PR.KL=DELAY3(INR.JK,PD)                                      9, R
PD=12                                                        9.1, C
    PR      - PRODUCTION RATE (UNITS/MONTH)
    INR     - INITIATION RATE (UNITS/MONTH)
    PD      - PRODUCTION DELAY (MONTHS)
```

DYNAMO equations for commodity model (cont'd)

```
CONSUMPTION SECTOR

PCC.K=TABLE(PCCT,P.K,0,100,20)                            10, A
PCCT=7/6/4/2/1/0                                          10.1, T
    PCC    - PER CAPITA CONSUMPTION (UNITS/MONTH/PERSON)
    PCCT   - PER CAPITA CONSUMPTION TABLE
    P      - MARKET PRICE (DOLLARS/UNIT)

CR.KL=POP*PCC.K                                           11, R
POP=200                                                   11.1, C
    CR     - CONSUMPTION RATE (UNITS/MONTH)
    POP    - POPULATION (PEOPLE)
    PCC    - PER CAPITA CONSUMPTION (UNITS/MONTH/PERSON)
CONTROL STATEMENTS

DT=.5
LENGTH=120
PLTPER=3
PLOT DI=I(2000,8000)/P=P(20,80)/CR=C,PR=*(0,900)
```

FIGURE E10-3

Figure E10-4 indicates the relationship between the producers' desired production capacity and the market price they expect. Producers form their current expected market price by exponentially smoothing actual price over the past five months. Having determined their desired production capacity, producers either increase or lower their actual production capacity to eliminate one-fourth of the discrepancy between actual capacity and desired capacity in one month. The initiation of new units of commodity is determined solely by production capacity. A third-order exponential production delay connects initiation of a unit and its receipt in the distributors' inventory. The delay time equals twelve months.

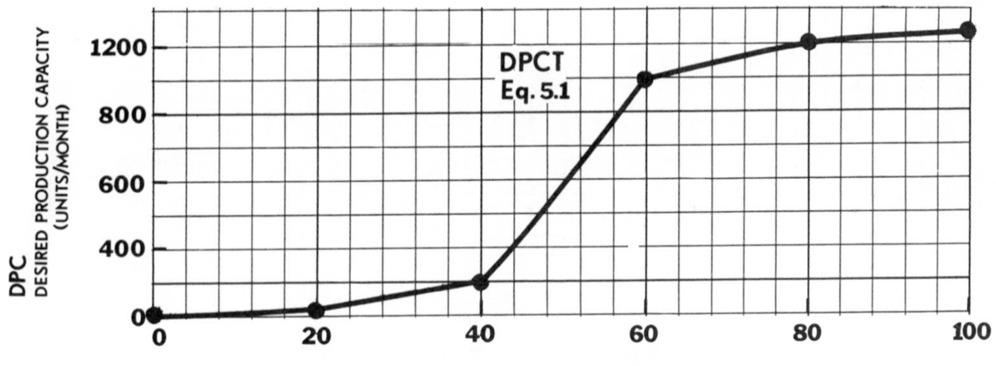

FIGURE E10-4

Whenever the inventory fluctuates, distributors adjust actual market price to bring their inventory within an acceptable range. Figure E10-5 depicts this decision rule.

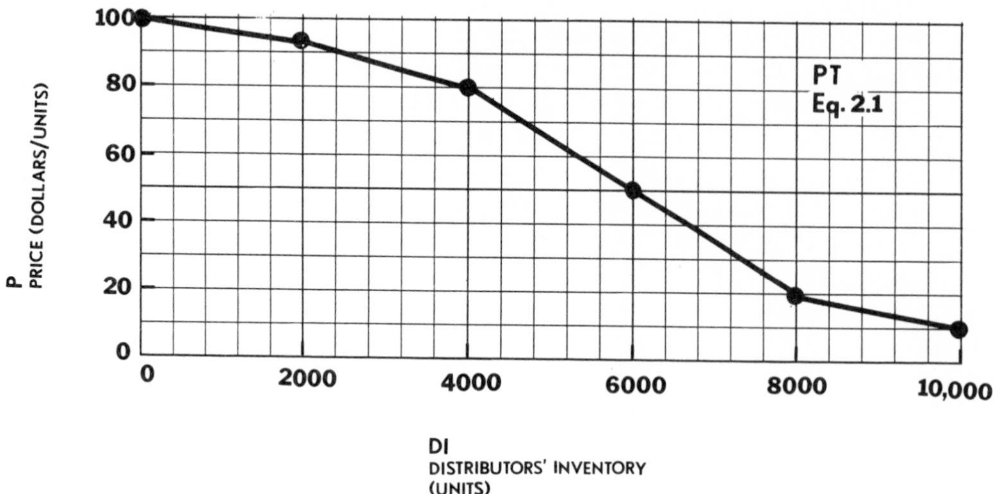

E10-5 Price versus inventory relationship

Consumers obtain their units from the distributors' inventory. Figure E10-6 indicates the relationship between the per capita consumption rate and actual market price. Population remains constant at 200 people.

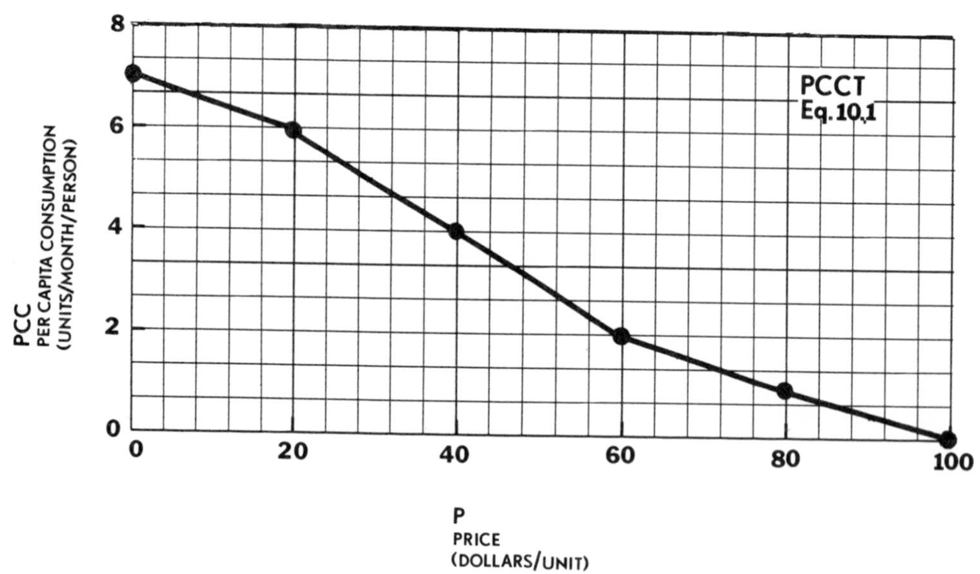

E10-6 Consumption versus price relationship

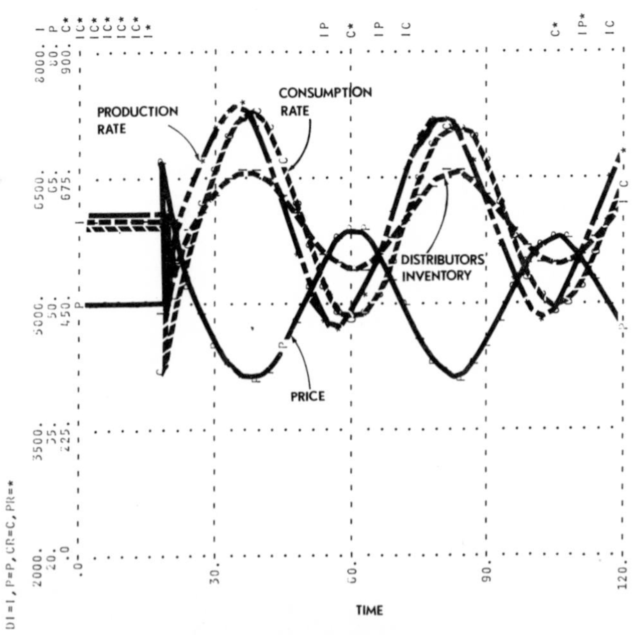

FIGURE E10-7

System response to an initial disequilibrium caused by lowering inventory from the equilibrium value at month 15

STRUCTURAL ADDITIONS

System dynamics requires the ability to perceive and model structure essential to observed behavior in the real world. The preceding commodity model flow diagram and equations incorporate the basic structure for explaining long-term commodity fluctuations. However, the highly simplified model does not include several real-world features which might have importance for understanding commodity cycles. We will describe several of those structural features and then incorporate the new structure into the existing model.

For each description, redraw the appropriate portion of the flow diagram and write necessary equation changes or additions to integrate the new structure into the overall model. Keep the structural additions as simple as possible. Treat each description separately. Provide initial values and parameter values where appropriate. Sketch all table functions. Precise values are not important.

E10.1 Production capacity PC does not last indefinitely, but depreciates gradually until no longer useful. Assume a useful life of ten years. Represent depreciation in the model structure.

E10.2 Population does not, in general, remain constant. Provide structure and equations to test the impact on commodity cycles of different rates of exponential population growth.

E10.3 In some commodities the production delay PD depends on price P. For example, holding livestock longer before slaughter may become desirable. For pigs, the delay between initiation (i.e., breeding) and production (i.e., addition to the inventory) may vary in length from 10-14 months when the price varies between 30 and 70. Represent this in the model.

E10.4

a) Inventory may not directly determine price, but simply determine the percent and the direction of price change. Offer a new structure to study the importance of this difference. For your table of percentage change in price, supply values consistent with the original model.

b) Represent the case where inventory determines an "equilibrium" value for price. Actual price adjusts toward the equilibrium with an adjustment delay of two months.

E10.5 Some seasonal factors often influence production rate. For example, hog breeders do not evenly breed their sows through the year. The figure below summarizes the empirical data on birth rate (and by implication the breeding or initiation rate INR) for 1962. The figure indicates the percent of the total year's farrowings which took place in each month. In the basic model, initiation rate INR (i.e., breeding rate) equaled production capacity PC. Make a revised diagram and equation set to study the effect of the relationship shown in the figure on total system stability. The change requires no additional levels.

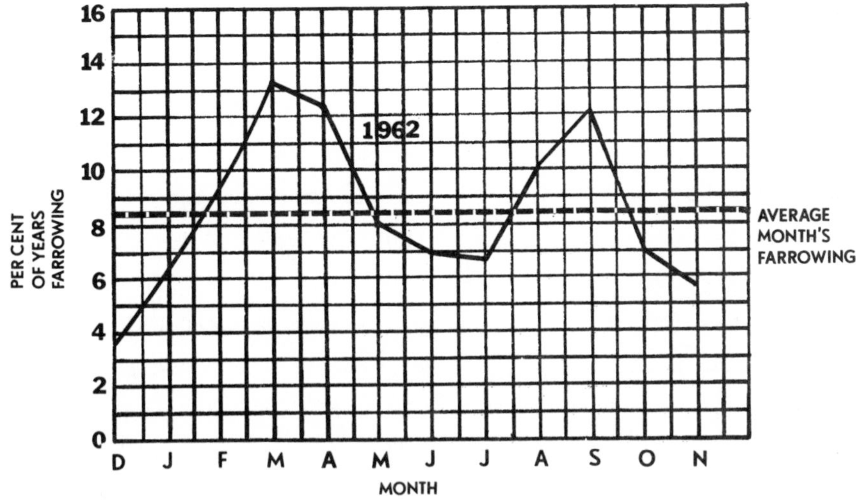

Seasonal Farrowing

Source: Meadows, Dynamics of Commodity Production Cycles, p. 40.

Price variability forces distributors and manufacturers to bear some risk for some metal and fiber commodities: tin, copper, lead, jute and cotton. Investment in the development of synthetic substitutes is proportional to the average risk over the past few years. The more substitutes available, the lower per capita consumption PCC at any given price level. This reformulation will probably require an additional level (synthetic substitutes) and two smoothing functions (average risk and average price). **E10.6**

For some commodities, such as copper, we cannot treat consumption as a sink. About 40 percent of all copper eventually finds its way into a scrap inventory. As price increases, more and more of this scrap returns to distributors' inventory DI. About a twenty-year delay connects consumption rate and scrap accumulation rate. Represent this in the model. **E10.7**

DELAY ANALYSIS

Response to Price Variation.

E10.8 Imagine an exogenous influence in Figure E10-2 which acts on price P as a sine wave with a 1-month period, an amplitude of 5(dollars/unit) and mean of zero. The equation for price P is:

```
A P.K = TABLE(PT,DI.K,0,10000,20000)+5*SIN(6.28*TIME.K/1)
```

How will this factor influence the initiation rate INR? Determine the magnitude and phase lag of the impact. The phase and gain charts of Figure E9-11 in Exercise 9 may be useful for your calculations. Assume the system is open between DI and P.

Response to Seasonal Farrowing.

E10.9 What influence does seasonal farrowing in the figure in E10.5 have on the production rate PR in Figure E10-2? On the price P? Determine the magnitude and phase lag. Remember that farrowing directly affects the initiation rate. Assume the system is in equilibrium and open between P and EPCR. Approximate the farrowing seasonal with a sine wave period of six months and amplitude of 0.5 as in E10.5.

Solution 10

STRUCTURAL ADDITIONS

S10.1

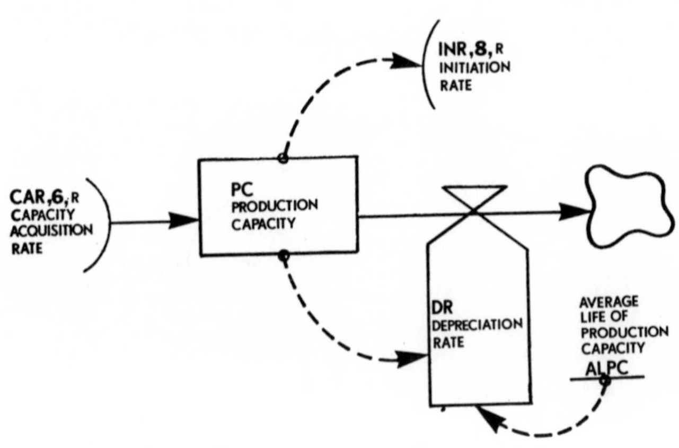

L	PC.K = PC.J+(DT)(CAR.JK-DR.JK)	Production Capacity (Units/month)
R	DR.KL = PC.K/ALPC	Depreciation Rate (Units/month/month)
C	ALPC = 120	Average Life of Production Capacity (Months)

S10.2

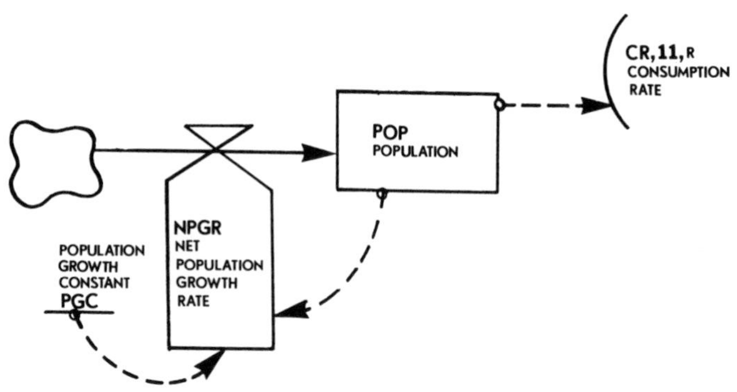

L	POP.K = POP.J+(DT)(NPGR.JK)	Population (People)
N	POP = 200	Initial Population (People)
R	NPGR.KL = (POP.K)(PGC)	Net Population Growth Rate (People/month)
C	PGC = Constant	Population Growth Constant (Fraction/month)

S10.3

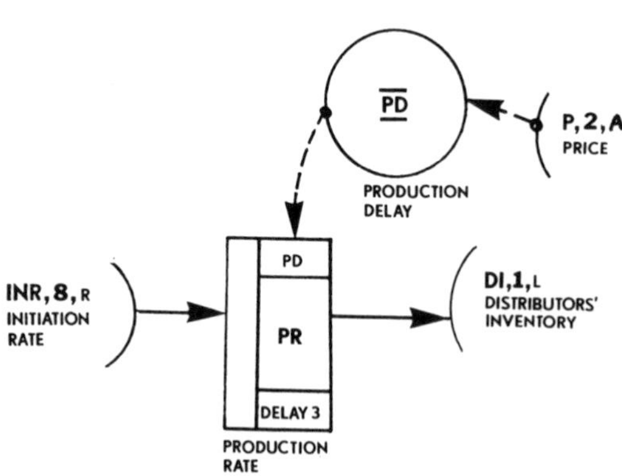

R	PR.KL = DELAY3(INR.JK,PD.K)	Production Rate (Units/month)
A	PD.K = TABHL(PDT,P.K,30,70,20)	Production Delay (Months)
T	PDT = 10/12/14	Production Delay Table

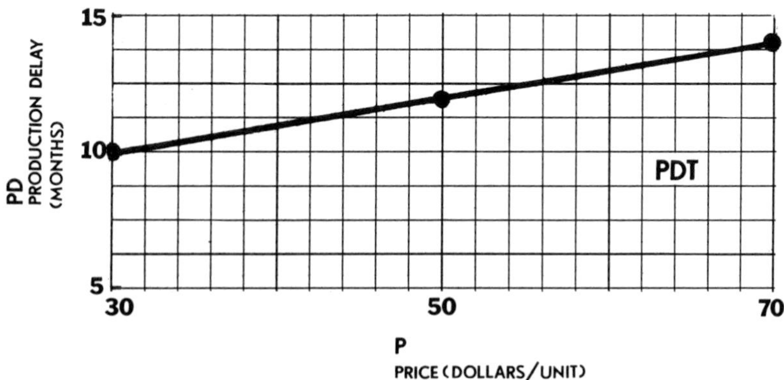

a) **S10.4**

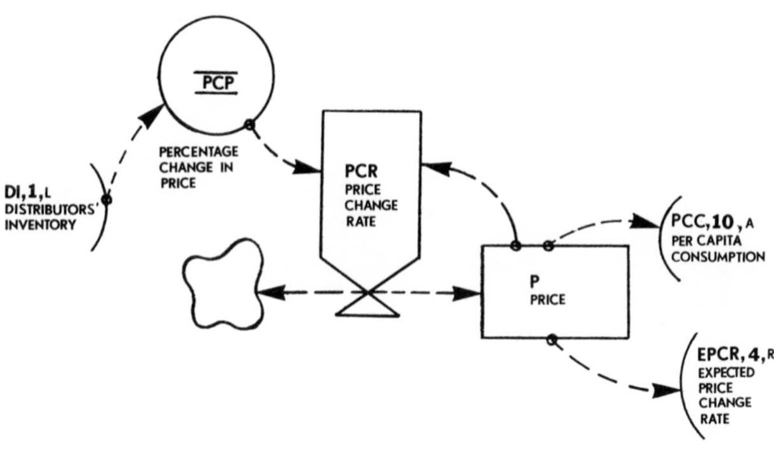


```
L  P.K = P.J+(DT)(PCR.JK)                  Price (Dollars/unit)
N  P = 80                                  Initial Price (Dollars/unit)
R  PCR.KL = (PCP.K)(P.K)                   Price Change Rate (Dollars/
                                             unit/month)
A  PCP.K = TABHL(PCPT,DI.K,0,12000,2000)   Percentage Change in Price
                                             (Fraction/month)
T  PCPT = .1/.1/.08/0/-.08/-.1/-.1         Percentage Change in Price
                                             Table
```

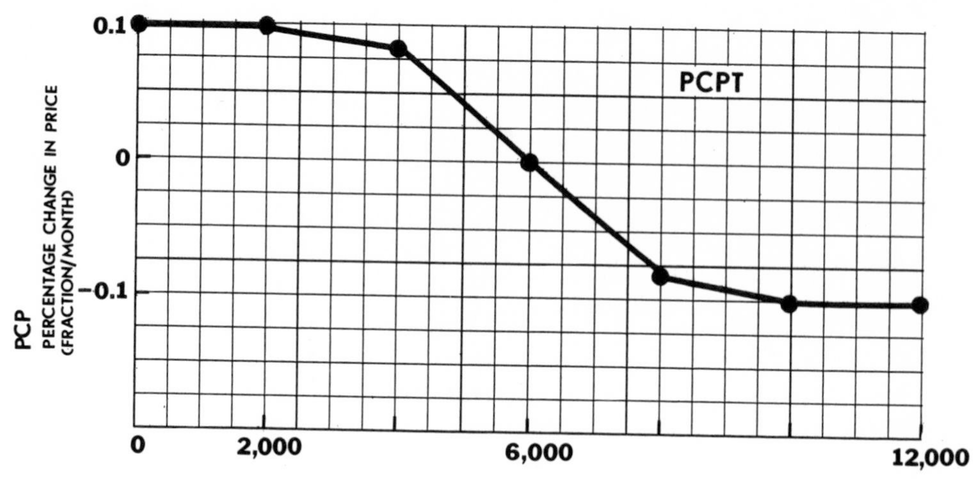
PCPT
0.1
0
-0.1
PCP
PERCENTAGE CHANGE IN PRICE
(FRACTION/MONTH)
0
2,000
6,000
12,000
DI
DISTRIBUTORS'
INVENTORY (UNITS)

b)

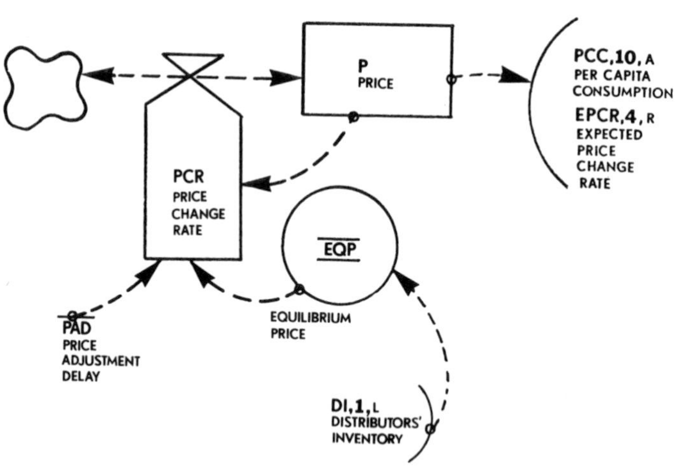
P
PRICE
PCC,10,A
PER CAPITA
CONSUMPTION
EPCR,4,R
EXPECTED
PRICE
CHANGE
RATE
PCR
PRICE
CHANGE
RATE
EQP
EQUILIBRIUM
PRICE
PAD
PRICE
ADJUSTMENT
DELAY
DI,1,L
DISTRIBUTORS'
INVENTORY

L	P.K = P.J+(DT)(PCR.JK)	Price (Dollars/unit)
N	P = 80	Initial Price (Dollars/unit)
R	PCR.KL = (EQP.K-P.K)/PAD	Price Change Rate (Dollars/unit/month)
C	PAD = 2	Price Adjustment Delay (Months)
A	EQP.K = TABLE(EQPT,DI.K,0,10000,2000)	Equilibrium Price (Dollars/unit)
T	EQPT = 100/94/80/50/20/10	Equilibrium Price Table

S10.5 Approximate seasonal farrowing with a sine wave of period six months and amplitude of 0.5.

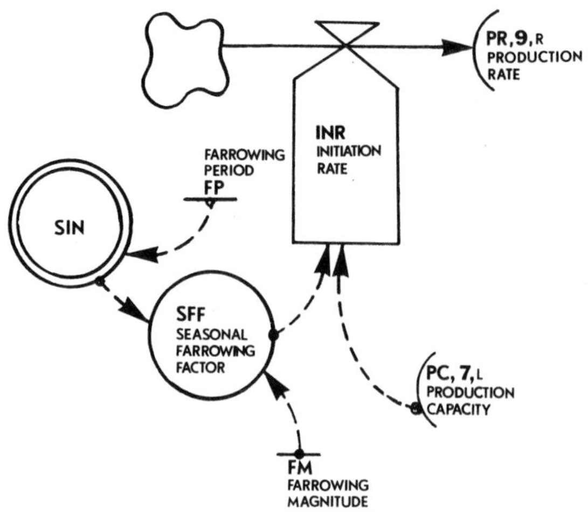

R	INR.KL = (PC.K)(SFF.K)	Initiation Rate (Units/month)
A	SFF.K = 1+(FM)(SIN((6.28*TIME.K)/FP))	Seasonal Farrowing Factor (Dimensionless)
C	FM = .5	Farrowing Magnitude (Dimensionless)
C	FP = 6	Farrowing Period (Months)

S10.6

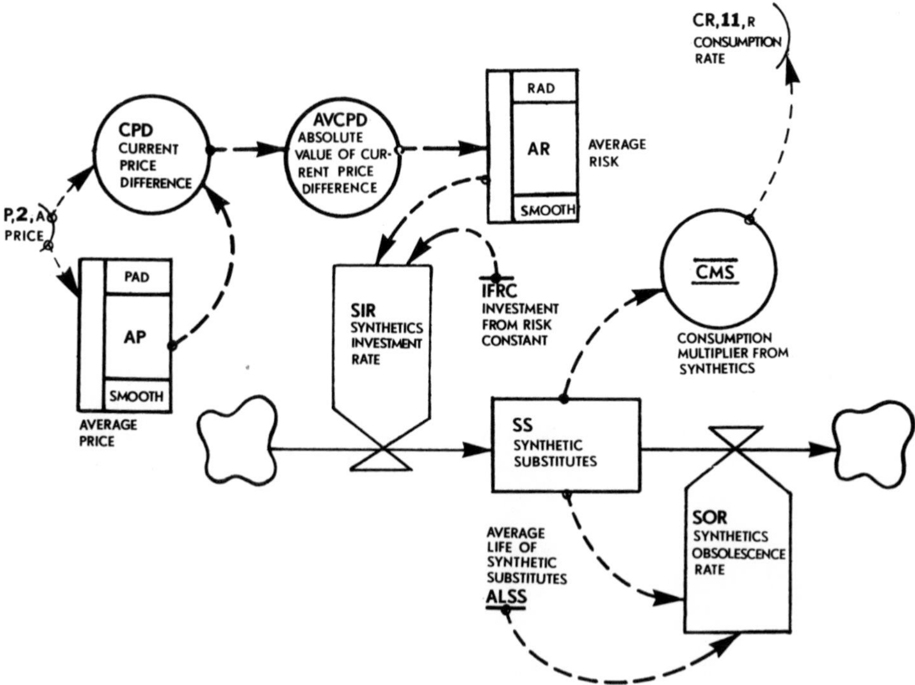
CR,11,R
CONSUMPTION RATE
CPD
CURRENT PRICE DIFFERENCE
AVCPD
ABSOLUTE VALUE OF CURRENT PRICE DIFFERENCE
RAD
AR
SMOOTH
AVERAGE RISK
P,2,A
PRICE
PAD
AP
SMOOTH
AVERAGE PRICE
SIR
SYNTHETICS INVESTMENT RATE
IFRC
INVESTMENT FROM RISK CONSTANT
CMS
CONSUMPTION MULTIPLIER FROM SYNTHETICS
SS
SYNTHETIC SUBSTITUTES
SOR
SYNTHETICS OBSOLESCENCE RATE
AVERAGE LIFE OF SYNTHETIC SUBSTITUTES
ALSS

```
A  AP.K = SMOOTH(P.K,PAD)                  Average Price (Dollars/unit)
C  PAD =                                   Price averaging delay (Months)
A  CPD.K = P.K-AP.K                        Current Price Difference
                                             (Dollars/unit)
A  AVCPD.K = MAX(CPD.K,-CPD.K)             Absolute Value of Current Price
                                             Difference (Dollars/unit)
L  AR.K = SMOOTH(AVCPD.K,RAD)              Average Risk (Dollars/unit)
C  RAD = 24                                Risk Averaging Delay (Months)
R  SIR.KL = AR.K*IFRC                      Synthetics Investment Rate
                                             (Units/month)
C  IFRC =                                  Investment From Risk Constant
                                             ((Units/month)/(dollars/unit))
L  SS.K = SS.J+(DT)(SIR.JK-SOR.JK)         Synthetics Substitutes (Units)
N  SS =                                    Initial Value (Units)
R  SOR.KL = SS.K/ALSS                      Synthetics Obsolescence Rate
                                             (Units/month)
C  ALSS =                                  Average Life of Synthetic
                                             Substitutes (Months)
A  CMS.K = TABLE(CMST,SS.K,0,100,25)       Consumption Multiplier from
                                             Synthetics (Dimensionless)
T  CMST = 1/1/.6/.2/.2                     Consumption Multiplier from
                                             Synthetics Table
R  CR.KL = POP*PCC.K*CMS.K                 Consumption Rate (Units/month)
```

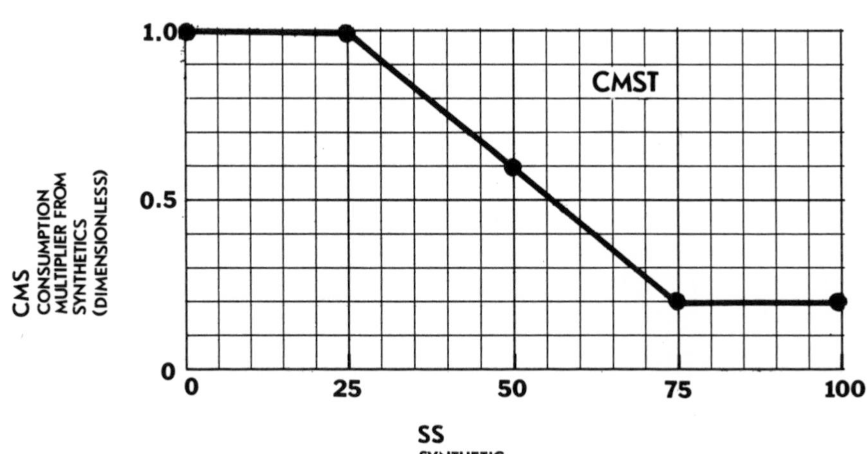

S10.7

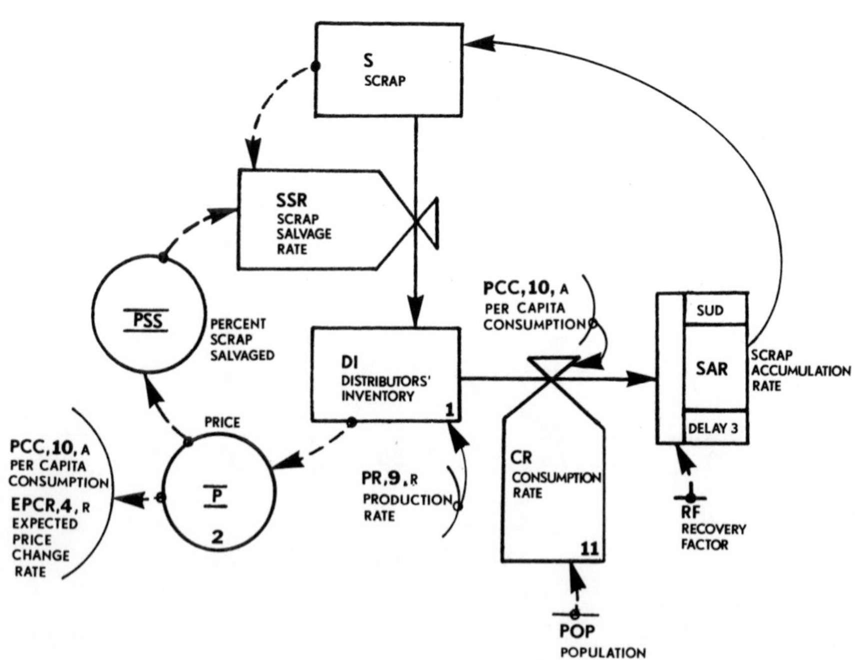

R	SAR.KL = DELAY3(CR.JK,SUD)*RF	Scrap Accumulation Rate (Units/month)
C	SUD = 240	Scrap Usage Delay (Months)
C	RF = .4	Recovery Factor (Dimensionless)
L	S.K = S.J+(DT)(SAR.JK-SSR.JK)	Scrap (Units)
N	S = 4000	Initial Value
R	SSR.KL = (PSS.K)(S.K)	Scrap Salvage Rate (Units/month)
A	PSS.K = TABLE(PSST,P.K,0,100,25)	Percent Scrap Salvaged (Fraction/month)
T	PSST = 0/0/.1/.2/.4	Percent Scrap Salvaged Table

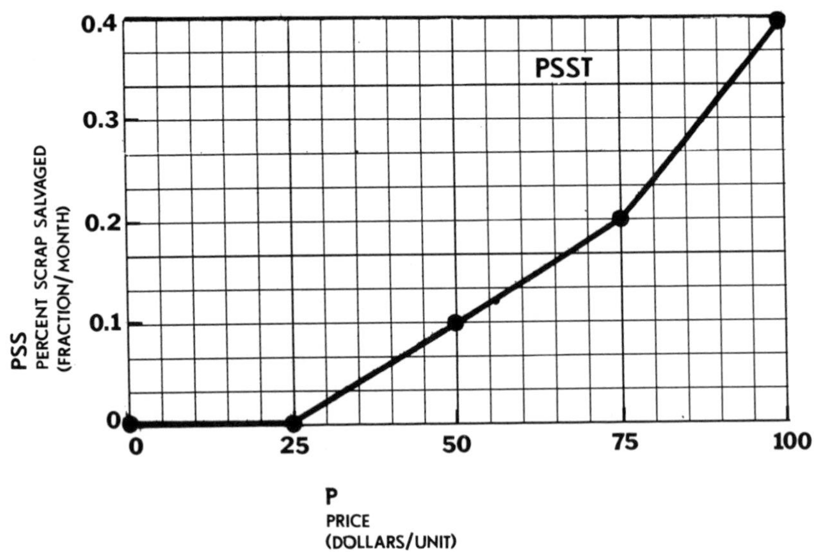

DELAY ANALYSIS

Response to Price Variation.

S10.8 The exogenous influence on P has a period of 1 month and an amplitude of 5 (dollars/unit).

a) EP and EPCR constitute a first-order information delay of 5 months. The ratio (T/P) of the time constant (delay time) and the input period equals 5. The phase and gain charts in Exercise 9 show that the phase shift equals 90° and gain equals 0.03.

b) DPC introduces a multiplier (gain) which depends on the value of EP from the DPCT table. Initially, DI = 6000, so P = 50 and EP = 50. Consequently DPC = 600. In this region of the table function DPC = -1400+40*EP. The gain equals +40 and, since neither sign reversal nor integration occurs, the phase shift equals 0°.

c) The equations for CAR and PC constitute a first-order delay of DPC with delay time CAD = 4. T/P = 4 and the gain equals 0.04 with a phase shift of 90°.

d) Finally, INR = PC so that the gain between these two elements equals 1.0 and the phase shift equals 0°.

The table below summarizes the computations. The gains from P to EP, from DPC to PC, and from PC to INR are dimensionless, while the multiplier (gain) between EP and DPC has the dimension (units^2/ month*dollars). This difference occurs because we measure P and EP in dollars per unit, while we measure DPC, PC and INR in units per month. Since the amplitude of the oscillation in P equals 5(dollars/unit), the amplitude of the oscillation in INR equals the product of 0.048(units^2/month*dollars) and 5(dollars/unit) or 0.24(units/month), shifted 180° behind P. The 10 percent variation in price produces only a 0.03 percent variation in the initiation rate. See simulation below.

SUMMARY

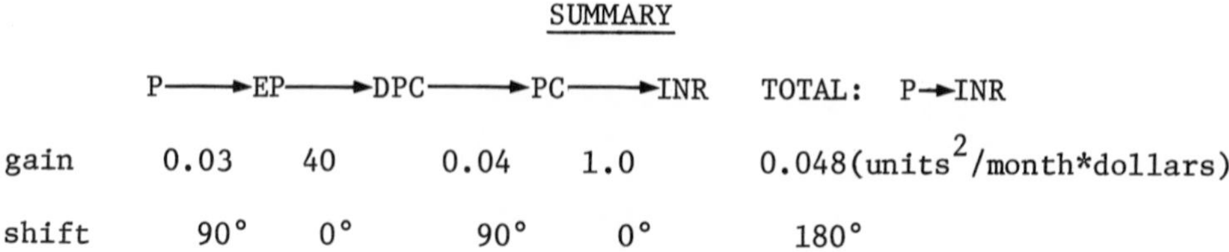

	P→EP	EP→DPC	DPC→PC	PC→INR	TOTAL: P→INR
gain	0.03	40	0.04	1.0	0.048(units^2/month*dollars)
shift	90°	0°	90°	0°	180°

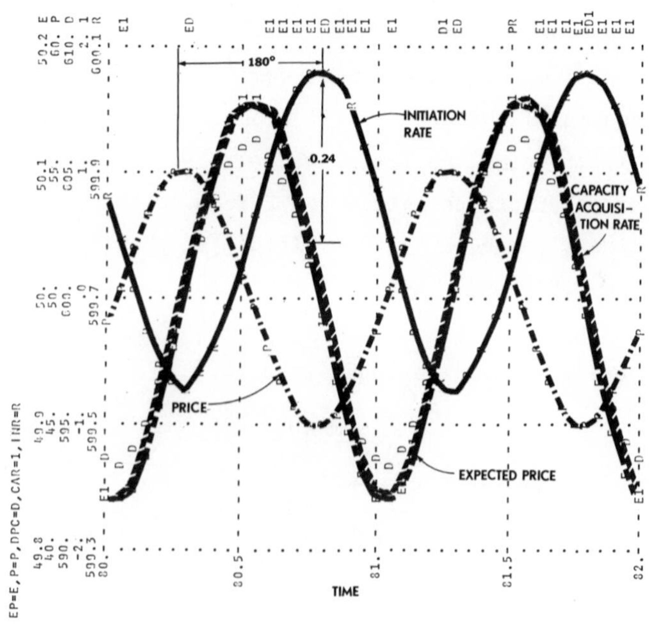

Response to price variation

Response to Seasonal Farrowing.

S10.9 The exogenous influence on INR has a period of 6 months and an amplitude of 300 units per month.

a) The effect on PR involves a third-order delay of INR. T/P = 2, so the gain = 0.014 and phase shift = 230°.

b) The effect on CR involves a first-order delay of PR. The delay time remains implicit in the table-function relationships connecting DI, P, PPC, and CR. Near the equilibrium point, these table functions become linear, and can be expressed by:

P = 140-0.015*DI

PCC = 13-0.2*P

CR = POP*PCC = 200*PCC.

Combine these equations:

CR = 200(13-0.2*(140-0.015*DI))

= 0.6*DI-3000 = (DI/1.67)-3000

CR, therefore, consists of a constant outflow of 3000 (units/month) plus a first-order delay with time constant of 1.67 (months). In this delay, T/P = 0.28 and the gain = 0.49 with the phase shift = 60°.

c) Since CR = DI/1.67-3000, near equilibrium, an oscillation imposed on DI through PR will produce an oscillation in CR only 1/1.67 times as large.

d) We can calculate the effect on P from the table function. In the linear range, P = 140-0.015*DI, so that from DI to P a gain of 0.015 and a shift of 180° (for the change of sign) both occur.

The table below summarizes these results.

SUMMARY

	INR→PR	PR→CR	CR→DI	DI→P	TOTAL: INR→P
gain	0.014	0.49	1.67	0.015	$1.8*10^{-4}$(dollars*month/units^2)
shift	230°	60°	0°	180°	470° (110°)

Since the magnitude of oscillation in INR equals 300(units/month), we expect that PR oscillates by roughly 4(units/month), 230° behind INR while P oscillates by roughly 0.054(dollars/month), (an insignificant amount), 110° behind INR. See simulation below.

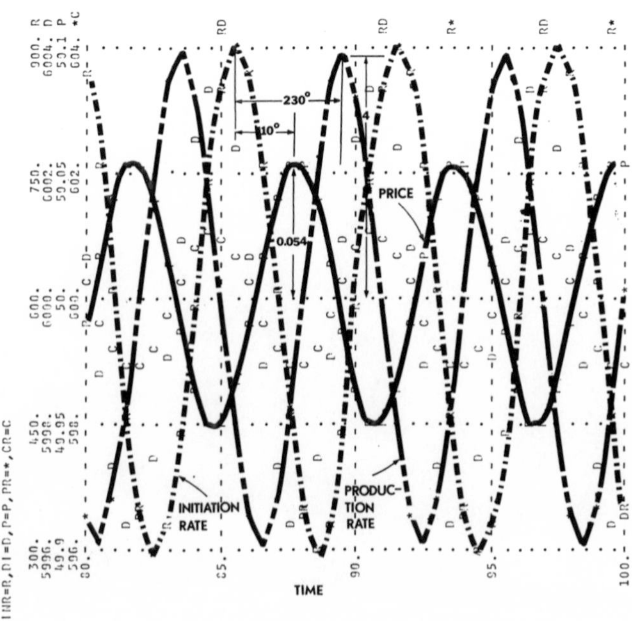

Response to seasonal farrowing

Exercise 11
Analysis of Market Growth Model

by Narendra K. Patni

This exercise investigates in some detail the structure and behavior of feedback loops described in Jay W. Forrester's "Market Growth as Influenced by Capital Investment." Step-by-step analysis of the system loops brings out interesting, but frequently overlooked, behavioral aspects. The analysis centers on simplifying the loops and examining the effects of parameters and table functions on behavior. The reader may wish to study the "Market Growth" paper to obtain maximum benefit from this exercise. However, the exercise does include the relevant flow diagrams, equation listing, and computer runs.*

Some students may encounter difficulty with this exercise. For instance, the reader must apply open-loop, steady-state analysis, previously introduced in Exercise 9, to systems affected by step inputs. The reader should closely monitor his own answers against the provided solutions since many questions build on previous answers.

Figure E11-1 contains a simplified, causal-loop diagram of three major loops in the market growth model. We will investigate only loops 1 and 2 here. Figure E11-2 presents the flow diagram of loop 1; Figure E11-5, the flow diagram for loop 2. The flow diagram for loop 3 appears in Figure E11-11. Figures E11-3 and E11-4 depict two computer simulation runs for loop 1. Figure E11-8 displays the run for loop 2; Figures E11-9 and E11-10, runs for both loops. Figures E11-6 and E11-7 graph the two table functions used in loop 2. A complete DYNAMO equation listing follows the figures.

---

*Jay W. Forrester, "Market Growth as Influenced by Capital Investment," Industrial Management Review 9, no.2 (Winter 1968): 83-105. Reprinted in Jay W. Forrester, Collected Papers of Jay W. Forrester (Cambridge: Wright-Allen Press, 1975).

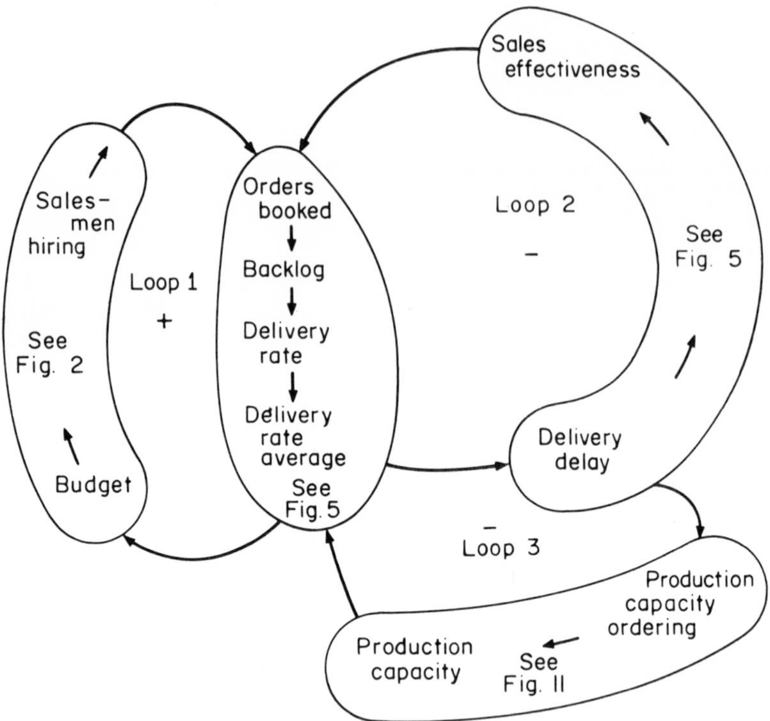

E11-1

Loop structure for sales growth, delivery delay, and capacity expansion

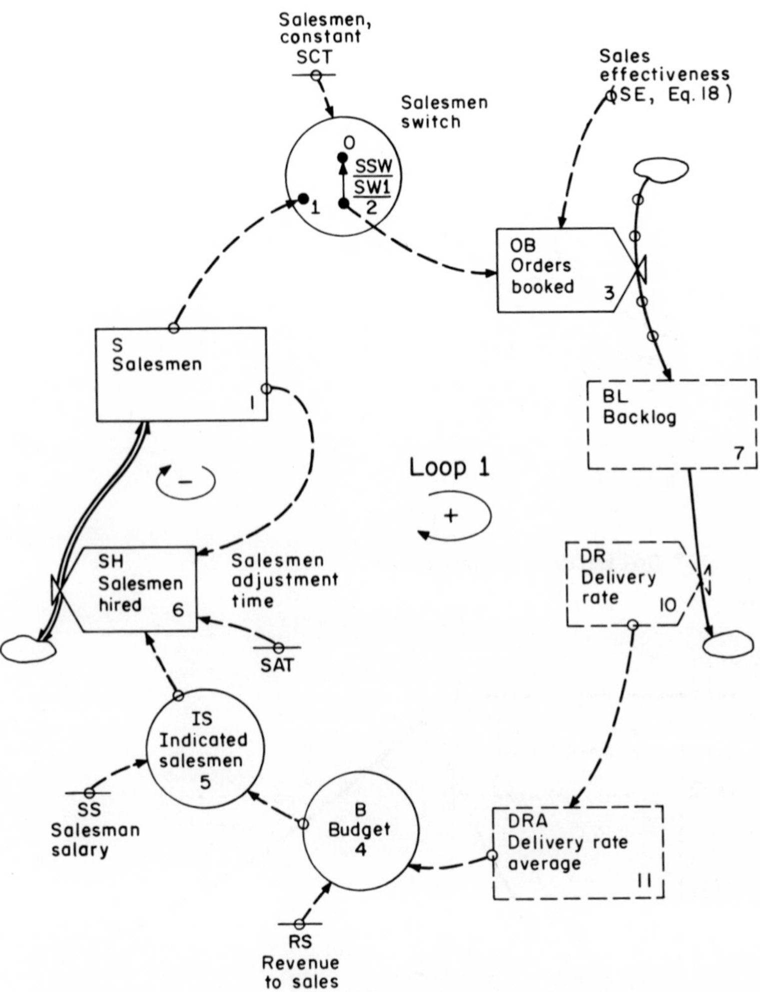

E11-2

Salesmen-hiring loop with sales generating revenue to support selling effort

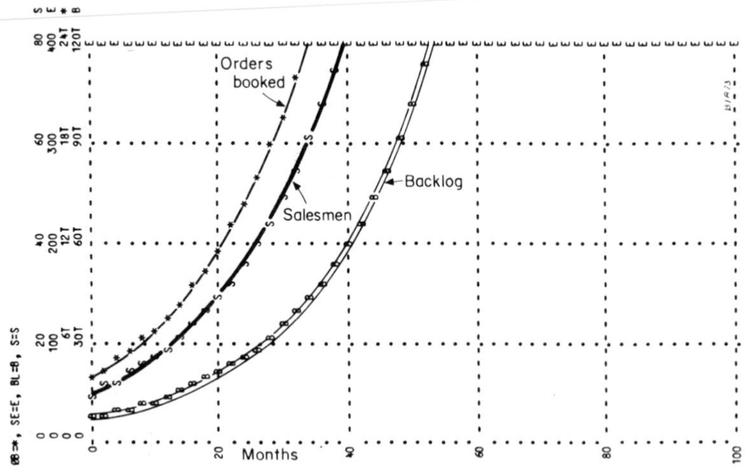

E11-3

Unlimited exponential growth in loop 1

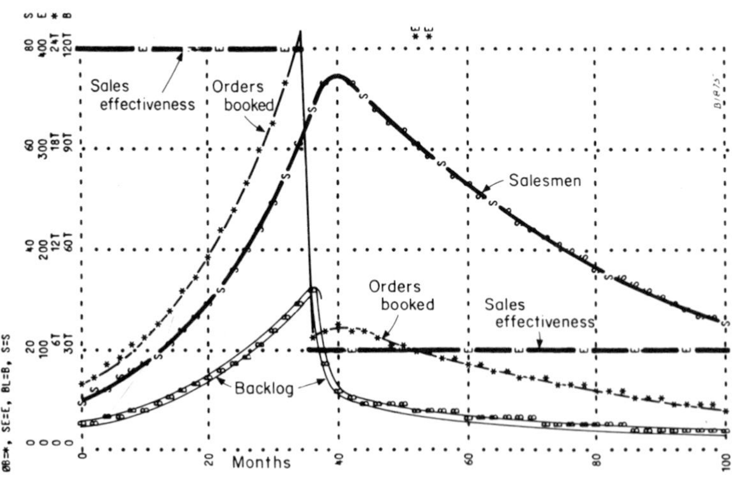

E11-4

Initial growth in loop 1 followed by decline
when sales effectiveness is reduced

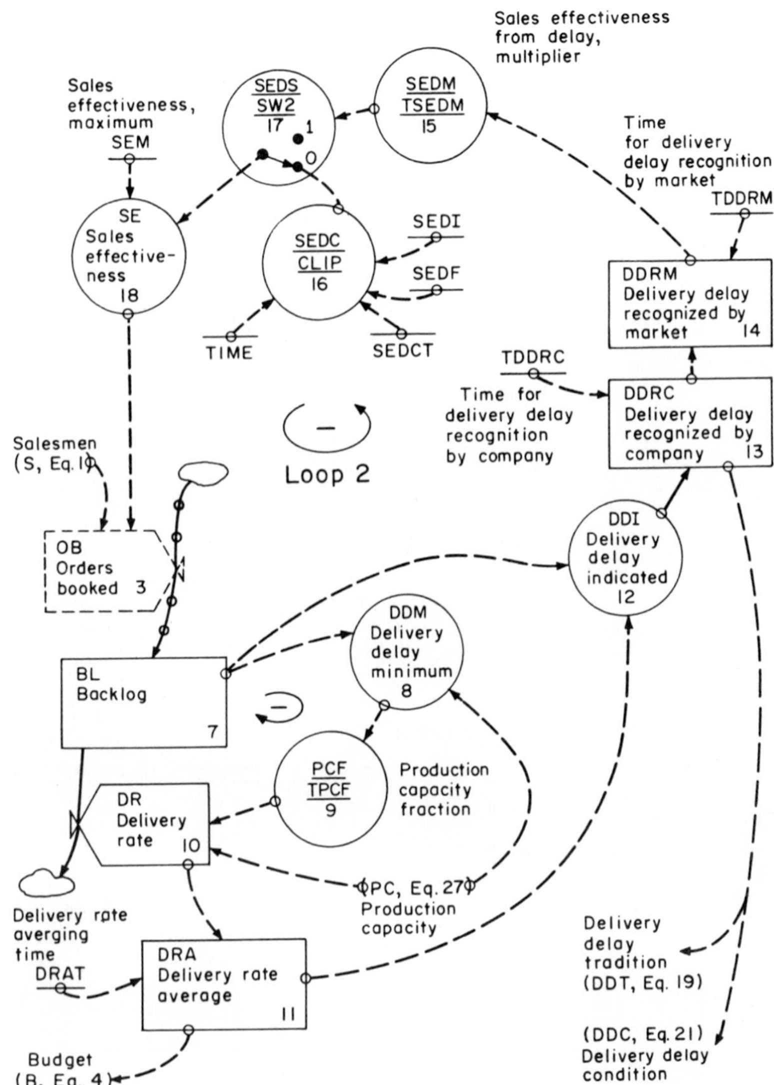

E11-5

Market loop with delivery delay determining product attractiveness

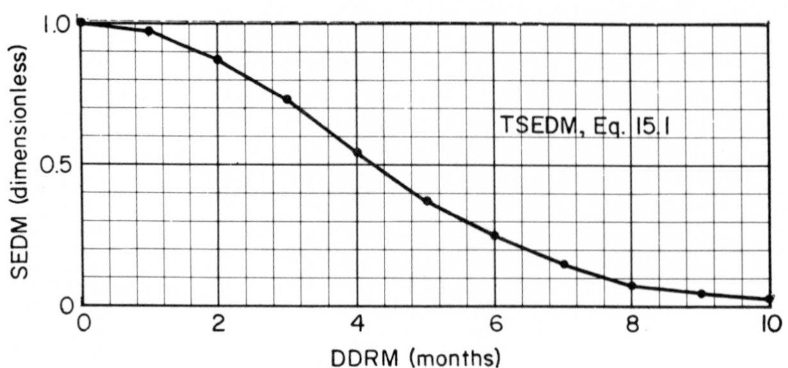

E11-6

Table for sales effectiveness from delay multiplier as it depends on DDRM

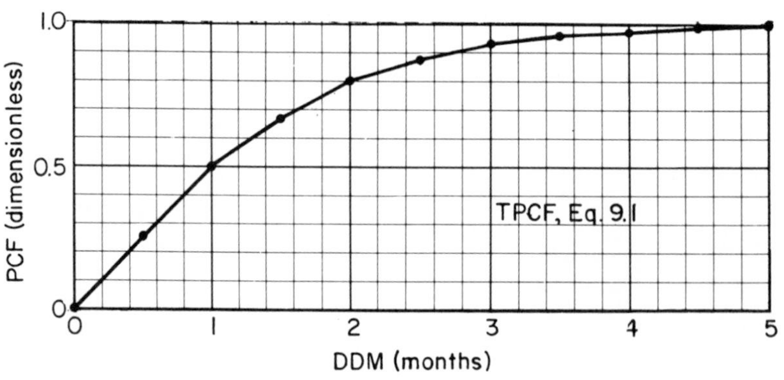

E11-7

Table for production capacity fraction as it depends on delivery delay minimum

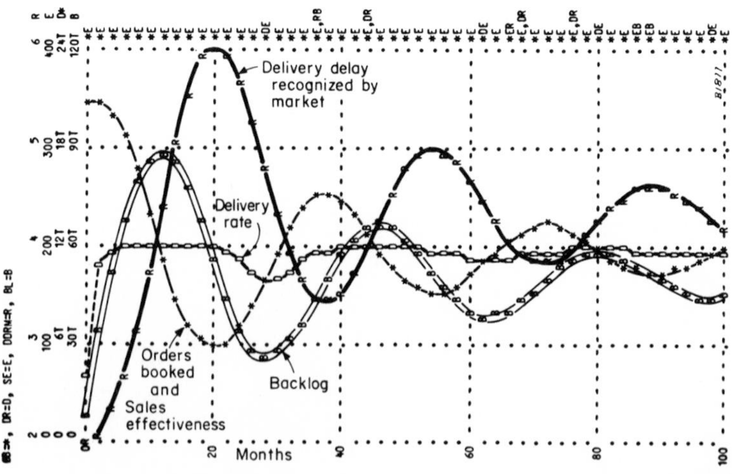

E11-8

Fluctuation caused by delayed responses within loop 2

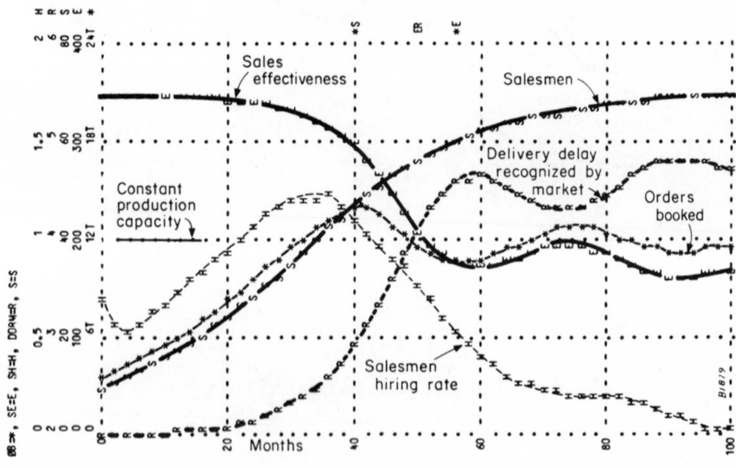

E11-9

Sales growth and stagnation caused by interaction of loops 1 and 2

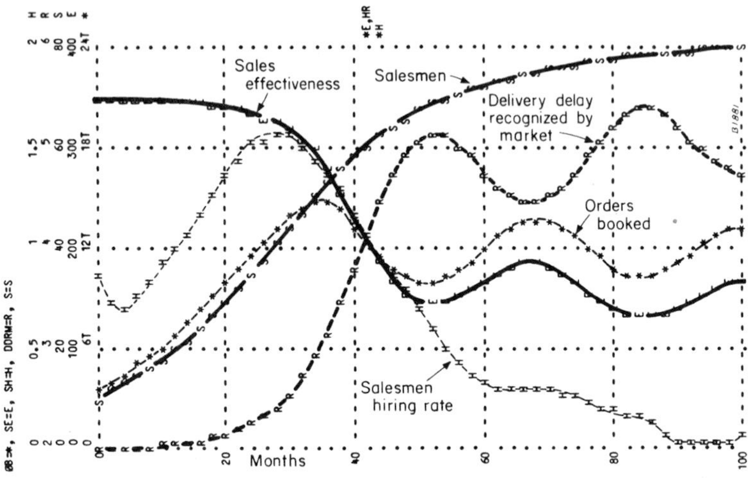

E11-10

Higher budget to sales effort

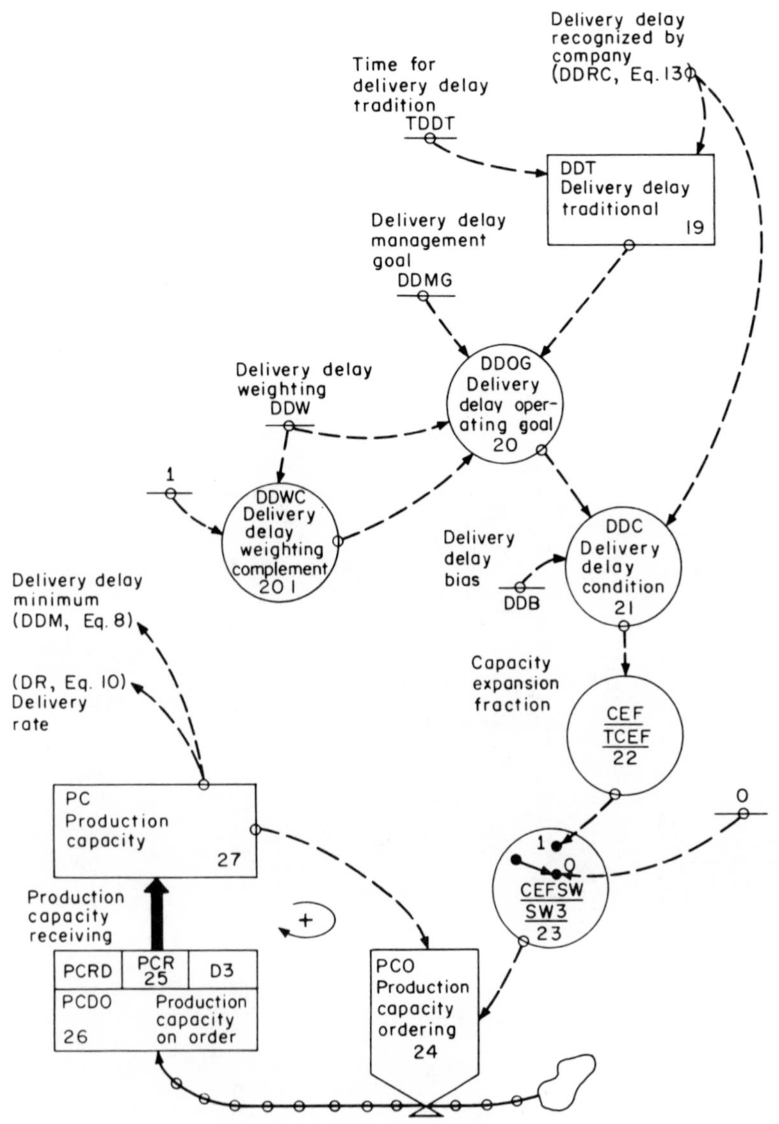

E11-11

Capacity expansion loop

DYNAMO Equations:

```
0.1            MARKET LOOPS
0.2     RUN   STD
0.3     NOTE  U. OF ILL.--CONVERSE AWARDS SYMPOSIUM, APRIL 13, 1967
0.4     NOTE
0.5     NOTE  POSITIVE LOOP--SALESMEN
0.6     NOTE
1       1L    S.K=S.J+(DT)(SH.JK+0)
1.1     6N    S=10
2       49A   SSW.K=SWITCH(SCT,S.K,SW1)
2.1     C     SCT=60
2.2     C     SW1=0
3       12R   OB.KL=(SSW.K)(SE.K)
4       12A   B.K=(DRA.K)(RS)
4.1     C     RS=12
5       20A   IS.K=B.K/SS
5.1     C     SS=2000
6       21R   SH.KL=(1/SAT)(IS.K-S.K)
6.1     C     SAT=20
6.4     NOTE
6.5     NOTE  NEGATIVE LOOP--MARKET
6.6     NOTE
7       1L    BL.K=BL.J+(DT)(OB.JK-DR.JK)
7.1     6N    BL=8000
8       20A   DDM.K=BL.K/PC.K
9       58A   PCF.K=TABHL(TPCF,DDM.K,0,5,.5)
9.1     C     TPCF*=0/.25/.5/.67/.8/.87/.93/.95/.97/.98/1
10      12R   DR.KL=(PC.K)(PCF.K)
11      3L    DRA.K=DRA.J+(DT)(1/DRAT)(DR.JK-DRA.J)
11.1    6N    DRA=DR
11.2    C     DRAT=1
12      20A   DDI.K=BL.K/DRA.K
13      3L    DDRC.K=DDRC.J+(DT)(1/TDDRC)(DDI.J-DDRC.J)
13.1    6N    DDRC=DDI
13.2    C     TDDRC=4
14      3L    DDRM.K=DDRM.J+(DT)(1/TDDRM)(DDRC.J-DDRM.J)
14.1    6N    DDRM=DDRC
14.2    C     TDDRM=6
15      58A   SEDM.K=TABHL(TSEDM,DDRM.K,0,10,1)
15.1    C     TSEDM*=1/.97/.87/.73/.53/.38/.25/.15/.08/.03/.02
16      51A   SEDC.K=CLIP(SEDF,SEDI,TIME.K,SEDCT)
16.1    C     SEDF=1
16.2    C     SEDI=1
16.3    C     SEDCT=36
17      49A   SEDS.K=SWITCH(SEDC.K,SEDM.K,SW2)
17.1    C     SW2=0
18      12A   SE.K=(SEDS.K)(SEM)
18.1    C     SEM=400
18.4    NOTE
18.5    NOTE  CAPITAL INVESTMENT
18.6    NOTE
19      3L    DDT.K=DDT.J+(DT)(1/TDDT)(DDRC.J-DDT.J)
19.1    6N    DDT=DDRC
19.2    C     TDDT=12
20      15A   DDOG.K=(DDT.K)(DDW)+(DDMG)(DDWC)
20.1    7N    DDWC=1-DDW
20.2    C     DDW=0
20.3    C     DDMG=2
21      27A   DDC.K=(DDRC.K/DDOG.K)-DDB
21.1    C     DDB=.3
22      58A   CEF.K=TABHL(TCEF,DDC.K,0,2.5,.5)
22.1    C     TCEF*=-.07/-.02/0/.02/.07/.15
23      49A   CEFSW.K=SWITCH(0,CEF.K,SW3)
23.1    C     SW3=0
24      12R   PCO.KL=(PC.K)(CEFSW.K)
25      39R   PCR.KL=DELAY3(PCO.JK,PCRD)
25.1    C     PCRD=12
26      1L    PCOO.K=PCOO.J+(DT)(PCO.JK-PCR.JK)
26.1    12N   PCOO=(PCO)(PCRD)
27      1L    PC.K=PC.J+(DT)(PCR.JK+0)
27.1    6N    PC=PCI
27.2    C     PCI=12000
```

```
27.5    NOTE
27.6    NOTE   CONTROL CARDS
27.7    NOTE
27.8    PLOT   OB=*,PC=C(0,24000)/SE=E(0,400)/S=S(0,80)
27.9    NOTE   B42, RERUNS OF B41
28      NOTE
28.1    RUN    A
28.2    NOTE   UNLIMITED EXPONENTIAL GROWTH
28.3    SPEC   DT=.5/LENGTH=100/PRTPER=100/PLTPER=2
28.4    PRINT  1)S
29      C      SW1=1
29.1    C      PCI=100000
29.4    PLOT   OB=*(0,24000)/SE=E(0,400)/BL=B(0,120000)/S=S(0,80)
29.5    RUN    B
29.6    NOTE   GROWTH AND DECLINE
30      C      SW1=1
30.1    C      SEDF=.25
30.2    C      PCI=100000
30.5    RUN    C
30.6    NOTE   NEGATIVE LOOP OSCILLATION
31      C      SW2=1
31.3    PLOT   OB=*,DR=D(0,24000)/SE=E(0,400)/DDRM=R(2,6)/BL=B(0,120000)
31.4    RUN    D
31.5    NOTE   SALES STAGNATION
32      C      SW1=1
32.1    C      SW2=1
32.4    PLOT   OB=*(0,24000)/SE=E(0,400)/SH=H(0,2)/DDRM=R(2,6)/S=S(0,80)
32.5    RUN    E
32.6    NOTE   INCREASED SALES BUDGET ALLOCATION
33      C      SW1=1
33.1    C      SW2=1
33.2    C      RS=13.6
33.5    RUN    F
33.6    NOTE   CAPACITY EXPANSION
34      C      SW1=1
34.1    C      SW2=1
34.2    C      SW3=1
34.5    PLOT   OB=*,PC=C(0,24000)/SE=E(0,400)/DDRM=R,DDOG=G(2,6)/S=S(0,80)/CEF=F(
34.6    X1     -.06,.18)
34.7    RUN    G
34.8    NOTE   GOAL=TRADITION WITH DELIVERY DELAY BIAS PRESSURE
35      C      SW1=1
35.1    C      SW2=1
35.2    C      SW3=1
35.3    C      DDW=1
```

Analysis of loop 1 (Figure E11-2) and its behavior (Figures E11-3 and E11-4).

E11.1 What loops are active in the run shown in Figure E11-3?
What characteristics do these loops display?
What does the run imply about PC?

E11.2 Identify the delay times (or time constants) around loop 1.
Note whether the values remain constant during the model run.

E11.3 Assume that a broken link between delivery rate DR and delivery rate average DRA makes loop 1 an open loop.

a) What is the equilibrium ratio (gain) of delivery rate DR to orders booked OB when a step input in OB occurs?

b) What is the numerical value of the time constant between backlog BL and DR? Assume that production capacity PC still equals 100,000.

E11.4 Assume loop 1 is open. Put in a step input of 10,000 units/month in orders booked OB. Plot BL and DR as functions of time on the grid below. BL is initially zero.

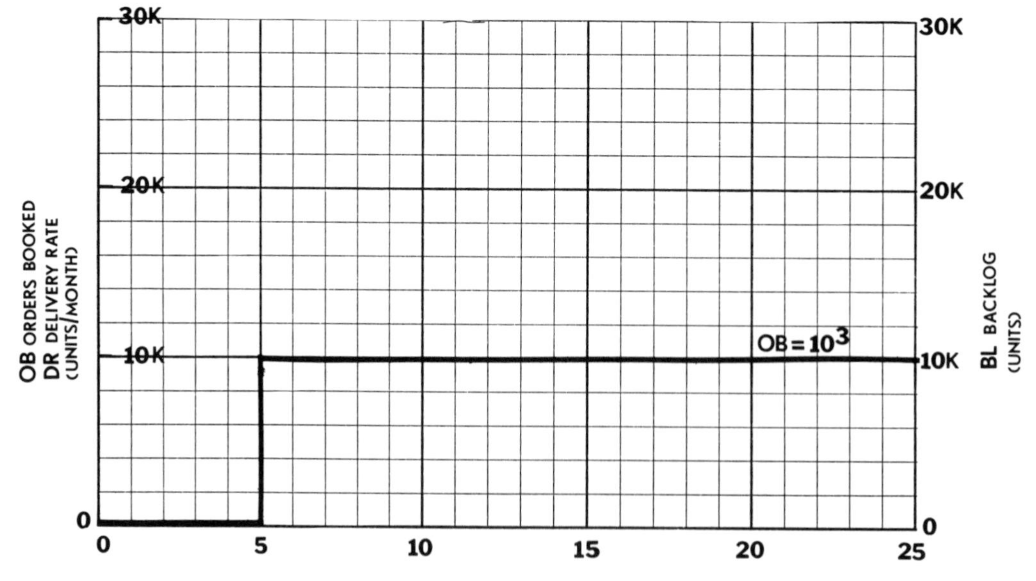

Now plot DR over time in Figure E11-3. E11.5

Assume that loop 1 remains open between delivery rate DR and delivery rate average DRA. E11.6

a) What parameters and numerical values determine the total open-loop gain from DRA to DR?
b) What is the numerical value of the total gain as computed from the above values?

What happens when the open-loop gain is greater than 1? Less than 1? Equal to 1? E11.7

What effect does a changed salesmen adjustment time SAT have on the behavior in Figure E11-3? E11.8

Is there any limitation to growth in Figure E11-3? What behavior and terminal values would DR, BL, OB, and S have if the run continued indefinitely? E11.9

What factors can cause the decay in Figure E11-4? E11.10

a) Why do the variables in loop 1 decline in value in the latter portion of Figure E11-4? E11.11
b) At what value of sales effectiveness SE will system variables neither grow nor decay?

How does the minor negative loop involving salesmen S and salesmen hiring SH affect the behavior in Figure E11-4? Explain. E11.12

What effect does the delivery rate averaging time DRAT have on the behavior in loop 1? E11.13

<u>Analysis of loop 2 (Figure E11-5) and its behavior (Figure E11-8).</u>

What is the goal of the minor negative loop involving the backlog BL and the delivery rate DR? E11.14

E11.15 In Figure E11-8, why does the delivery rate DR flatten out at 12,000?

E11.16 Find an algebraic expression for the delivery delay indicated DDI in terms of delivery delay minimum DDM and production capacity fraction PCF. Plot DDI versus DDM and DDI versus BL on the graph below. Ignore delivery rate average DRA.

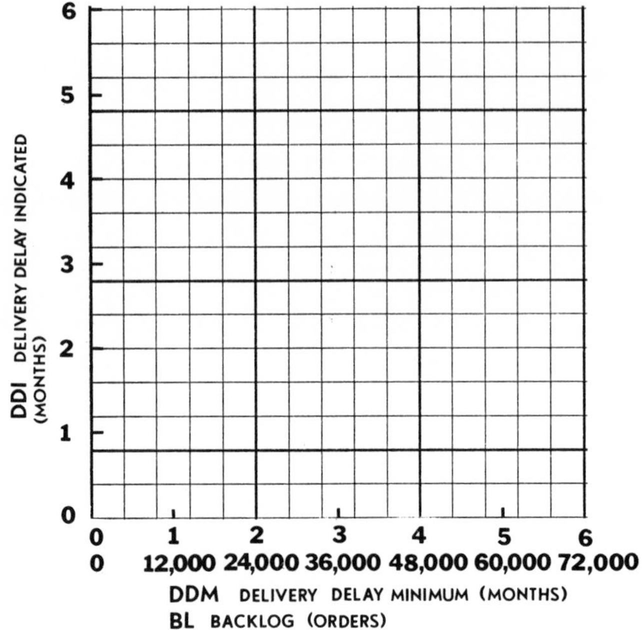

E11.17 Offer a real world interpretation of the relationships developed in the previous question.

E11.18 Given the initial values of the run in Figure E11-8, explain why sales effectiveness SE starts at a value of 350.

E11.19

a) Consider loop 2 with 80 salesmen S generating enough orders to keep the plant operating at full capacity at equilibrium. Calculate equilibrium values of SE, BL, DDI, and DDM. Assume PC remains at 12,000.

b) Compute the equivalent time constant of the system. Assume a step increase in salesmen S from 80 to 160. Ignore the recognition delays DDRC and DDRM. Also, assume PC remains at 12,000.

E11.20

Sketch the behavior of SE, BL, and DDI with the assumptions stated in question E11.19 b). Use the grids below.

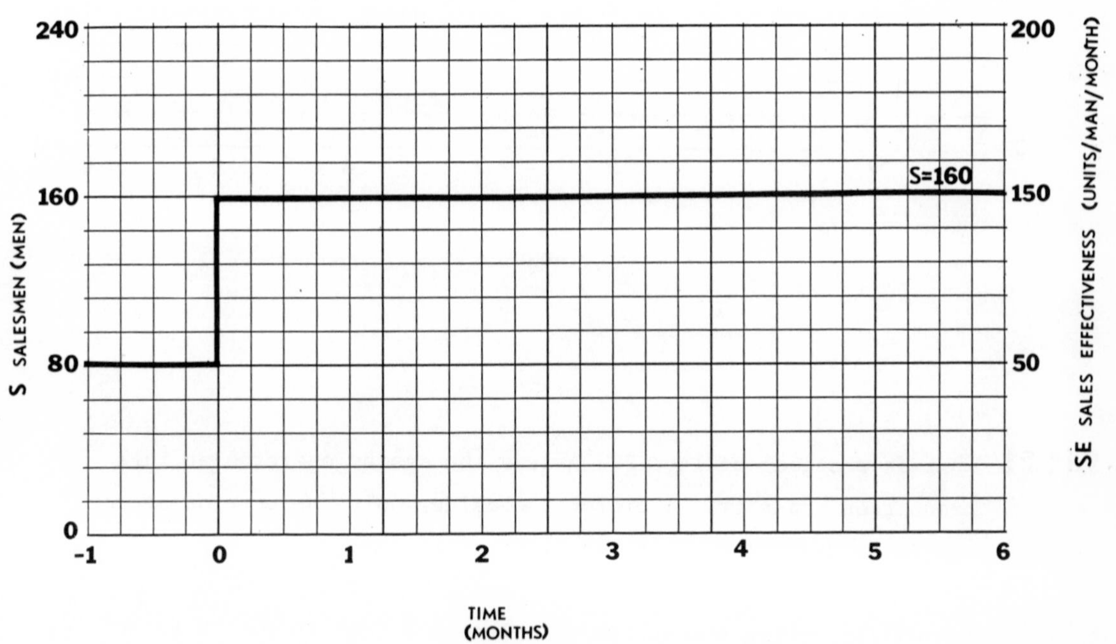

(a)

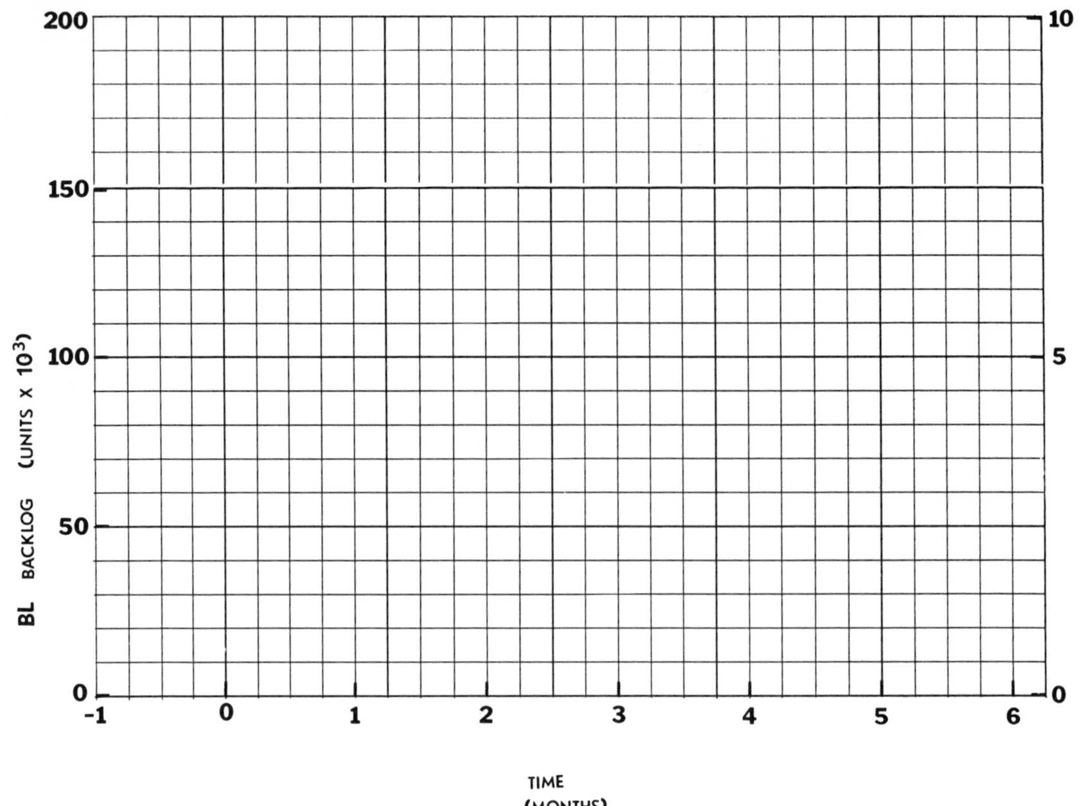

(b)

E11.21 What is the goal of loop 2? Assume the system operates at full production capacity. Therefore, consider only the portion of the run in Figure E11-8 after the first 40 months.

E11.22 How will doubling the number of salesmen S from 60 to 120 affect the steady-state values implied by the run in Figure E11-8? Assume a constant delivery rate DR at production capacity PC. Identify the new steady-state values for DDRM and SE.

E11.23 How does changing the shape of the SEDM curve shown in Figure E11-6 affect the behavior of the negative loop? Examine the three curves in the figure below. What new steady-state value will SE have:

a) When SEM equals 400 (unchanged)?

b) When SEM equals 300?

Assume a constant delivery rate DR at the production capacity PC.

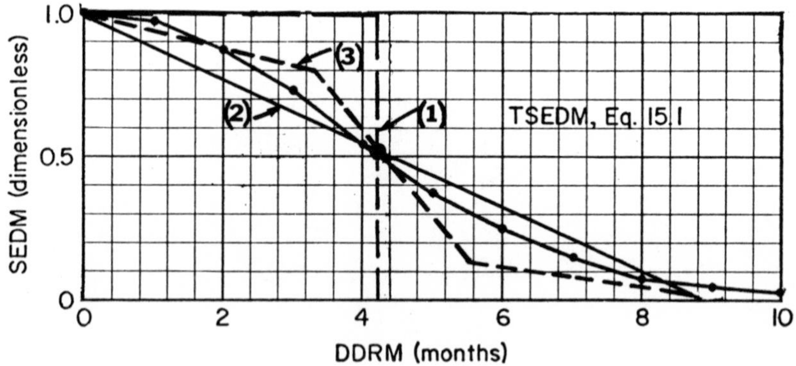

Solution 11

Analysis of loop 1 and its behavior.

Loop 1 is an active positive loop that produces exponential growth. The small (minor) negative loop within loop 1 has a goal of equating S to IS. Therefore, the negative loop acts like a first-order delay. Other system loops become inactive when we set all switches except SW1 at zero. We have set the production capacity at a high initial value of 100,000 to remove the constraint from production capacity PC. **S11.1**

The three loop delays have the following respective time constants: **S11.2**

1. salesmen adjustment time SAT = 20;
2. delivery rate averaging time DRAT = 1 month; and
3. variable delay time between orders booked OB and delivery rate DR. From Figure E11-5, we note that DR equals the product of production capacity fraction PCF and production capacity PC. Thus,

   DR.KL = PC.K*PCF.K

   PC is constant in the run in Figure E11-3 at a value of 100,000. PCF is a function of delivery delay minimum DDM according to the table functions in Figure E11-7. Since DDM is the ratio of backlog BL to PC, we can express DR completely as a function of BL. The time constant associated with DR then depends on the slope of the PCF curve in Figure E11-7. Question E11.4 derives the delay time value.

S11.3 a) $\text{Gain} = \frac{\text{Equilibrium output}}{\text{Equilibrium input}} = \frac{DR}{OB} = 1$

b) DR.KL = PC.K*PCF.K

PCF.K = function of DDM

$\text{DDM.K} = \frac{\text{BL.K}}{\text{PC.K}}$

At PC = 100,000 and BL less than 100,000, the operating range for DDM is from 0 to 1. We can approximate the table function in Figure E11-7 by:

PCF.K = K*(BL.K/PC.K)

where K = slope of the curve in Figure E11-7 in the operating range. Hence, K = 0.5 and

DR.KL = PC.K*K*(BL.K/PC.K)

= K*BL.K

= 0.5*BL.K (11.1)

The time constant, therefore, equals 1/0.5 or 2 months. This time constant changes only when the operating point shifts to the non-linear portion of the PCF curve. Given a fixed time constant, the structure acts as a pure first-order delay. The equilibrium or steady-state gain across a delay always equals 1 in the presence of a step input. This delay structure is similar to the pollution absorption structure in Chapter 3 (section 3.12).

From S11.3 b) we plot BL and DR using a time constant of 2 months. **S11.4**

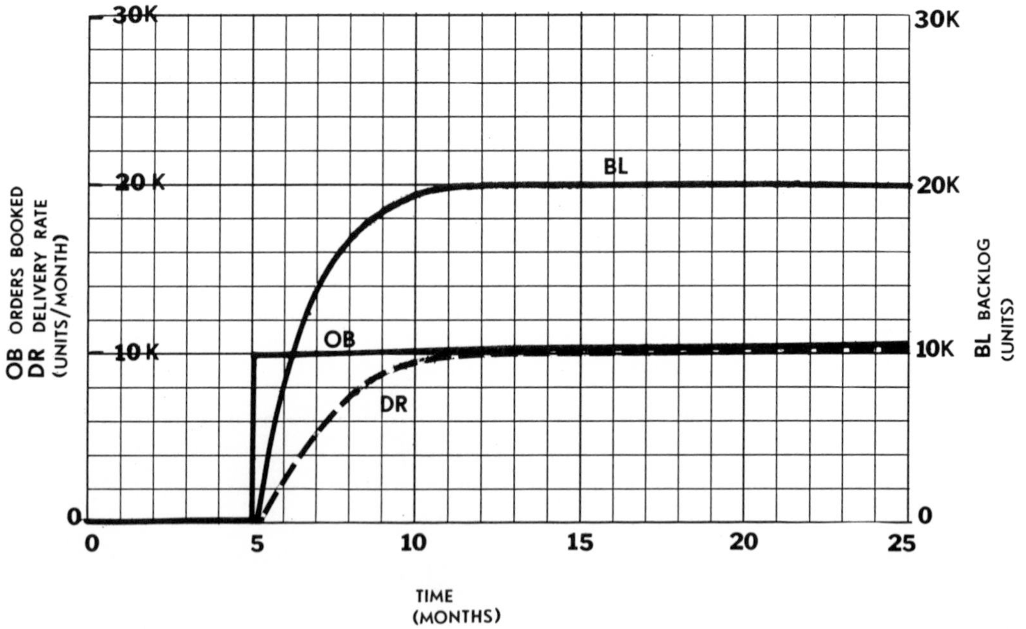

DR will lag orders by roughly two months and run parallel to OB. We can plot DR by using equation (11.1) above. **S11.5**

a) Gain parameters and their numerical values: **S11.6**
The steady-state gain across the two delays (S-SH and BL-DR) in loop 1 equals 1 (see question E11.3). The gain between budget B and delivery rate average DRA, therefore, equals revenue to sales RS which has a value of 12 (dollars/unit). The gain between indicated salesmen IS and budget B equals 1/SS or 1/2,000 (dollars/man/month). Finally, the gain between orders booked OB and salesmen S equals sales effectiveness SE which has a value of 400 (units/salesmen/month).

b) Total Gain value:
To determine the total gain, we multiply the steady-state gain values among the elements or:

$$G_{total} = (RS)(SE)(1/SS) \quad (11.2)$$
$$= (12 \text{ dollars/unit})(400 \text{ units/man/month}) / (2{,}000 \text{ dollars/man/month})$$
$$= 2.4$$

S11.7 When the open-loop gain exceeds 1, the loop exhibits the exponential growth characteristic of positive feedback. If the loop gain falls below 1, the system exhibits the exponential decay, seeking zero as a goal, characteristic of negative feedback. A gain value of 1 implies a system in equilibrium.

S11.8 SAT affects the growth rate of variables in the loop, but does not affect the mode of behavior. A larger SAT would produce slower growth; a smaller SAT, faster growth.

S11.9 Growth is limited by production capacity PC. DR will finally approach the PC value of 100,000 unit/month. This trend will generate enough revenue to support 600 salesmen, who in turn will generate 240,000 orders/month. Therefore, the backlog will continue to increase linearly at the rate of 140,000 per month while DR, S, and OB remain constant.

S11.10 Any factor which tends to reduce loop gain to less than one can bring about decay. As seen in equation (11.2), increasing SS, lowering RS, and/or lowering SE can cause this change.

S11.11 a) The loop gain drops from 2.4 to 0.6 by decreasing SE. Salesmen S do not generate enough revenue to support their numbers. With a gain less than 1, salesmen S decrease exponentially.
b) When SE = 166.6, the loop gain will equal 1 and the system will stabilize.

S11.12 The minor loop has a goal of equating salesmen S with indicated salesmen IS. When the gain around the outside positive loop exceeds one, indicated salesmen IS will always <u>exceed</u> actual salesmen S and both IS and S will grow without limit. When SE = 100, the gain drops to 0.6. Indicated salesmen IS never rises to equal actual salesmen; therefore salesmen S are continually released. Furthermore, the rate of release is proportional to the absolute difference between salesmen S and indicated salesmen IS. Since S and IS maintain a ratio of 0.6 if we ignore the effects of averaging and delays in the loop, the absolute difference declines with time. Salesmen S, indicated salesmen IS, orders booked OB, and delivery rate DR all approach zero asymptotically.

DRAT, like SAT, only affects the rate of growth or rate of decline, but cannot change the mode of behavior. **S11.13**

Analysis of loop 2 and its behavior.

This loop has a goal of equating DR to OB. The minor loop attains this goal by allowing backlog BL to adjust the production capacity fraction PCF and, thereby, the delivery rate DR. **S11.14**

In Figure E11-8, production capacity PC has a constant value of 12,000. Since the delivery rate DR cannot exceed PC, DR levels off at 12,000. This leveling off occurs when backlog BL forces delivery delay minimum DDM to a value of 5. A production capacity fraction PCF of 1 results. **S11.15**

Dropping the DYNAMO subscripts, we get: **S11.16**

DDI = BL/DR

DDM = BL/PC

DR = PCF*PC

Therefore,

$$DDI = BL/DR = (BL/PC)/(PC/DR)$$

$$= \frac{DDM}{(PC*PCF/PC)}$$

$$= DDM/PCF.$$

From Figure E11-7, as DDM increases, PCF approaches 1. As shown above, DDI and DDM then have the same value. Thus, after DR reaches PC, DDI and DDM become equal and interchangeable.

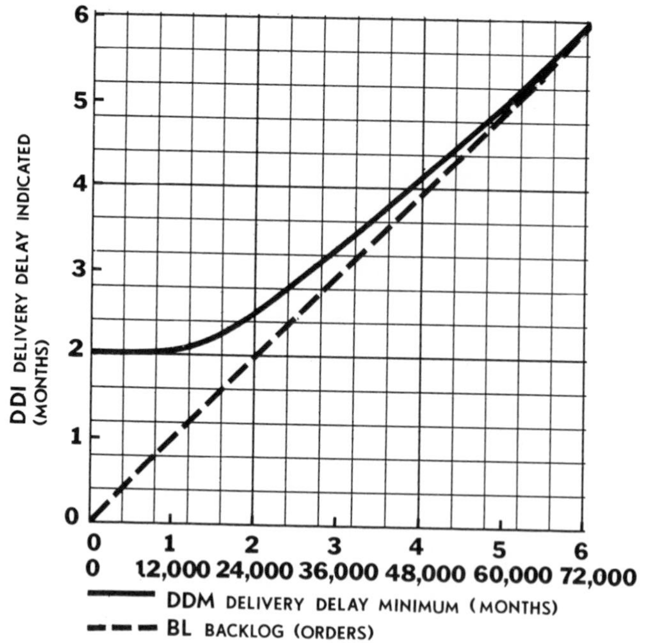

S11.17 Plant capacity can become full only when the order backlog and, hence, delivery delay substantially increase. Any organization may experience inherent delays in scheduling and planning production. The two month minimum delivery delay reflects these normal obstructions.

S11.18 In Figure E11-8, backlog BL has an initial value of 8,000, while salesmen S has a value of 60. Production capacity PC equals 12,000 units/month. Therefore, the delivery delay minimum DDM equals 0.66 month. From the table in Figure E11-7, PCF equals 0.33. This PCF permits a delivery rate DR of 4,000. Therefore, DDI equals 2 months. DDI initializes DDRM and DDRC at 2 months. From the table in Figure E11-6, SEDM initially equals 0.87. Sales effectiveness SE then equals 400*(0.87) or 350.

S11.19 a) Equilibrium values are:

DDI = DDM = 5 months

BL = 5*12,000 = 60,000

SE = 12,000/80 = 150

b) The single negative-feedback loop shown below approximates the system described above.

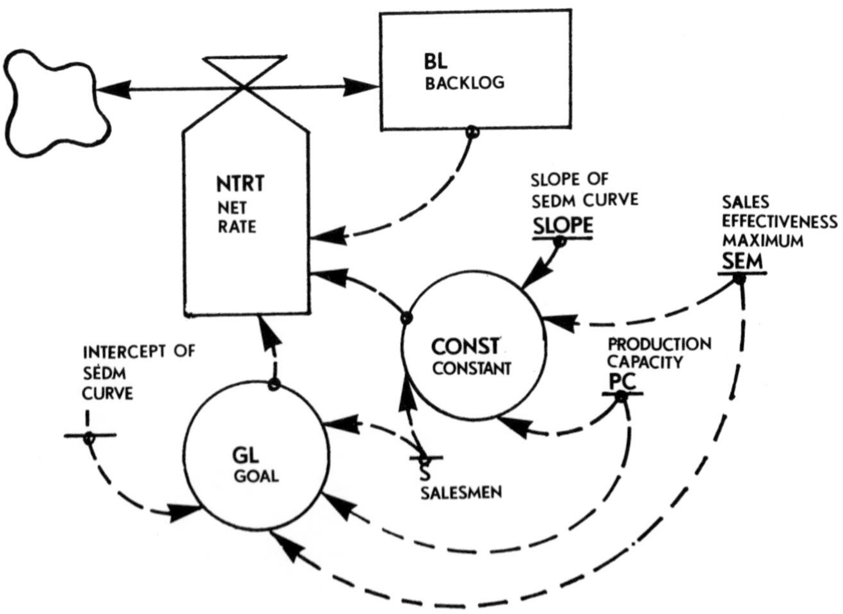

From Figure E11-5 we have the following equations:

```
L  BL.K = BL.J+(DT)(OB.JK-DR.JK)
N  BL = 60000
R  OB.KL = S*SE.K
C  S = 160
A  SE.K = SEM*SEDM.K
C  SEM = 400
A  SEDM.K = I-SLOPE*DDI.K
C  SLOPE = 0.11
C  I = 0.94
A  DDI.K = BL.K/DR.JK
R  DR.KL = PC
C  PC = 12,000
```

SEDM is a linearized algebraic representation of the table in Figure E11-6. The slope value of 0.11 derives from the linear portion of the SEDM curve between DDRM equal to 5 and 7. The intercept I, then, has a value of approximately 0.94. By substitution and by defining (OB-DR) as the net rate NTRT, we can reduce the preceding equations to:

```
L  BL.K = BL.J+(DT)(NTRT.JK)
N  BL = 60000
R  NTRT.KL = GL.K-CONST.K*BL.K
A  GL = S*SEM*I-PC
A  CONST = SLOPE*SEM*S/PC
C  S = 160
C  SEM = 400
C  I = 0.94
C  PC = 12000
C  SLOPE = 0.11
```

CONST, the inverse of the time constant, has a value of (0.11/months) *(400 unit/man/month)*(160 men)/(12,000 units/month) or (0.58/months). The approximate time constant therefore equals 1.7 months.

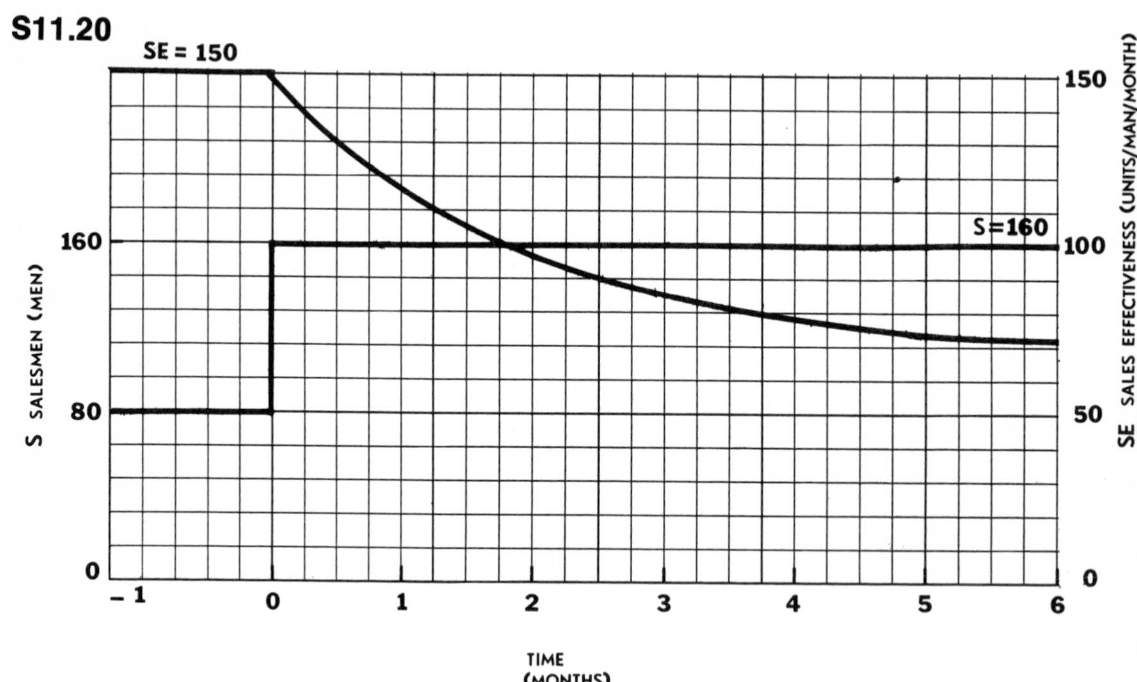

(a)

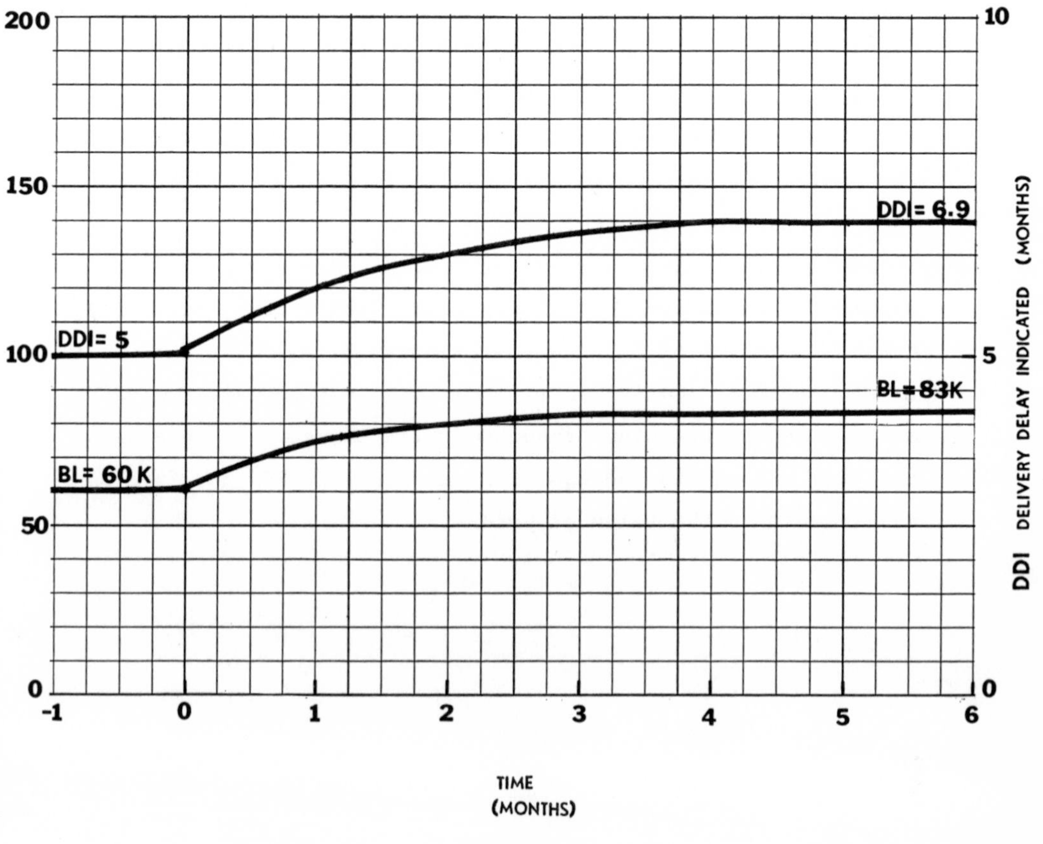

(b)

The goal of the system is to adjust the value of sales effectiveness SE so that orders booked OB (for a constant number of salesmen S) equals delivery rate DR which in turn equals production capacity PC. Algebraically, we can state that the goal is to move the system toward the following equality: **S11.21**

$$OB = DR = SE*S \tag{11.3}$$

According to equation (11.3) the system will stabilize itself with sales effectiveness SE reduced from 200 to 100. Thus, the effect of increasing the number of salesmen S is offset by a decrease in the value of sales effectiveness SE. OB must still equal DR. From the SEDM table in Figure E11-6, we find DDRM would equal 6 months in equilibrium. **S11.22**

S11.23 a) In equilibrium SE must equal 200. For SEM at 400, this implies SEDM must equal 0.5 and DDRM must equal 4.2. All three curves produce the same value of SEDM when DDRM equals 4.2. The curves do not affect the steady-state values. However, the shape of the table-function curves will have a significant effect on the loop's transient response.

b) SE must still achieve a value of 200. An SEM value of 300 implies that SEDM must equal 0.66. At SEDM = 0.66, curve (1) has a DDRM value of 4.2, curve (2) a value of 2.8, and curve (3) a value of 3.8. Consequently, the steady-state values of DDRM, as well as the transient response, will change. In general, changing SEM shifts the equilibrium value of SEDM and forces adjustments in the value of DDRM.

Exercise 12
Residential Community Model

by Michael R. Goodman

Exercise 12 is the first of four model conceptualization exercises included in this book. It simplifies and highlights some important population-housing interactions found in Urban Dynamics.[1] It also builds upon and extends many of the concepts introduced in the land-use model in Chapter 5. Exercise 12 helps the reader gain experience with various stages of model construction and analysis. The exercise contains a verbal description of the basic structural relationships likely to be found in an area such as a resort or retirement village where population growth primarily depends upon housing availability and such natural qualities as location and weather. The model, when completed, should be able to trace the area's growth from an early undeveloped stage, through a growth stage, and into an equilibrium stage within an 80-year time horizon.

DESCRIPTION--RESIDENTIAL COMMUNITY

Consider a fixed, geographical area which might represent a large retirement village, a resort town, or perhaps a suburb. Aside from attractive location, climate, and recreational facilities, housing availability is the primary determinant of population growth in the community. As long as housing supply equals housing desired, which is proportional to population, people move into the area. The community's attractive qualities bring people in at a normal rate of 14.5 percent per year of the resident population. Under these conditions, area

---

[1] Familiarity with Urban Dynamics, however, is not a prerequisite for this exercise.

residents are also leaving at a normal rate of 2 percent per year for various personal reasons. An abundant housing situation in the area tends to attract people at a greater than normal 14.5 percent per year while discouraging the normal 2 percent outflow. The excess housing causes purchase and rental prices to decline, provides a greater selection of available housing, and forces developers to escalate promotion of the area. When a tight housing situation appears, the opposite actions result. Prospective migrants are deterred from moving into the area. Unable to find desirable housing, residents leave at an accelerated rate. Prospective migrants perceive changes in housing availability only after a 5-year delay. Besides migration flows into and out of the area, the population experiences an annual net death rate of 2.5 percent in consequence of the largely elderly character of the local population.

The housing construction industry responds both to housing availability and land availability within the area. New construction continues as long as plenty of choice land remains available. Under these conditions, the housing construction rate will equal 12 percent per year of the existing housing stock just to keep up with normal population growth. When surplus housing exists, builders cut back on construction of new units. When tight housing market conditions prevail, the construction rate increases to meet demand. As the land zoned for residential development fills up, construction ceases. Since the average lifetime of housing units equals approximately 50 years, the annual demolition rate equals 2 percent.

E12.1 CAUSAL-LOOP DIAGRAM

From the verbal description of a hypothetical residential community, develop a simple causal-loop diagram. Complete your diagram with only those variables necessary to close the indicated loops.

E12.2 FLOW DIAGRAM

Convert your causal-loop diagram to a system dynamics flow diagram.

DYNAMO EQUATIONS

E12.3

Convert your flow diagram to DYNAMO equations and briefly explain each equation. Draw and explain each table function.

MODEL BEHAVIOR AND ANALYSIS

E12.4

Run the model and analyze the runs. Perform a limited sensitivity analysis of your table functions and any key parameters. Does the model behavior seem reasonable? Briefly explain the behavior and how your model assumptions lead to observed behavior. Give a brief critique of model usefulness.

Solution 12

This detailed solution follows the outline suggested in the exercise. S12.1 contains a statement-by-statement development of the multi-loop structure presented in the verbal description. S12.2 translates the causal-loop diagram into a flow diagram. S12.3 contains a detailed explanation of model equations, including table functions. S12.4 investigates model behavior and explores various table function and parametric alternations. A conclusion and critique of the model completes the exercise.

We must keep in mind that the model developed in this exercise is extremely simplified. At most, the model offers an initial conceptualization that could form the groundwork for a more substantive model. This exercise aims primarily to introduce a more complex model and to illustrate how structural assumptions determine model behavior. In the process the exercise exposes the reader to some implications of important assumptions found in Urban Dynamics. These assumptions include the "housing-migration" loop and the "land fraction occupied-construction" loop. The model demonstrates how interaction of these two loops can alone produce the growth and stagnation life-cycle typical of many urban areas. The exercise also demonstrates the relative insensitivity of this simple multi-loop structure to parametric changes.[1]

CAUSAL-LOOP DIAGRAM **S12.1**

We can sketch the individual causal relationships suggested in the exercise description. Chapter 1 presents the conventions and mechanics for causal-loop diagramming.

---

[1]For a complete step-by-step analysis of the urban system see Louis E. Alfeld, Introduction to Urban Dynamics, (Cambridge: Wright-Allen Press, 1975). Alfeld's textbook evolves a series of ten simple urban models. Chapter 6, for example, presents a population-housing model similar to the residential community model developed in this exercise.

Consider a fixed geographical area which might represent a large retirement village, resort town, or perhaps a suburb. Aside from attractive location, climate, and recreational facilities, housing availability is the primary determinant of population growth in the community.

Only two factors control migration streams in the fixed area: natural qualities which we assume remain unchanged throughout the area's life cycle, and housing availability which changes as population and housing stocks vary. Figure (a) diagrams the appropriate variables.

Since the determination of the supply of housing has not yet been presented, we temporarily show housing as an exogenous variable. The natural attractiveness, an externally determined variable, also lies outside the loop. The closed loop contains three elements: population, housing availability, and migration. These elements produce a negative feedback path. Migration keeps population in balance with housing.

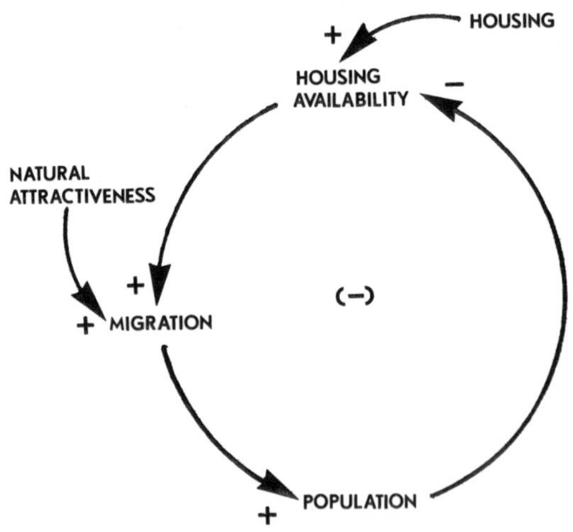

(a)

Basic migration-housing availability loop

As long as housing supply equals housing desired, which is proportional to the population, people move into the area. The community's attractive qualities bring people in at a normal rate of 14.5 percent per year of the resident population.

This statement simply adds a positive link from population directly to migration and indicates that housing availability depends on housing desired and housing as shown in Figure (b).

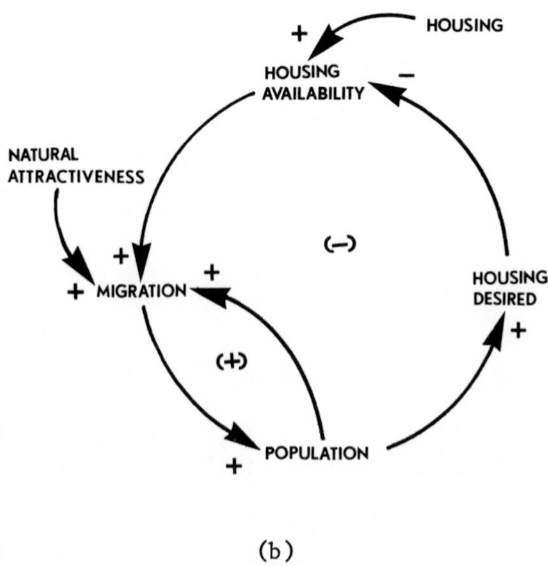

(b)
Population-migration link added

Under these conditions, area residents are also leaving at a normal rate of 2 percent per year for various personal reasons.

That statement indicates that we must disaggregate migration into its two components: in-migration and out-migration. Figure (c) depicts the two streams. The model now contains an additional negative loop between population and out-migration. We have not yet established a causal link from housing availability to out-migration.

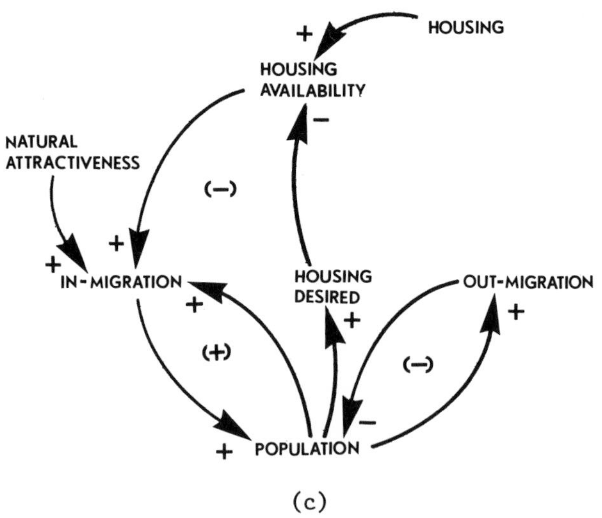

(c)

Out-migration loop added

An abundant housing situation in the area tends to attract people at a greater than normal 14.5 percent per year while discouraging the normal 2 percent out-flow. The excess housing causes purchase and rental prices to decline, provides a greater selection of available housing, and forces developers to escalate promotion of the area.

The first statement links out-migration, as well as in-migration, to housing availability through the attractiveness variable from housing availability introduced in the second statement. That is, the statements indicate that not excess housing alone, but the consequences of excess housing as well, make the area more attractive. As shown in Figure (d), the aggregate effects on migration form a single attractiveness variable which depends in a positive way on housing availability.

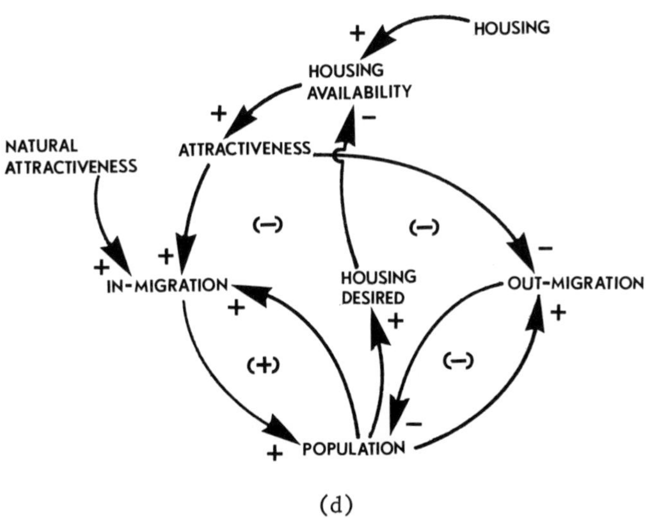

(d)

Attractiveness variable added,
out-migration linked to attractiveness

When a tight housing situation appears, the opposite actions result. Prospective migrants are deterred from moving into the area. Unable to find desirable housing, residents in the area leave at an accelerated rate.

This statement does not require any refinement of the last diagram because the structure can already account for either a glutted or tight housing market in the area.

Prospective migrants perceive changes in housing availability only after a five-year delay.

We insert a perception delay between the attractiveness and in-migration variables. Since we assume that residents in the area have more current information about housing, no perception delay appears in the out-migration-attractiveness link. Figure (e) shows the delay.

Besides migration flows into and out of the area, the population experiences an annual net death rate of 2.5 percent in consequence of the largely elderly character of the local population.

This statement indicates the need for an additional net death rate loop to complete the population sector of the model.

Figure (e) shows four negative loops and one positive loop in the population sector.

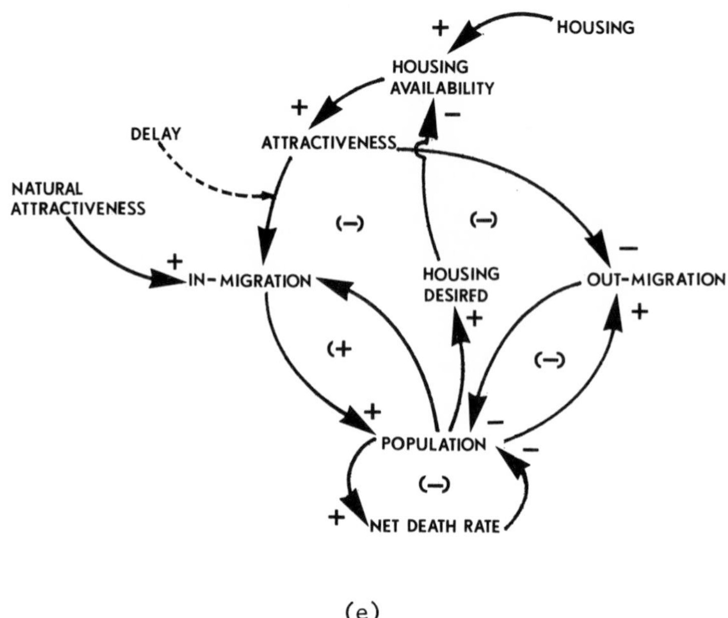

(e)
Complete population sector

The housing construction industry responds both to housing availability and land availability within the area.

The statement indicates that two factors, as shown in Figure (f), influence the housing construction rate: housing availability and land availability. By inference, both housing availability and lack of available land (or increase in land occupied) depress construction.

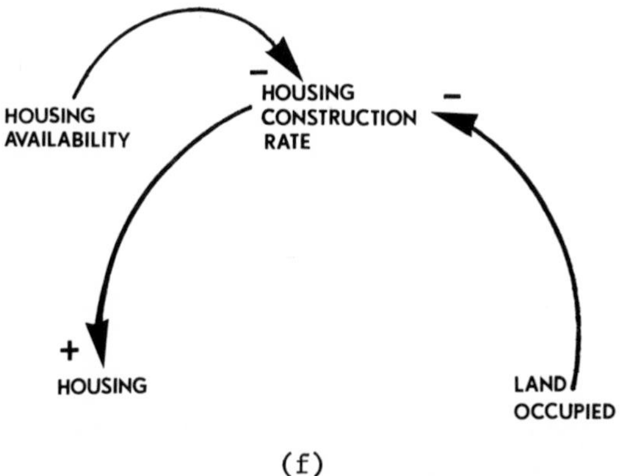

(f)

Basic factors in housing construction

New construction continues as long as plenty of choice land remains available. Under these conditions, the housing construction rate will equal 12 percent per year of the existing housing stock just to keep up with normal population growth.

We add a positive link from housing to the construction rate. Figure (g) contains the updated causal-loop diagram. Figure (g) also contains the positive link from housing to housing availability.

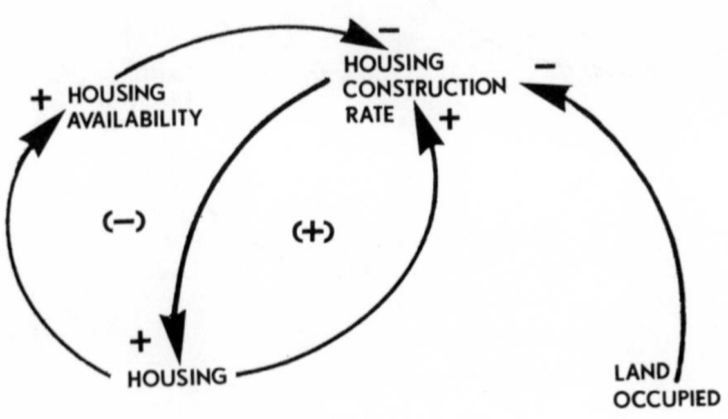

(g)

Link between housing and
housing construction added

When surplus housing exists, builders cut back on construction of new units. When tight housing market conditions prevail, the construction rate increases to meet demand.

The two statements simply describe the decision process of the construction industry in the area. The diagram needs no modification.

As the land zoned for residential development fills up, construction ceases.

We add the obvious positive link between housing and land occupied. However, in order to account explicitly for the total land available, the diagram must be slightly altered. We introduce the land fraction occupied which simply equals the ratio of land occupied by housing to total land available. Figure (h) shows this.

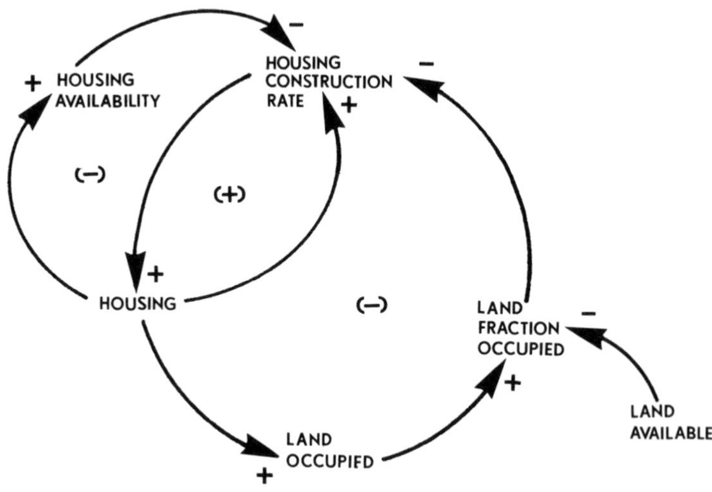

(h)

Land fraction occupied loop added

Since the average lifetime of housing units equals approximately 50 years, the annual demolition rate equals 2 percent.

This statement calls for a demolition loop in the housing sector. The complete housing sector, shown in Figure (i), contains one positive loop and three negative loops.

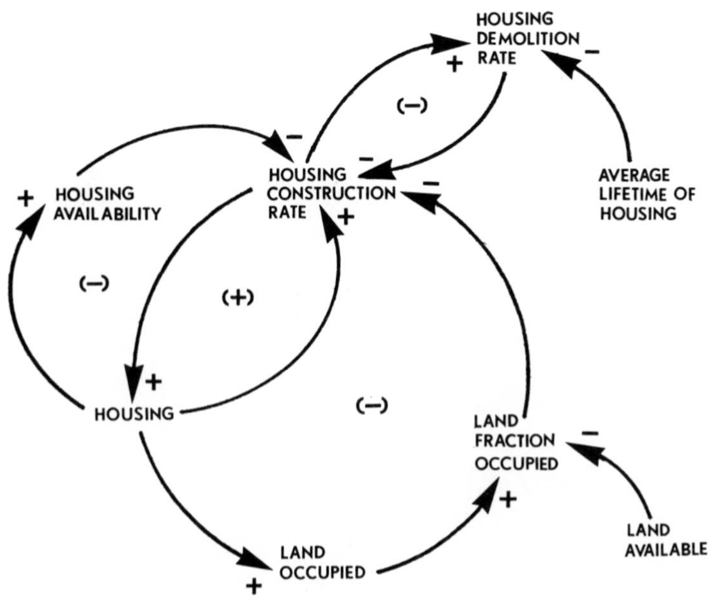

(i)

Demolition loop added

The complete causal-loop diagram appears in Figure (j). The population sector and housing sector are linked exclusively through the housing availability variable. The loops and their respective polarities represent the model assumptions. The following flow diagram and equations spell out the specifics of those assumptions and their implications.

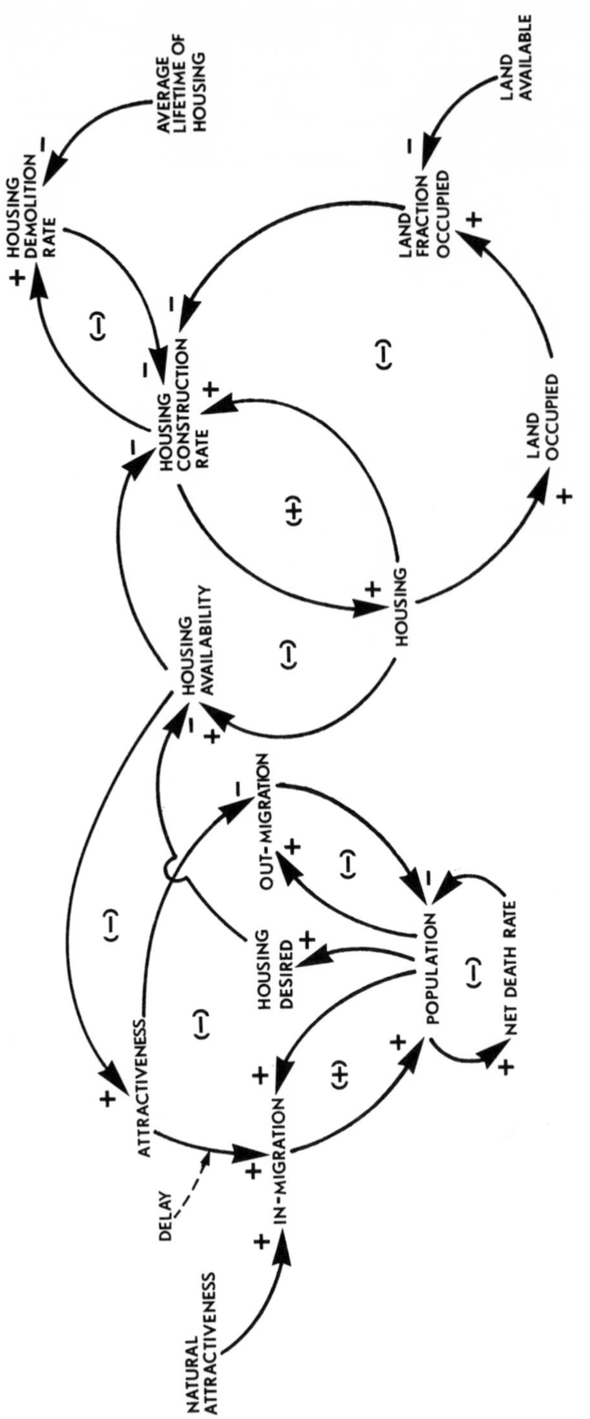

(j)

Complete causal diagram

FLOW DIAGRAM

S12.2

A few guidelines can aid in converting causal-loop diagrams into DYNAMO flow diagrams. We can identify levels by deciding which variables in the causal-loop diagram embody the process of accumulation. Population and housing easily qualify under this test. Alternatively, imagine the system instantly frozen; all action and activity halted. Since all rates would disappear, those variables which we could still measure or count are either levels, auxiliaries (functions of levels) or constants.[2] Freezing the community system would leave the measurable physical quantities of people and houses. Such computable variables as land fraction occupied, a function of the number of houses and the constant land area, would be modeled as an auxiliary.

Once we have isolated the levels, auxiliaries, and constants, then only the rates remain. We can apply an easy check on whether rates have been correctly identified. Only rates can alter levels. Hence, any variable directly affecting an identified level must be a rate. The rates in this model are fairly obvious because we have already identified them as such.

Population Sector. The causal-loop diagram indicates that the level of population POP is increased by the in-migration rate IMR and decreased by the out-migration rate OMR and the net death rate NDR. The causal-loop diagram further indicates that all three rates depend directly on the population level (ignoring other influences for the moment). According to the verbal description, as long as housing keeps up with population growth, the three population flow rates experience the following percentage growth rates:

In-Migration	14.5 percent per year
Out-Migration	2.0 percent per year
Net Death Rate	2.5 percent per year

---

[2]Constants are quantities that do not change over time and are not affected by other variables within the boundary of the system. They should be obvious from the causal-loop diagram.

These parameters, respectively defined as the normal in-migration NIM, normal out-migration NOM, and death rate fraction DRF, appear as constants in the flow diagram of Figure (a). For consistency, we will use the same procedure when specifying the parameters associated with the housing construction and demolition rates. When housing desired remains equal to the housing available, an in-migration rate of 14.5 percent per year and a simultaneous outflow rate of 4.5 percent per year from out-migration and deaths produces a net population growth of 10 percent per year. Thus population doubles in size every 7 years. The net population growth rate arises from the natural attractiveness of the area, all else being equal.

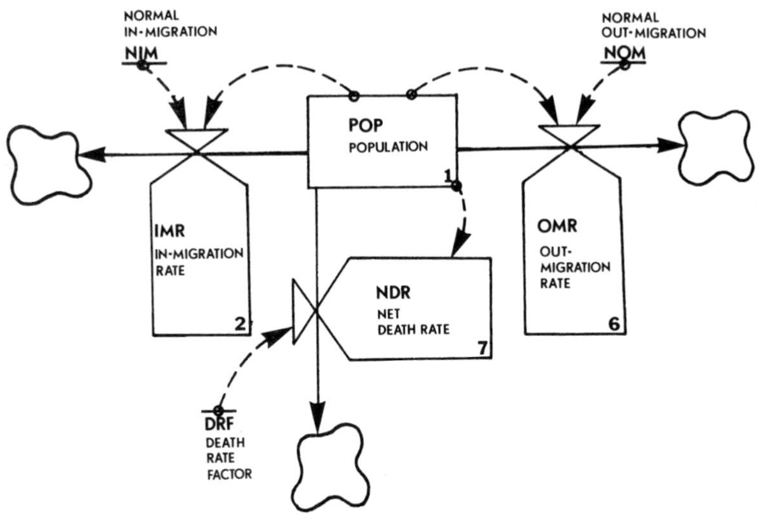

(a)

Population sector

Housing desired HD, simply proportional to the population, appears in the model as an auxiliary. For this model, we assume that three people on the average can comfortably occupy each housing unit or that each person desires one-third of a unit. A housing unit occupies a fixed amount of land and includes not only the housing structure but all necessary service facilities, access roads, and sidewalks. Figure (b) includes the housing desired auxiliary.

For simplicity, housing availability, which links the two levels of the model, equals the ratio of the housing level to housing desired and is redefined as the housing ratio HR. HR provides a rough index of the condition of the housing market in the area. Other indices such as rental/purchase prices, occupancy ratios, or the delay time in acquiring suitable housing would require additional structure. For the level of aggregation in this model, the housing ratio formulation is sufficient. A ratio less than one reflects a shortage condition; a ratio greater than one reflects an excess condition. A unity value indicates an equilibrium condition between housing and housing desired. The housing ratio HR (housing availability) appears in Figure (b).

From the causal-loop diagram of the population sector, we can see that the attractiveness for migration variable AMM depends upon the housing ratio HR. The attractiveness variable requires a table function to describe its nonlinear relationship with the housing ratio. The next section of this exercise presents the details of the relationship. For the moment, simply assume that, as the housing ratio HR varies, the area becomes more or less than "normally" attractive. The normal 14.5 percent per year in-migration flow varies upward or downward in response to the attractiveness variable or multiplier as housing conditions change. A perception delay AMMP intervenes in this process. The completed loop appears in Figure (b).

The flow diagram representation for the population sector of the model now lacks only the information link from the attractiveness for migration multiplier AMM to the out-migration rate OMR. The inverse of the attractiveness multiplier, the departure migration multiplier DMM, modulates out-migration rate OMR. That is, as the area becomes more attractive because of ample housing availability, more people arrive and fewer depart than normally in a given period. The opposite tendencies result when housing becomes tight. Figure (c) contains the entire population sector flow diagram.

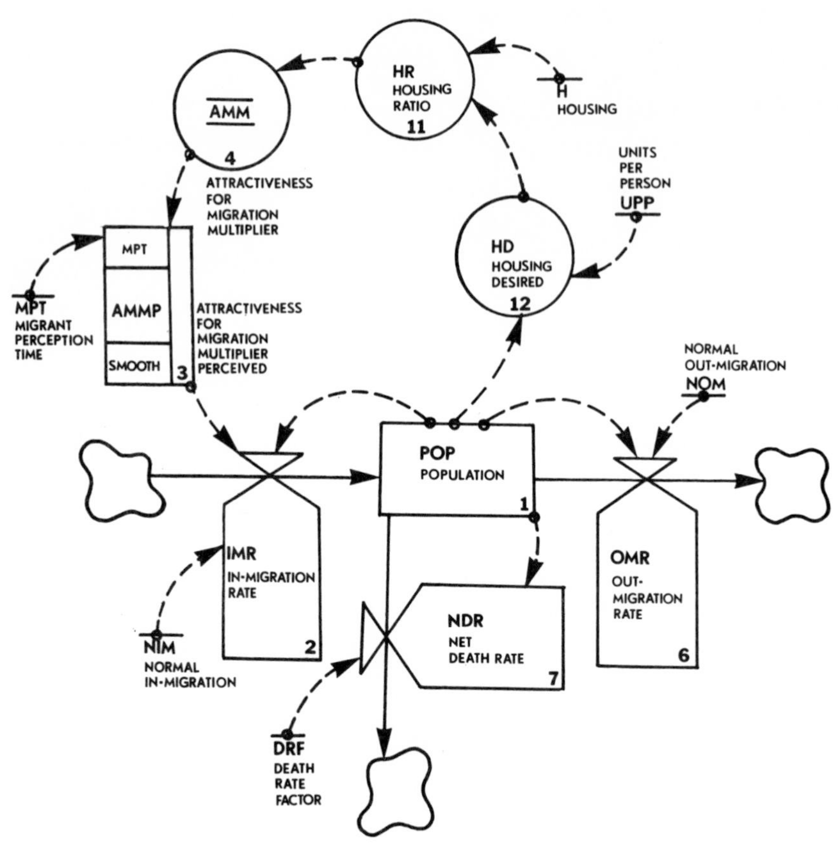

(b)

Population sector with housing desired loop

(Note: Housing H is shown as an exogenous constant within the population sector.)

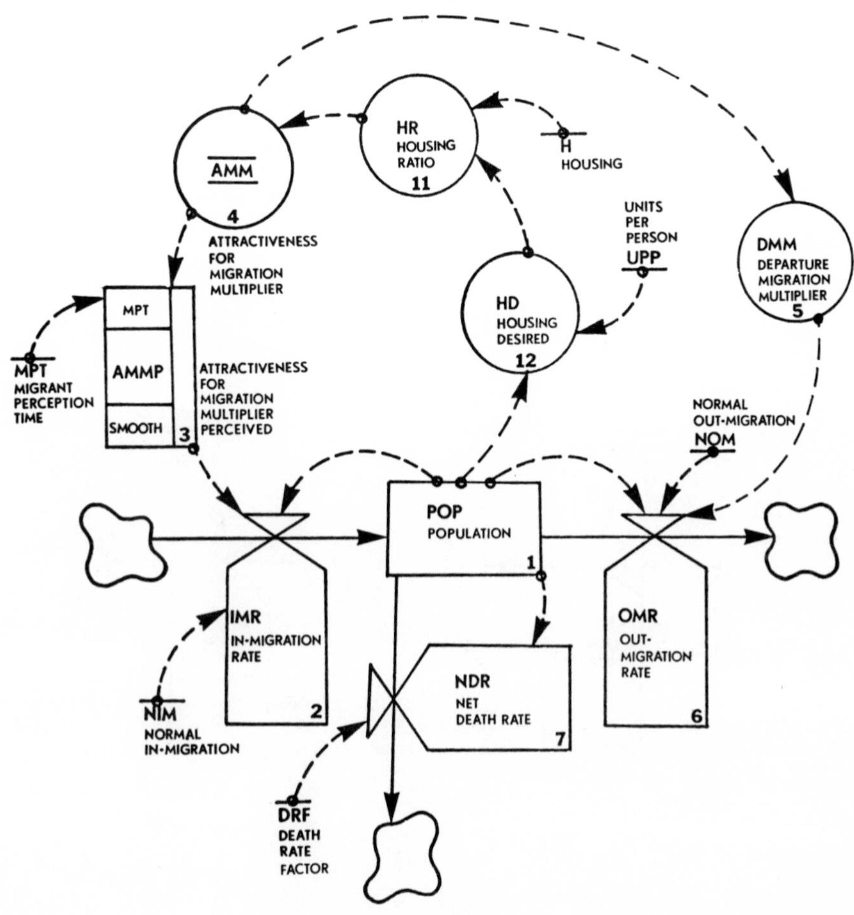

(c)

Complete population sector

Housing Sector. The level of housing H is increased by the housing construction rate HCR and decreased by the housing demolition rate HDR. Like the population flow rates, both housing flow rates are functions of a level and other variables. As indicated in the verbal description, when the amount of housing just equals the housing desired (HR equals one), construction proceeds at 12 percent per year and demolition proceeds at 2 percent per year. Normal housing construction NHC and average lifetime of housing ALTH, respectively, denotes these two model parameters in the flow diagram of Figure (d). The level of housing H under these normal circumstances grows at 10 percent per year, the same growth rate as in the population sector. We have defined the five rate parameters--NIM, NOM, DRF, NHC, and ALTH--in the model in a manner consistent with the definitional point of a housing ratio HR of one. As long as HR remains at one, the area experiences a simultaneous 10 percent per year growth in population and housing.

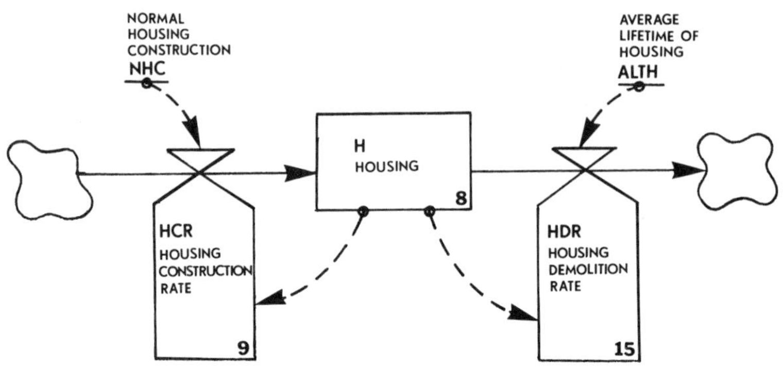

(d)

Housing sector

The negative land availability loop imposes an important constraint on housing construction. The loop contains two auxiliaries: the land fraction occupied LFO and the land availability multiplier LAM. LFO is simply a function of the housing level and two constants: land area LAND and average amount of land per unit LPU. The land availability multiplier LAM, represented by a table, has a nonlinear relationship with LFO. LAM modifies the normal yearly construction rate NHC in a manner analogous to the attractiveness multiplier. As long as LFO is much less than one, LAM has a value of unity and hence, does not affect the normal construction rate. The next section discusses the shape of the LAM table in detail. The LFO loop appears in Figure (e).[3]

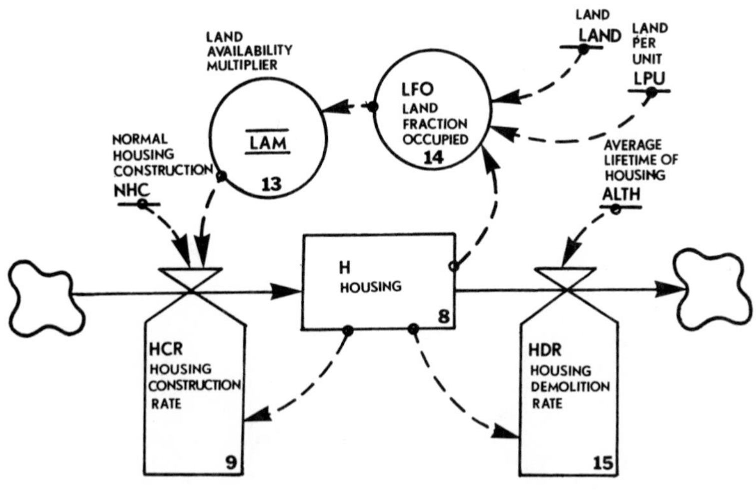

(e)

Housing sector with land availability

Housing construction is linked to the housing ratio HR through a housing construction multiplier HCM. The multiplier simply adjusts the normal construction rate upward or downward to reflect housing market conditions represented by HR. We use a table function because the multiplier is not linearly related to the housing ratio over the entire range of possible housing conditions. The details of the HCM table are discussed in the next section. Figure (f) contains the entire housing sector flow diagram.

---

[3]Note this structure is similar to the land-use model in Chapter 5. S-shaped growth in housing results.

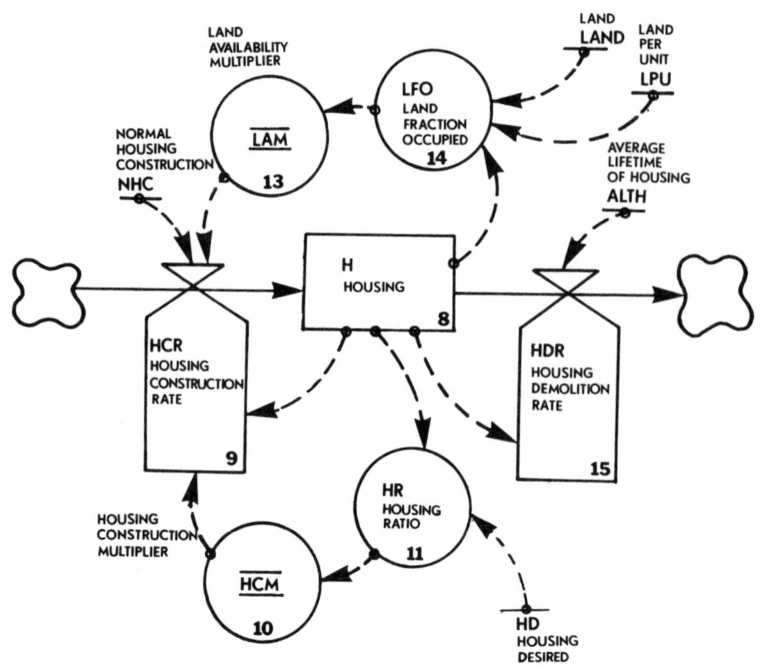

(f)

Complete housing sector

(Note: Housing desired HD is shown as an exogenous constant within the housing sector.)

The complete flow diagram appears on the following page. From this relatively simple third-order system, containing many of the relationships found in Urban Dynamics, some interesting behavior results. S12.3 provides the detailed equation descriptions necessary for understanding how the model fits together.

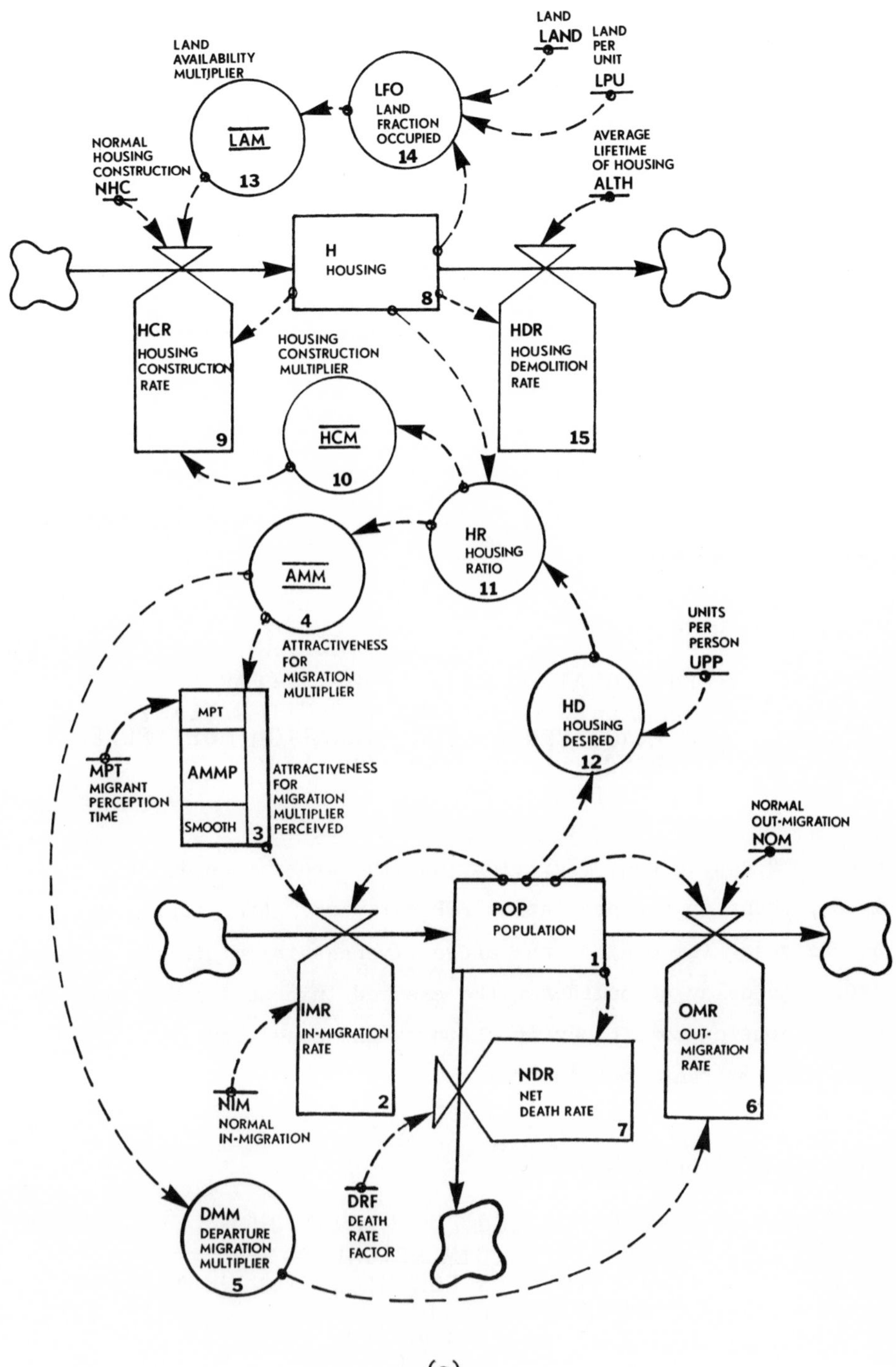

(g)

Complete flow diagram

S12.3 DYNAMO EQUATIONS

Population Sector. The level of population is altered by three flow rates: in-migration, out-migration, and deaths.

```
POP.K=POP.J+(DT)(IMR.JK-OMR.JK-NDR.JK)                    1, L
POP=POPI                                                  1.1, N
POPI=30.3                                                 1.2, C
    POP     - POPULATION (PEOPLE)
    IMR     - IN-MIGRATION RATE (PEOPLE/YEAR)
    OMR     - OUT-MIGRATION RATE (PEOPLE/YEAR)
    NDR     - NET DEATH RATE (PEOPLE/YEAR)
    POPI    - INITIAL VALUE OF POPULATION (PEOPLE)
```

The in-migration rate IMR is the product of population POP, the attractiveness for migration multiplier perceived AMMP, and the normal in-migration NIM.

```
IMR.KL=NIM*AMMP.K*POP.K                                   2, R
NIM=.145                                                  2.1, C
    IMR     - IN-MIGRATION RATE (PEOPLE/YEAR)
    NIM     - NORMAL IN-MIGRATION (FRACTION/YEAR)
    AMMP    - ATTRACTIVENESS FOR MIGRATION MULTIPLIER
                PERCEIVED (DIMENSIONLESS)
    POP     - POPULATION (PEOPLE)
```

When AMMP equals one, migration into the area proceeds at a rate of 14.5 percent of the population POP per year. AMMP represents a 5-year delay (smooth) of the attractiveness for migration multiplier AMM. The delay accounts for the assumed information lag between actual housing conditions in the area and conditions perceived by potential migrants outside the area.

```
AMMP.K=SMOOTH(AMM.K,MPT)                                  3, A
MPT=5                                                     3.1, C
    AMMP    - ATTRACTIVENESS FOR MIGRATION MULTIPLIER
                PERCEIVED (DIMENSIONLESS)
    AMM     - ATTRACTIVENESS FOR MIGRATION MULTIPLIER
                (DIMENSIONLESS)
    MPT     - MIGRANT PERCEPTION TIME (YEARS)
```

AMM relates in-migration to housing conditions in the area.

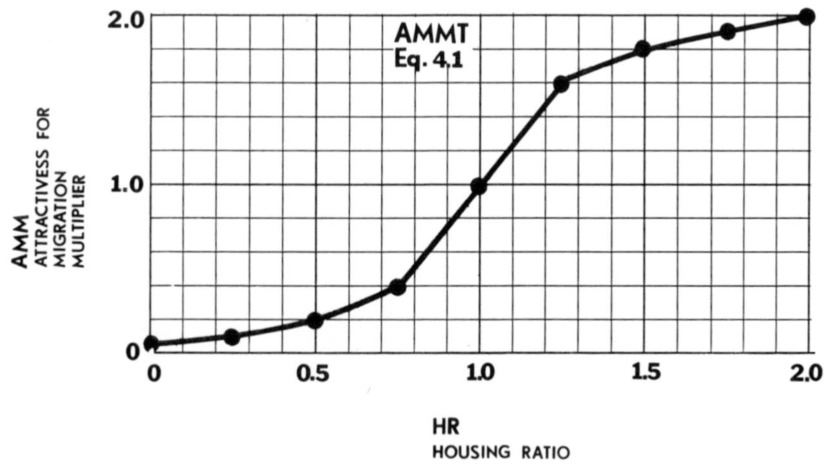


```
AMM.K=TABLE(AMMT,HR.K,0,2,.25)                                4, A
AMMT=.05/.1/.2/.4/1/1.6/1.8/1.9/2                             4.1, T
     AMM     - ATTRACTIVENESS FOR MIGRATION MULTIPLIER
                 (DIMENSIONLESS)
     AMMT    - ATTRACTIVENESS FOR MIGRATION MULTIPLIER
                 TABLE
     HR      - HOUSING RATIO (DIMENSIONLESS)
```

(a)

When HR equals one, AMM equals one, and in-migration occurs at 14.5 percent per year. When housing desired exceeds housing available, AMM begins to decline toward zero because of housing scarcity. An excess supply of housing induces migration, presumably in response to lower prices and a wider selection of housing units. We assume a saturation phenomenon however, once housing attains a 25 percent excess. That is, additional excess housing provides a smaller and smaller additional incentive to migrate. The slope of the table function becomes steepest through the definitional point (HR = 1, AMM = 1) because in this region we assume migrants are most sensitive to changes in housing availability.

The departure migration multiplier DMM equals the reciprocal of AMM. That is, we assume that the same housing conditions inducing people to move into the area would also tend to keep residents in the area from moving out.

```
DMM.K=1/AMM.K                                              5, A
    DMM      - DEPARTURE MIGRATION MULTIPLIER (DIMENSIONLESS)
    AMM      - ATTRACTIVENESS FOR MIGRATION MULTIPLIER
                  (DIMENSIONLESS)
```

The out-migration rate OMR is the product of the departure migration multiplier DMM, population POP, and the normal out-migration NOM.

```
OMR.KL=NOM*DMM.K*POP.K                                     6, R
NOM=.02                                                    6.1, C
    OMR      - OUT-MIGRATION RATE (PEOPLE/YEAR)
    NOM      - NORMAL OUT-MIGRATION (FRACTION/YEAR)
    DMM      - DEPARTURE MIGRATION MULTIPLIER (DIMENSIONLESS)
    POP      - POPULATION (PEOPLE)
```

The net death rate NDR is the product of the death rate factor DRF and the population POP.

```
NDR.KL=POP.K*DRF                                           7, R
DRF=.025                                                   7.1, C
    NDR      - NET DEATH RATE (PEOPLE/YEAR)
    POP      - POPULATION (PEOPLE)
    DRF      - DEATH RATE FACTOR (FRACTION/YEAR)
```

<u>Housing Sector</u>. The level of housing H is the net accumulation over time of the housing construction rate HCR and housing demolition rate HDR. We set the initial value of the housing level so that the housing ratio equals one.

```
H.K=H.J+(DT)(HCR.JK-HDR.JK)                                8, L
H=HI                                                       8.1, N
HI=10                                                      8.2, C
    H        - HOUSING (UNITS)
    HCR      - HOUSING CONSTRUCTION RATE (UNITS/YEAR)
    HDR      - HOUSING DEMOLITION RATE (UNITS/YEAR)
    HI       - INITIAL VALUE OF HOUSES (UNITS)
```

The housing construction rate HCR is a multiplicative function of the housing construction multiplier HCM, the land availability multiplier LAM, housing H, and the normal housing construction NHC.

```
HCR.KL=NHC*HCM.K*LAM.K*H.K                                  9, R
NHC=.12                                                     9.1, C
     HCR     - HOUSING CONSTRUCTION RATE (UNITS/YEAR)
     NHC     - NORMAL HOUSING CONSTRUCTION (FRACTION/YEAR)
     HCM     - HOUSING CONSTRUCTION MULTIPLIER
                 (DIMENSIONLESS)
     LAM     - LAND AVAILABILITY MULTIPLIER
                 (DIMENSIONLESS)
     H       - HOUSING (UNITS)
```

When both HCM and LAM equal one, then HCR equals 12 percent per year of the housing stock. The unity values of HCM and LAM define normal conditions in the model. Neither a land constraint nor an excess or surplus of housing prevails.

The housing construction multiplier HCM depends on the housing ratio HR. The table function reflects our assumptions about the response of the construction industry to housing availability. When HR equals one, the multiplier has essentially no effect and, assuming LAM also equals one, construction proceeds at 12 percent per year. When there is 25 percent more housing than desired, the multiplier declines substantially and forces construction to diminish. HCM approaches, but never achieves, zero on the assumption that someone will always want to build his own home. At the other extreme, when desired housing far exceeds available housing, the incentive for construction increases and we assume that builders construct at a maximum rate of 2.5 times the normal or 30 percent per year as long as land remains available.

```
HCM.K=TABLE(HCMT,HR.K,0,2,.25)                              10, A
HCMT=2.5/2.4/2.3/2/1/.37/.2/.1/.05                          10.1, T
     HCM     - HOUSING CONSTRUCTION MULTIPLIER
                 (DIMENSIONLESS)
     HCMT    - HOUSING CONSTRUCTION MULTIPLIER TABLE
     HR      - HOUSING RATIO (DIMENSIONLESS)
```

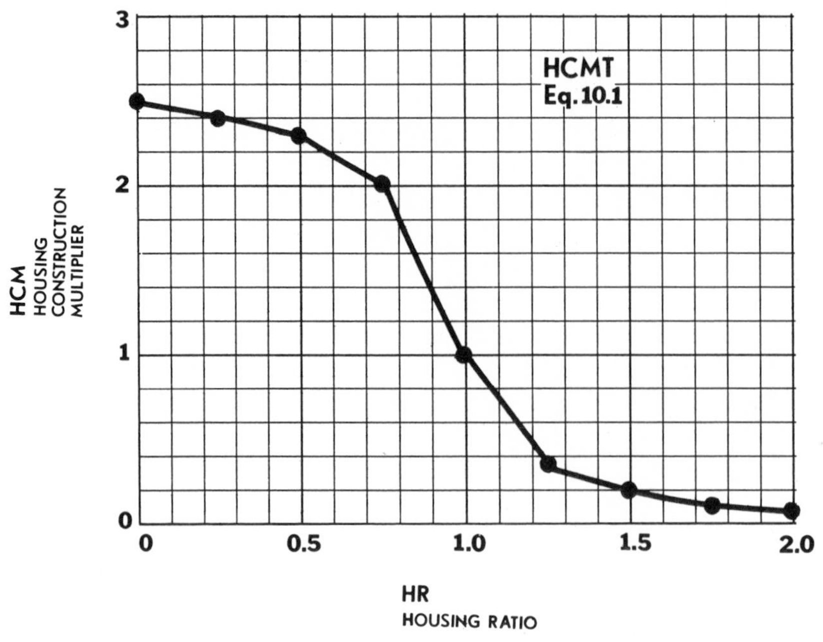

(b)

The housing ratio HR consists of the housing level H and housing desired HD.

```
HR.K=H.K/HD.K                                            11, A
     HR      - HOUSING RATIO (DIMENSIONLESS)
     H       - HOUSING (UNITS)
     HD      - HOUSING DESIRED (UNITS)
```

The housing desired HD is the product of the population POP and the units per person UPP, a constant equal to 1/3.

```
HD.K=POP.K*UPP                                           12, A
UPP=.33                                                  12.1, C
     HD      - HOUSING DESIRED (UNITS)
     POP     - POPULATION (PEOPLE)
     UPP     - UNITS PER PERSON (UNITS/PERSON)
```

The land availability multiplier LAM depends on the ratio of land area occupied by housing to total land area.

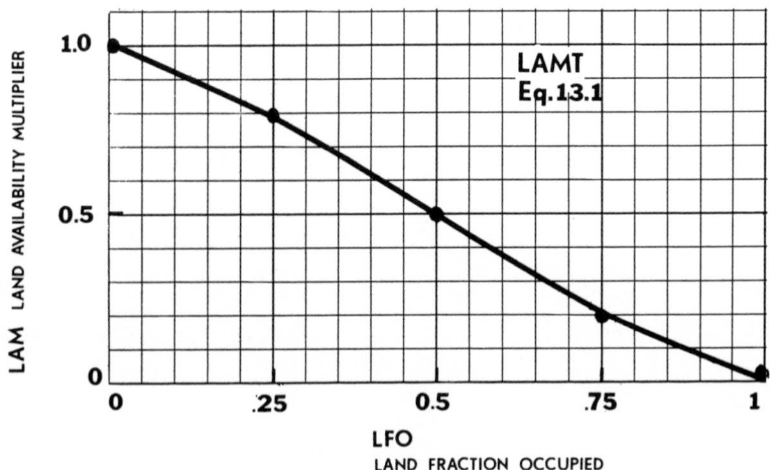


```
LAM.K=TABLE(LAMT,LFO.K,0,1,.25)                                 13, A
LAMT=1/.8/.5/.2/0                                               13.1, T
    LAM     - LAND AVAILABILITY MULTIPLIER
                  (DIMENSIONSIONLESS)
    LAMT    - LAND AVAILABILITY MULTIPLIER TABLE
    LFO     - LAND FRACTION OCCUPIED (DIMENSIONLESS)
```

(c)

The table function characterizes physical properties of the area. The first 25 percent of the land area used offers no prohibitive costs or restraints on construction. The next 50 percent requires additional development effort and generally has poorer quality than the first 25 percent. The final 25 percent is land developed only when demand becomes high enough to cover the development expenses.

In order to calculate LFO, we must specify the average amount of land per unit. We set the land per unit LPU in this model at one acre, and the total amount of land available for development at 1,500 acres. The land fraction occupied can then be computed by multiplying the number of houses by LPU and dividing by the land available LAND.

```
LFO.K=H.K*LPU/LAND                                          14, A
LPU=1                                                       14.1, C
LAND=1500                                                   14.2, C
    LFO     - LAND FRACTION OCCUPIED (DIMENSIONLESS)
    H       - HOUSING (UNITS)
    LPU     - LAND PER UNIT (ACRES/UNIT)
    LAND    - LAND (ACRES)
```

The housing demolition rate is a function of the average lifetime of housing ALTH, a constant, and the housing level H. A 50-year lifetime, used in the model, corresponds to a 2 percent yearly removal rate.

```
HDR.KL=H.K/ALTH                                             15, R
ALTH=50                                                     15.1, C
    HDR     - HOUSING DEMOLITION RATE (UNITS/YEAR)
    H       - HOUSING (UNITS)
    ALTH    - AVERAGE LIFETIME OF HOUSING (YEARS)
```

Control Statements. We set the model to run for 100 years, 25 years longer than the 75-year life cycle of the area, to insure that the area reaches equilibrium. The computer plots values every 2 years and makes computations each year. The two levels, the housing construction and demolition rates, the in-migration and out-migration rates, the housing ratio, and land fraction occupied are plotted in the print out.

```
PLOT H=H(0,2000)/POP=P(0,8000)/HCR=C,HDR=D(0,100)/OMR=0,
X IMR=M(0,600)/HR=R(.6,1.2)/LFO=L(0,1)
C DT=1
C LENGTH=100
C PLTPER=2
```

MODEL BEHAVIOR AND ANALYSIS

S12.4

Run 1 contains the standard run of the model. Three distinct phases characterize the life cycle of the area. The first 50 years are marked by rapid growth and development. The housing ratio during these years exceeds one, indicating a slight abundance of housing. In-migration and construction grow exponentially while out-migration is practically negligible. The positive loops involving population, housing construction, and migration dominate this phase. The next 20 years are marked by a transition phase. Construction reaches its maximum and begins to decline as the choice land fills up and the construction industry must restrict construction. Housing continues to grow, but at a decreasing rate. The housing ratio begins to decline, thereby retarding migration somewhat and accelerating out-migration. Because of land constraint, and in spite of excess demand (HR less than one), the construction rate continues to decline until construction just equals demolition (replacement). The negative loop involving land dominates. In the final equilibrium phase, housing levels off at 1,300 units, 200 units less than the 1,500 unit capacity of the area. The population levels at roughly 5,200 people.

Population Sector Sensitivity Analysis. The most striking behavior occurs in the housing ratio which indicates the degree of crowding as well as housing shortage. The housing ratio HR declines during the transition phase and eventually equilibrates at a value of 0.78. However, this apparent 22 percent overcrowded condition merely reflects the assumptions in the model. Once the housing level has equilibrated due to a land constraint, the area must become unattractive enough to inhibit in-migration and accelerate out-migration. Therefore, a net in-migration just compensates for the 2.5 percent per year death rate. This equilibrium occurs when the attractiveness for migration multiplier AMM falls below one. An AMM value less than one implies that the housing ratio HR must fall below one.

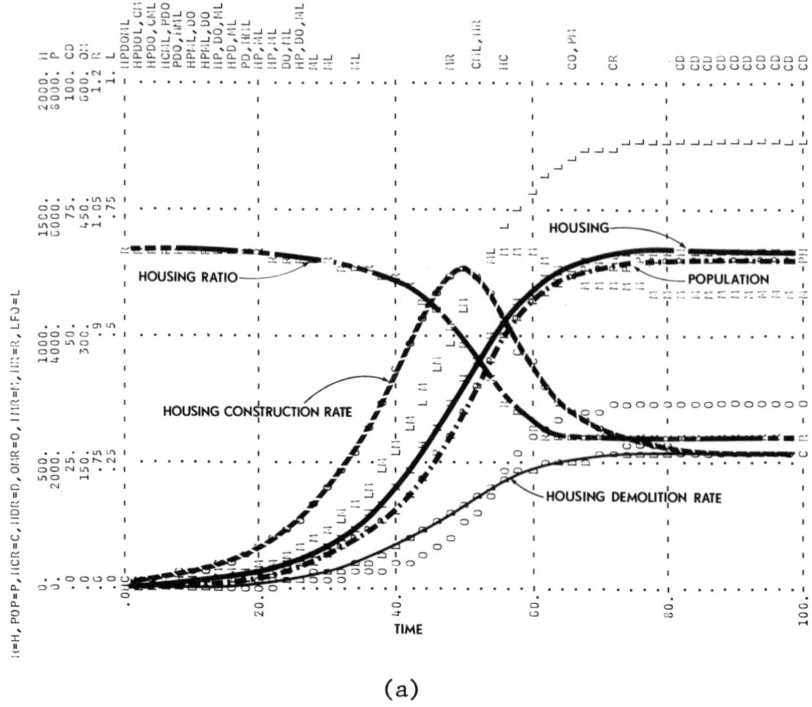

(a)

Run 1--standard run

People will continue to flow into the area until housing conditions become substantially unattractive because we assumed a positive inflow of people when HR equals one.

The preceding analysis permits us to make some observations about the influence of the attractiveness for migration AMM curve on model behavior. From Figure (b), a reproduction of the AMM table, we can see that the AMM value corresponding to an equilibrium HR value of 0.78 is 0.46. Regardless of the shape of the curve, for equilibrium to occur AMM must drop to 0.46. Should we hypothesize that housing availability only weakly affects migration, then the equilibrium housing ratio HR will become even smaller than 0.78. Curve B in Figure (b) illustrates this possibility. To achieve the necessary equilibrium AMM value, HR will decline to a value of 0.30. Curve C illustrates the effect of a very sensitive AMM curve. HR under such an assumption would closely approach 0.9 in equilibrium. The shape of the table function to the right of the definitional point does not affect the equilibrium HR value. The housing ratio must always decline to a value less than one to reach the equilibrium value of AMM as long as a net population inflow occurs when HR equals one.

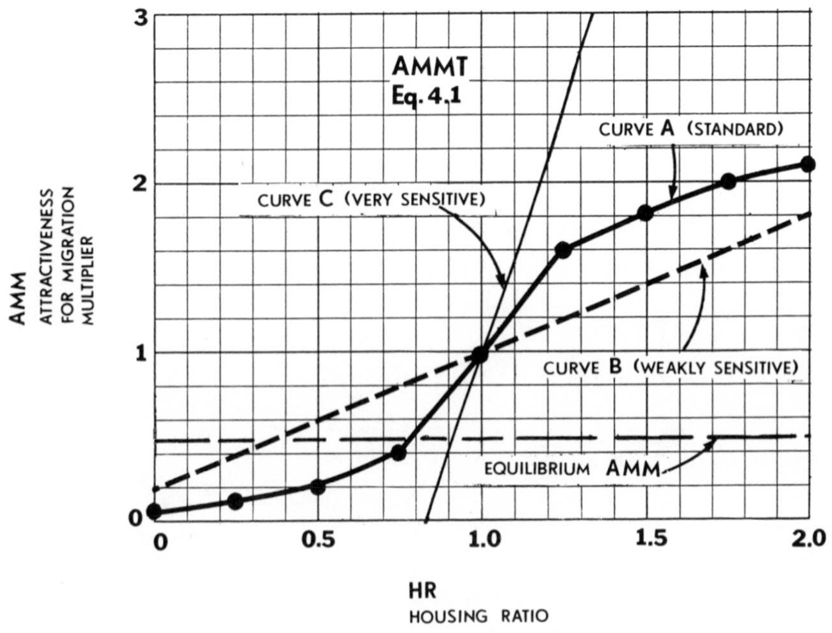

(b)

Alternative AMM curves

<u>Housing Sector Sensitivity Analysis</u>. Model behavior proves relatively insensitive to parameters and table functions in the housing sector. Changing the normal construction rate NHC or average lifetime of housing ALTH simply accelerates or slows down the net growth rate of housing, but the basic S-shaped behavior mode remains. The land constraint loop will always dominate the construction rate regardless of how much population pressure (higher demand) exists or how much optimism or pessimism the construction industry demonstrates.

Run 2 shows the result of doubling NHC from 12 percent per year to 24 percent per year. This change embodies an optimistic or speculative construction policy. That is, under normal conditions in which population growth equals 10 percent per year, housing stock growth will equal 22 percent per year. As shown in Run 2, the housing ratio remains above one for 30 years. The accelerated construction rate quickly reverses, however, as the land constraint begins to dominate growth. The migrant flow cannot reverse its growth trend as readily because of the perception delay. A slight overshoot in population results before the equilibrium housing ratio of 0.78 develops. The speculative normal construction rate simply contracts the 80-year growth cycle of the model to a 60 year cycle.

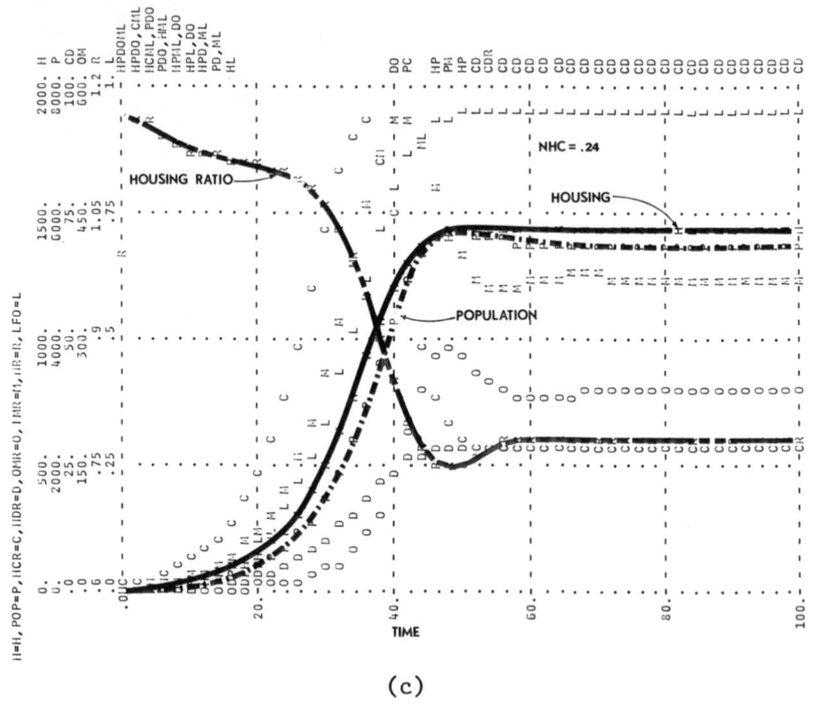

(c)

Run 2--speculative construction policy
(NHC = 0.24)

To test the effect of a relatively conservative housing industry, we change the housing construction multiplier table HCMT as shown in Figure (d). With surplus housing, the construction multiplier is depressed slightly more than in the standard curve. An excess demand condition, however, never increases the normal construction rate by more than 25 percent, as opposed to the potential 150 percent increase found in the standard curve.

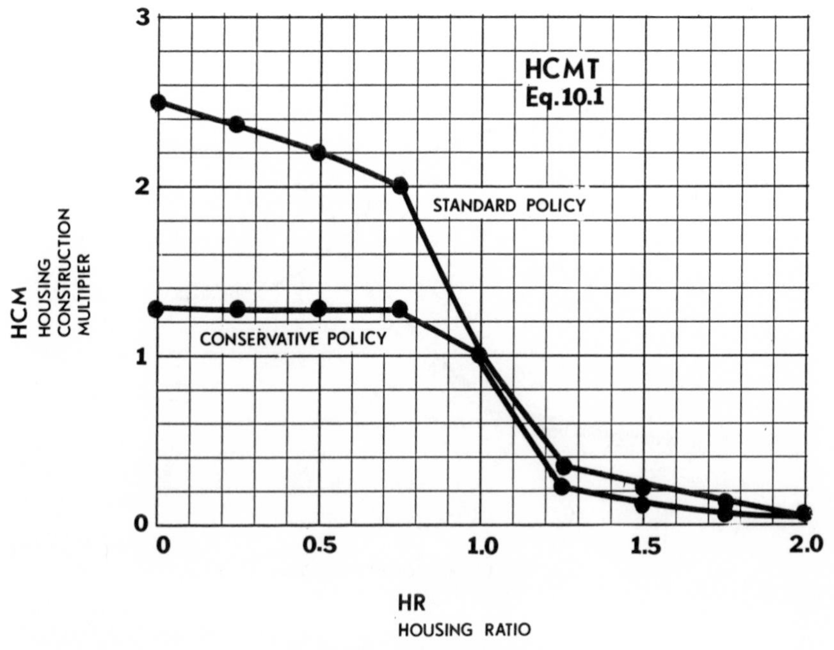

(d)

Alternative HCM curves

Run 3 simulates this construction policy. Construction proceeds more slowly than in the two previous runs and suppresses population growth, enough to eliminate any overshoot. The housing ratio HR remains above one for roughly 40 years before making a transition to its 0.78 equilibrium value. The equilibrium conditions of housing and population occurring after 80 years in this run mirror the standard run.

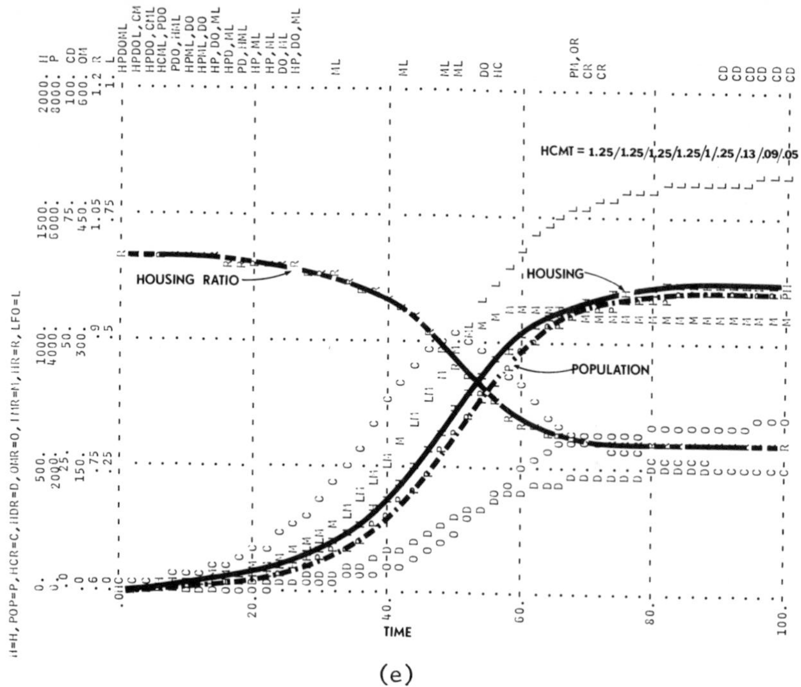

(e)

Run 3--conservative construction policy
(HCMT altered)

As a final test of model assumptions, assume that the land area zoned for residential development has a fairly uniform quality. Construction rapidly halts only when land becomes fairly scarce. Figure (f) illustrates the shape of the proposed land availability multiplier table LAMT.

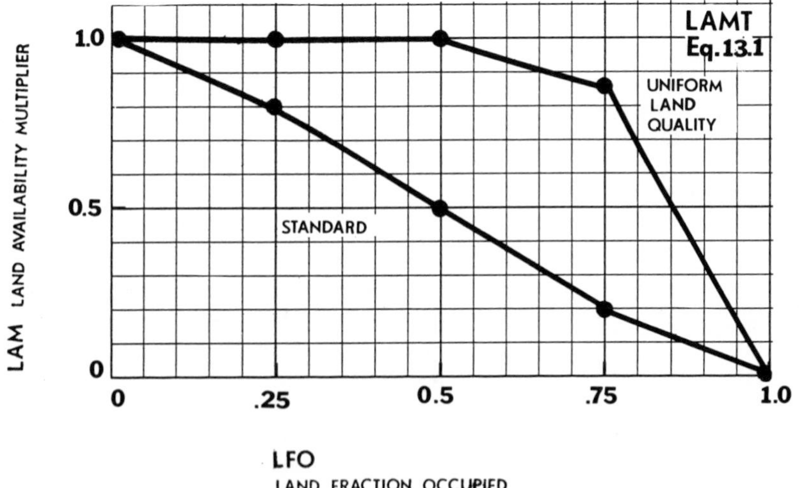

(f)

Alternative LAMT curves

Run 4 simulates model behavior with a uniform land hypothesis. Population and housing can freely grow at rates slightly higher than the normal 10 percent per year during the first 45 years of development because land imposes no constraint. Within the next 15 years, however, construction declines rather suddenly and brings the level of housing into equilibrium. Like Run 2, the migrant streams cannot quickly respond to the rapid housing equilibrium because they encounter the 5-year perception delay. The population overshoots the equilibrium housing ratio and comes to equilibrium some 20 years after the housing stock has ceased growing.

The equilibrium housing and population levels in Run 4 exceed those in the standard run (Run 1). Shifting the decline of the LAM curve to the right allows the construction industry to generate a larger number of houses before the construction rate drops enough to

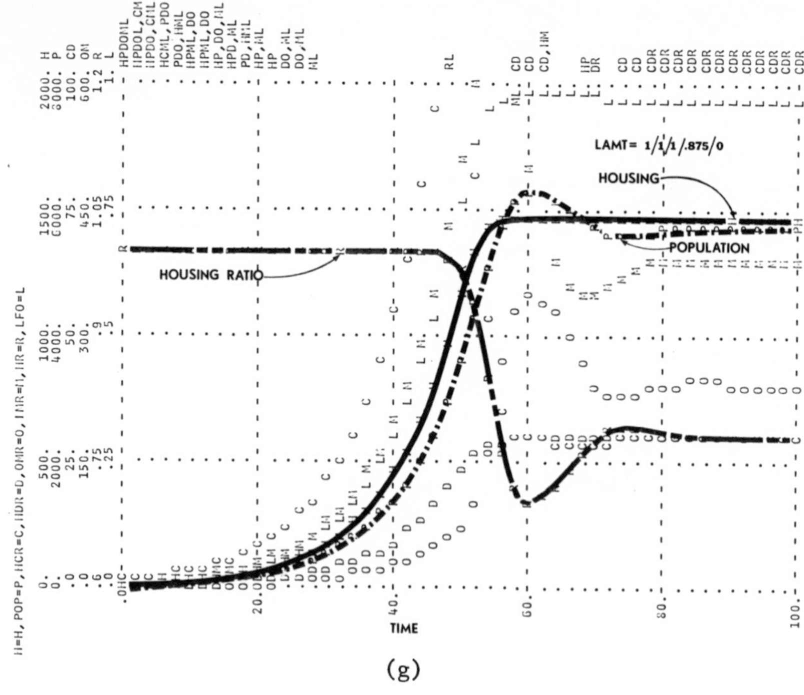

(g)

Run 4--uniform land quality
(LAMT altered)

balance the demolition rate. We could have obtained the same sort of equilibrium conditions by extending the average lifetime of housing ALTH to 75 or 100 years. Altering either the LAM table function or the ALTH in this fashion allows the community to initially develop more rapidly, but at the cost of a rather dramatic transition.

SUMMARY AND CRITIQUE

The residential community model, although a relatively simple structure, illustrates a number of important dynamic concepts. This third-order structure generates behavior much like the first-order structures in Chapters 4 and 5. During growth, the area attracts migrants. In fact, construction exceeds the net population growth slightly. Once the land constraint begins to depress construction, however, the community loses its attractive housing condition. Eventually, demolition matches construction and the level of housing stabilizes. The population stabilizes when housing conditions become sufficiently unattractive for potential migrants and area residents. The relatively unattractive housing condition sufficiently suppresses the net inflow of people to just offset the natural death rate.

The fundamental role of resource constraints in controlling growth and bringing about equilibrium clearly emerges in this exercise. The assumption of a finite amount of land available for housing construction dominates system behavior and generates the sigmoidal development pattern.

With a clear definition of purpose, we could modify or expand the community model into a viable urban management tool. At the present level of aggregation, sufficient for describing long-term community development, possible additions or modifications might include:

1. incorporating the average rate of population growth as a factor determining construction;
2. making demolition a function of land availability and housing availability; and
3. defining more specifically the housing information sources used by potential migrants in deciding whether to migrate.

An even more disaggregated model might include:

1. aging in the housing sector (filter-down) involving two or three more levels of housing;
2. such other migration factors as urban services, pollution levels, tax rates; and
3. an economic sector that includes pricing and land use.

Exercise 13
Future Electronics Model

by Edwin N. Jarmain

This exercise contains a brief description of a firm's management policies and its quality control problem.[1] *The reader is asked to construct a simulation model to analyze the described condition and suggest policies to improve performance. The exercise offers experience in model conceptualization and demonstrates how simple models can facilitate policy evaluation. The reader should review Chapters 5, 13, and Appendix O in Industrial Dynamics for background on model construction.*

DESCRIPTION--FUTURE ELECTRONICS COMPANY

The Future Electronics Company is a medium-sized firm producing a line of integrated circuits. Integrated circuits are single pieces of semi-conductor material that perform the functions of entire complex electronic circuits (amplifiers, oscillators, etc.). Because of the delicate production processes involved, only 30 to 50 percent of the items produced prove usable. Therefore, all units produced must undergo testing before sale.

The management of Future has for some time been worried about their quality image. From time to time they have heard such statements from customers as: "We are generally quite satisfied with your quality, and consider you one of our highest quality suppliers, but are bothered by some of the variations which occur. Every so often,

---

[1]This modeling exercise is based on Edwin Jarmain, Problems in Industrial Dynamics (Cambridge: The M.I.T. Press, 1963), pp. 29-30.

we receive a series of poor shipments from you. These create a disruption of our production, and we are forced to find a supplier whose quality is more dependable, even if their best is not as good as yours." While customers are not always so outspoken, Future has noticed that at times customers return many defective units, but at other times customers return very few defectives.

The management of Future is quite sensitive to this situation and upon noticing increased complaints and returns, they hire more people to increase the thoroughness of the testing procedure. They base their hiring decision on the number of testers presently employed and on the frequency of complaints.

The quite difficult testing procedure requires several months training, although some trainees learn faster than others. The testers in training do not test parts for shipment, since Future does not wish to take the chance that inexperienced testers might let bad units get through. The new people are trained by the experienced employees. An experienced tester assigned to the training of a new man must spend about half his time in this capacity, and thus take time away from actual testing.

Future has a policy against laying off testers, but will let natural attrition reduce an apparent excess. After becoming fully trained, a tester remains with Future an average of about three years.

At the present time, demand forces testers to attempt to keep up with production. Thus, the time spent on testing a unit depends on the volume of production. Future does not know a great deal about the policies of customers, but the company feels that customers take an appreciable amount of time to determine the quality of units which they receive.

E13.1 Study the described situation and make a concise statement of the problem or behavior which your model should explain. Identify such factors as delays and managerial policies that, in your opinion, could cause this behavior.

E13.2 Develop a causal-loop diagram based on your analysis.

From your causal-loop diagram, construct a flow diagram. **E13.3**

Develop DYNAMO equations. Supply parameters that correspond with the data given in the description or that you believe represent reasonable values. Sketch all table functions. **E13.4**

Run your model on the computer. Compare its behavior to the behavior expected from your analysis in question E13.1. **E13.5**

Experiment with structural and/or parametric changes that might alleviate the problem. What changes in management policy can bring about desirable system performance? **E13.6**

Solution 13

S13.1

<u>Problem Statement and Hypotheses</u>. Customers seem to have mixed feelings about the quality of Future's products. During some periods, customers return many defective units, while at other times customers return very few defective units.

The above implies that a good model should show a tendency toward fluctuations in the quality of products. When the testing rate per man rises because of increased orders, the observed quality drops. After some delay, additional testers are hired. The increase in trainees, however, initially reduces the number of effective testers since experienced testers must help train. This increased training effort further reduces observed quality. However, after training finishes, the testing rate per man declines and observed quality rises.

S13.2 Causal-loop Diagram.

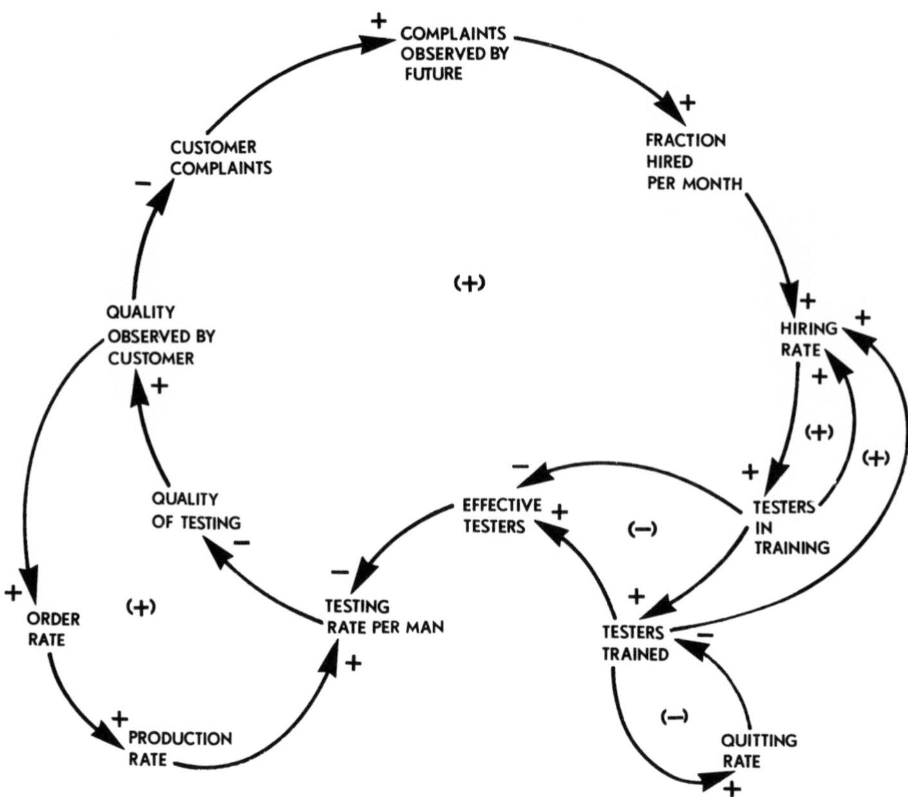

Simplified causal-loop diagram of tester hiring policy

S13.3

Flow Diagram.

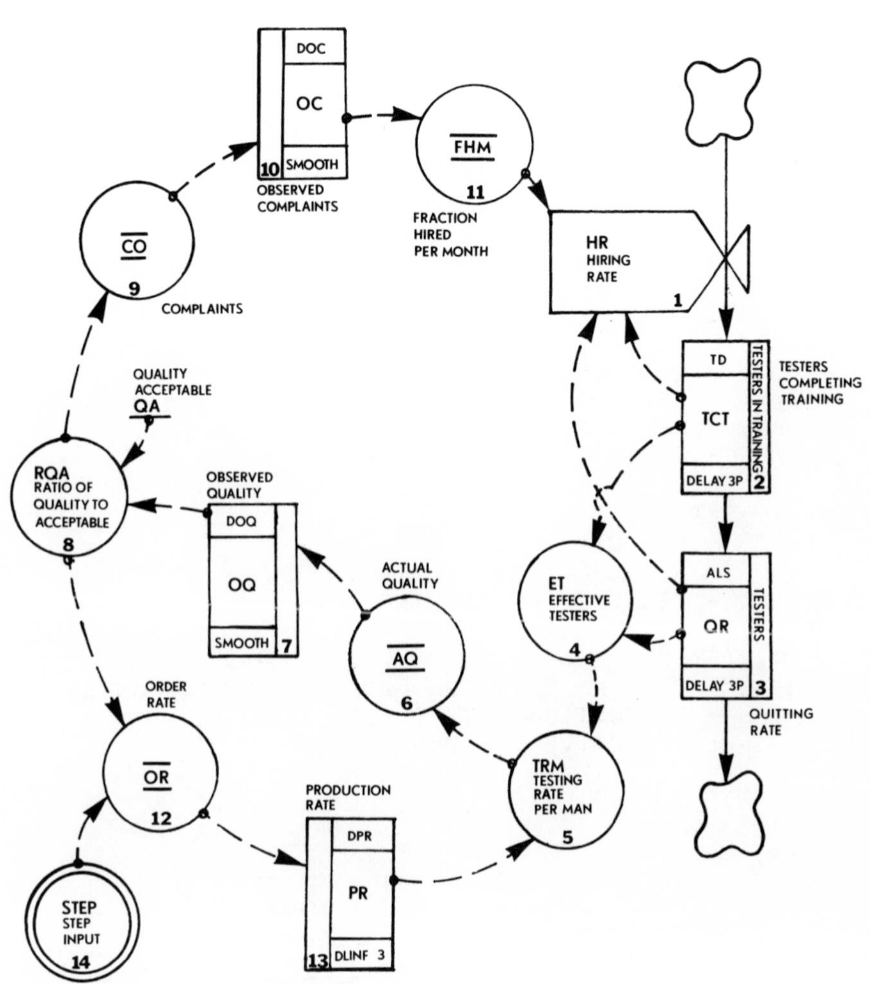

Flow diagram of tester hiring policy

S13.4 DYNAMO Equations.

```
HR.KL=(FHM.K)(TT.K+T.K)                              1, R
HR=20/69                                             1.1, N
     HR      - HIRING RATE (MEN/MONTH)
     FHM     - FRACTION HIRED PER MONTH (FRACTION/MONTH)
     TT      - TESTERS IN TRAINING (MEN)
     T       - TESTERS (MEN)

TCT.KL=DELAY3P(HR.JK,TD,TT.K)                        2, R
TD=3                                                 2.1, C
     TCT     - TESTERS COMPLETING TRAINING (MEN/MONTH)
     HR      - HIRING RATE (MEN/MONTH)
     TD      - TRAINING DELAY (MONTHS)
     TT      - TESTERS IN TRAINING (MEN)

QR.KL=DELAY3P(TCT.JK,ALS,T.K)                        3, R
ALS=36                                               3.1, C
     QR      - QUITTING RATE (MEN/MONTH)
     TCT     - TESTERS COMPLETING TRAINING (MEN/MONTH)
     ALS     - AVERAGE LENGTH OF SERVICE (MONTHS)
     T       - TESTERS (MEN)

ET.K=T.K-.5*TT.K                                     4, A
     ET      - EFFECTIVE TESTERS (MEN)
     T       - TESTERS (MEN)
     TT      - TESTERS IN TRAINING (MEN)

TRM.K=PR.K/ET.K                                      5, A
     TRM     - TESTING RATE PER MAN (UNITS/MAN/MONTH)
     PR      - PRODUCTION RATE (UNITS/MONTH)
     ET      - EFFECTIVE TESTERS (MEN)
```

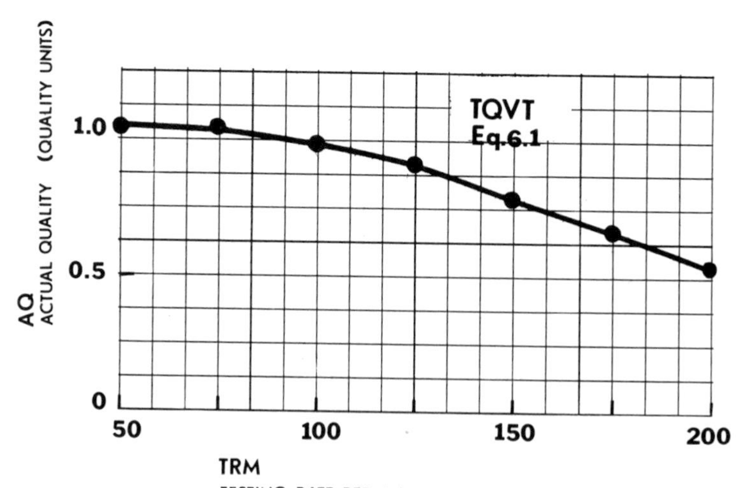

```
AQ.K=TABLE(TQVT,TRM.K,50,200,25)                        6, A
TQVT=1.1/1.1/1/.85/.7/.6/.55                            6.1, T
     AQ      - ACTUAL QUALITY (QUALITY UNITS)
     TQVT    - TABLE OF QUALITY VS. TESTING
     TRM     - TESTING RATE PER MAN (UNITS/MAN/MONTH)

OQ.K=SMOOTH(AQ.K,DOQ)                                   7, A
DOQ=6                                                   7.1, C
OQ=1                                                    7.2, N
     OQ      - OBSERVED QUALITY (QUALITY UNITS)
     AQ      - ACTUAL QUALITY (QUALITY UNITS)
     DOQ     - DELAY IN OBSERVING QUALITY (MONTHS)

RQA.K=OQ.K/QA                                           8, A
QA=1                                                    8.1, C
     RQA     - RATIO OF QUALITY TO ACCEPTABLE
                 (DIMENSIONLESS)
     OQ      - OBSERVED QUALITY (QUALITY UNITS)
     QA      - QUALITY ACCEPTABLE (QUALITY UNITS)
```

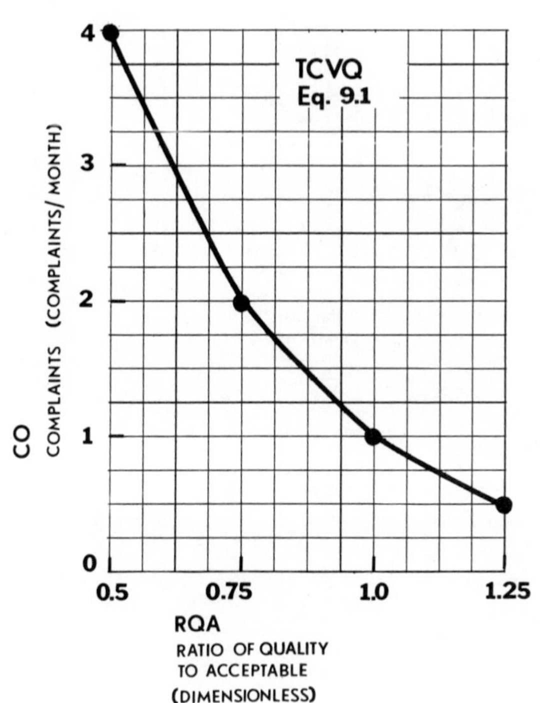

```
CO.K=TABLE(TCVQ,RQA.K,.5,1.25,.25)                                9, A
TCVQ=4/2/1/.5                                                     9.1, T
     CO       - COMPLAINTS (COMPLAINTS/MONTH)
     TCVQ     - TABLE OF COMPLAINTS VS. QUALITY
     RQA      - RATIO OF QUALITY TO ACCEPTABLE
                  (DIMENSIONLESS)

OC.K=SMOOTH(CO.K,DOC)                                             10, A
DOC=2                                                             10.1, C
OC=CO                                                             10.2, N
     OC       - OBSERVED COMPLAINTS (COMPLAINTS/MONTH)
     CO       - COMPLAINTS (COMPLAINTS/MONTH)
     DOC      - DELAY IN OBSERVING COMPLAINTS (MONTHS)
```

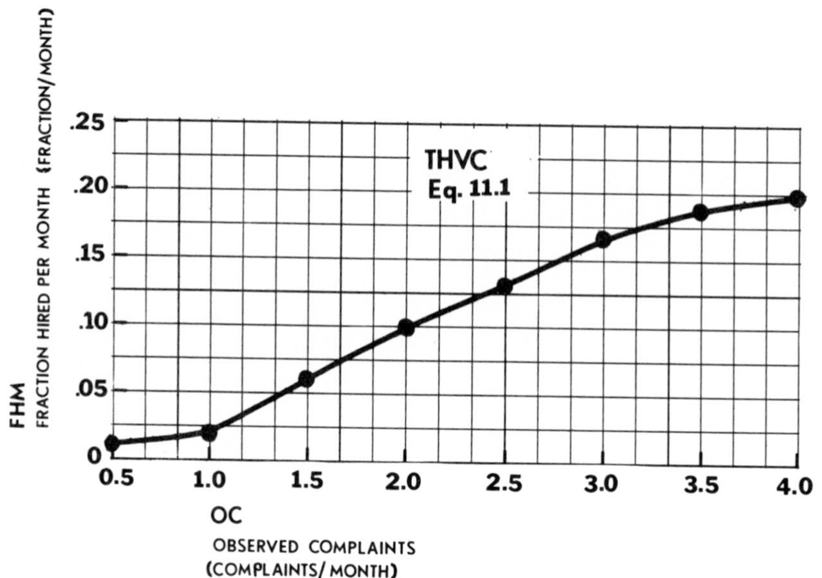


```
FHM.K=TABLE(THVC,OC.K,.5,4,.5)                                    11, A
THVC=.01/.0258/.06/.1/.135/.17/.185/.2                            11.1, T
     FHM      - FRACTION HIRED PER MONTH (FRACTION/MONTH)
     THVC     - TABLE OF HIRING VS. COMPLAINTS
     OC       - OBSERVED COMPLAINTS (COMPLAINTS/MONTH)
```

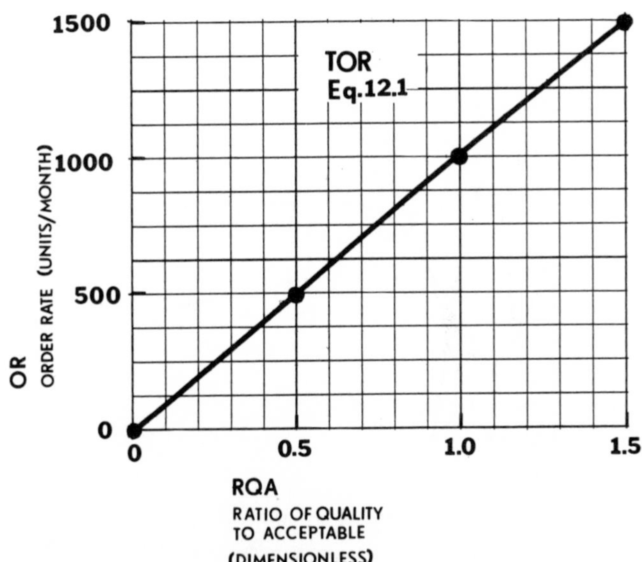


```
OR.K=TABLE(TOR,RQA.K,0,1.5,.5)(1+STEP.K)                12, A
TOR=0/500/1000/1500                                     12.1, T
     OR      - ORDER RATE (UNITS/MONTH)
     TOR     - TABLE OF ORDER RATE
     RQA     - RATIO OF QUALITY TO ACCEPTABLE
                  (DIMENSIONLESS)
     STEP    - STEP INPUT TO ORDER RATE (DIMENSIONLESS)

PR.K=DLINF3(OR.K,DPR)                                   13, A
DPR=3                                                   13.1, C
     PR      - PRODUCTION RATE (UNITS/MONTH)
     OR      - ORDER RATE (UNITS/MONTH)
     DPR     - DELAY FOR PRODUCTION (MONTHS)

STEP.K=STEP(STH,STT)                                    14, A
STH=.2                                                  14.1, C
STT=5                                                   14.2, C
     STEP    - STEP INPUT TO ORDER RATE (DIMENSIONLESS)
     STH     - STEP HEIGHT (DIMENSIONLESS)
     STT     - STEP TIME (MONTHS)

 CONTROL STATEMENTS

 C DT=.125
 C PLTPER=2.5
 C LENGTH=125
 PLOT HR=H(0,.8)/TT=T,T=*/TRM=R/OQ=Q/OR=O/OC=C
```

S13.5 Model Behavior. The figure below shows the behavior of the model when a 20 percent step increase in orders occurs in the fifth month.

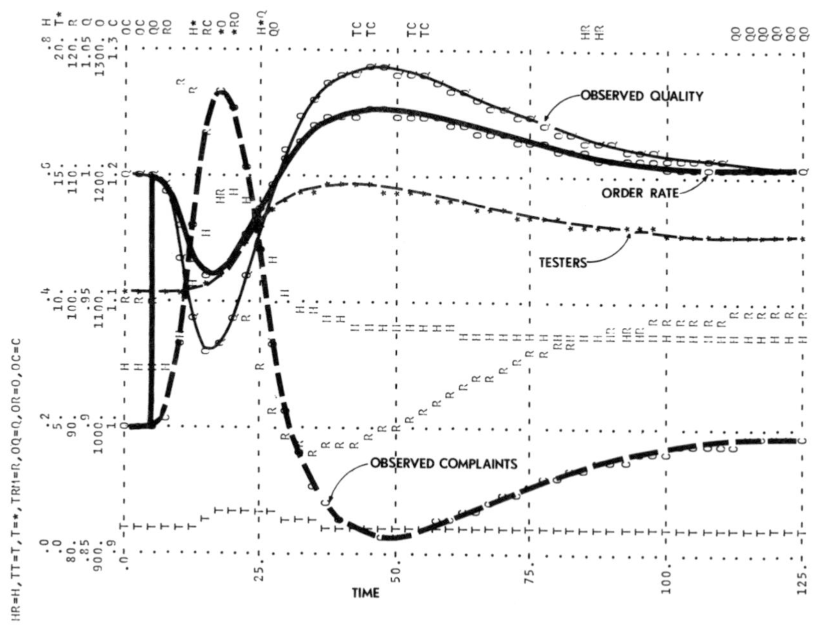

Response to 20 percent increase in orders

S13.6 Policy Analysis. We want to minimize the fluctuations in quality and maintain orders. A more effective policy might take explicit, planned action to meet foreseeable unfavorable conditions. The proposed new policy uses the order rate as an indication of the volume of units that need testing in the near future. Concurrently, information about the present number of testers, testers in training, and the expected turnover provides a means of estimating testers available to test forthcoming units. Hiring, then, is scheduled to provide for any discrepancy between testers needed and available testers. The following flow diagram and DYNAMO equations summarize the structural changes which, when added to the model in S13.3 and S13.4, reflect the new policy.

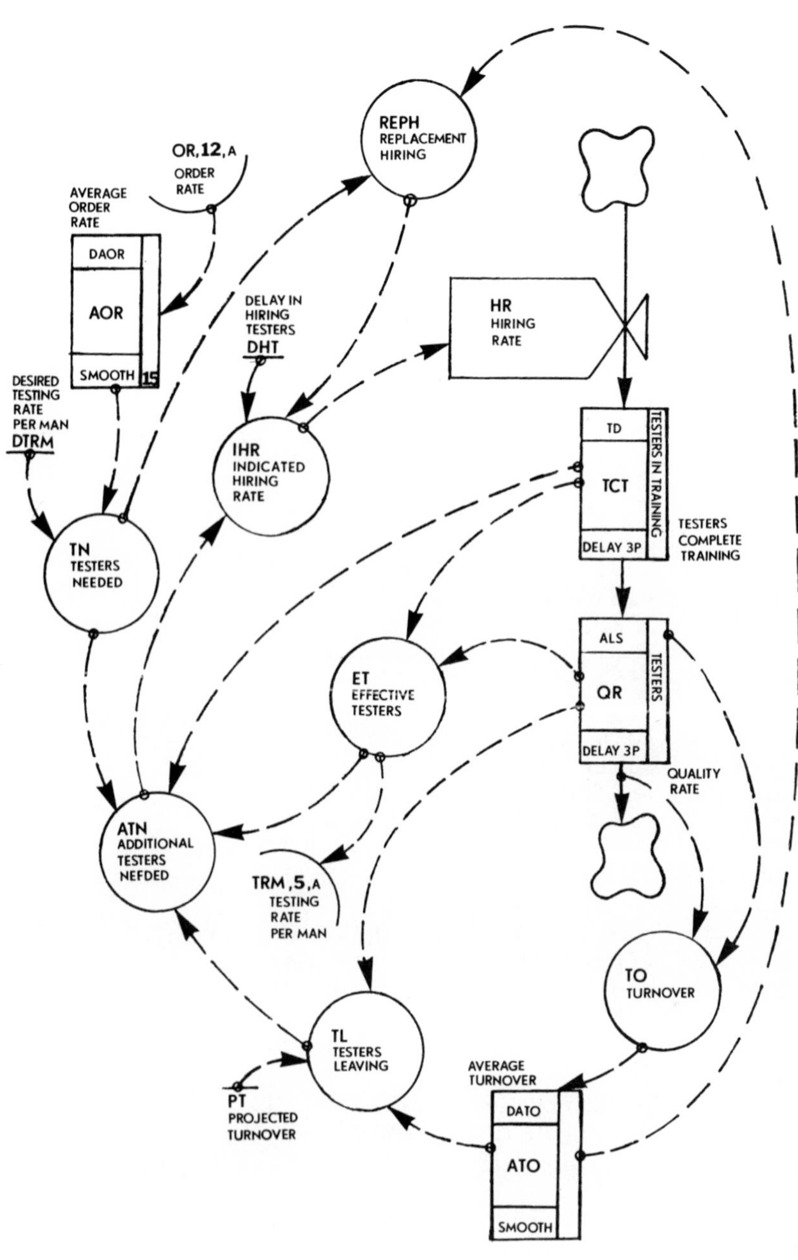

(a)

Flow diagram of new hiring policy

DYNAMO Equations of New Policy:

```
HR.KL=MAX(IHR.K,0)
HR=20/69
    HR       - HIRING RATE (MEN/MONTH)
    MAX      - MAXIMUM FUNCTION
    IHR      - INDICATED HIRING RATE (MEN/MONTH)

AOR.K=SMOOTH(OR.K,DAOR)
AOR=OR
DAOR=3
    AOR      - AVERAGE ORDER RATE (UNITS/MONTH)
    OR       - ORDER RATE (UNITS/MONTH)
    DAOR     - DELAY IN AVERAGING ORDERS (MONTHS)
TN.K=AOR.K/DTRM
DTRM=100
    TN       - TESTERS NEEDED (MEN)
    AOR      - AVERAGE ORDER RATE (UNITS/MONTH)
    DTRM     - DESIRED TESTING RATE PER MAN (UNITS/MAN/
                 MONTH)

ATN.K=TN.K-ET.K-TT.K+TL.K
    ATN      - ADDITIONAL TESTERS NEEDED (MEN)
    TN       - TESTERS NEEDED (MEN)
    ET       - EFFECTIVE TESTERS (MEN)
    TT       - TESTERS IN TRAINING (MEN)
    TL       - TESTERS LEAVING (MEN)

REPH.K=TN.K*ATO.K
    REPH     - REPLACEMENT HIRING (MEN/MONTH)
    TN       - TESTERS NEEDED (MEN)
    ATO      - AVERAGE TURNOVER (FRACTION/MONTH)

IHR.K=REPH.K+ATN.K/DHT
DHT=2
    IHR      - INDICATED HIRING RATE (MEN/MONTH)
    REPH     - REPLACEMENT HIRING (MEN/MONTH)
    ATN      - ADDITIONAL TESTERS NEEDED (MEN)
    DHT      - DELAY IN HIRING TESTERS (MONTHS)

TO.K=QR.JK/T.K
    TO       - TURNOVER (FRACTION/MONTH)
    QR       - QUITTING RATE (MEN/MONTH)
    T        - TESTERS (MEN)

ATO.K=SMOOTH(TO.K,DATO)
ATO=TO
DATO=4
    ATO      - AVERAGE TURNOVER (FRACTION/MONTH)
    TO       - TURNOVER (FRACTION/MONTH)
    DATO     - DELAY IN AVERAGING TURNOVER (MONTHS)
```

```
TL.K=T.K*ATO.K*PT
PT=3
     TL      - TESTERS LEAVING (MEN)
     T       - TESTERS (MEN)
     ATO     - AVERAGE TURNOVER (FRACTION/MONTH)
     PT      - PROJECTED TURNOVER (MONTHS)
```

Note: Equations 2 through 14 remain the same with the exception of equation 11 which is now omitted.

The following figure displays model response under the revised hiring policy. The fluctuations, including the variation in observed quality, are damped more quickly than under the former hiring policy.

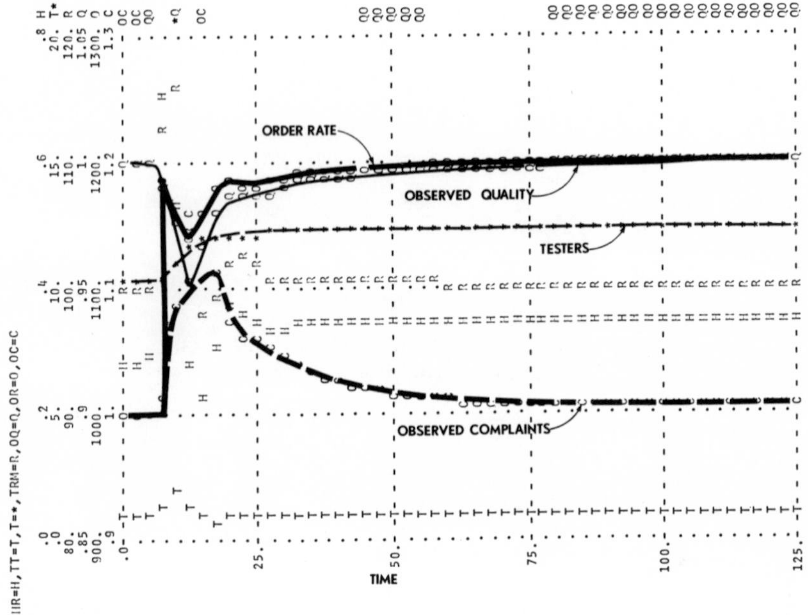

(b)

New policy response to 20 percent step increase in orders

Figure (c) details the first 25 months of the new response (PLTPER = 0.5).

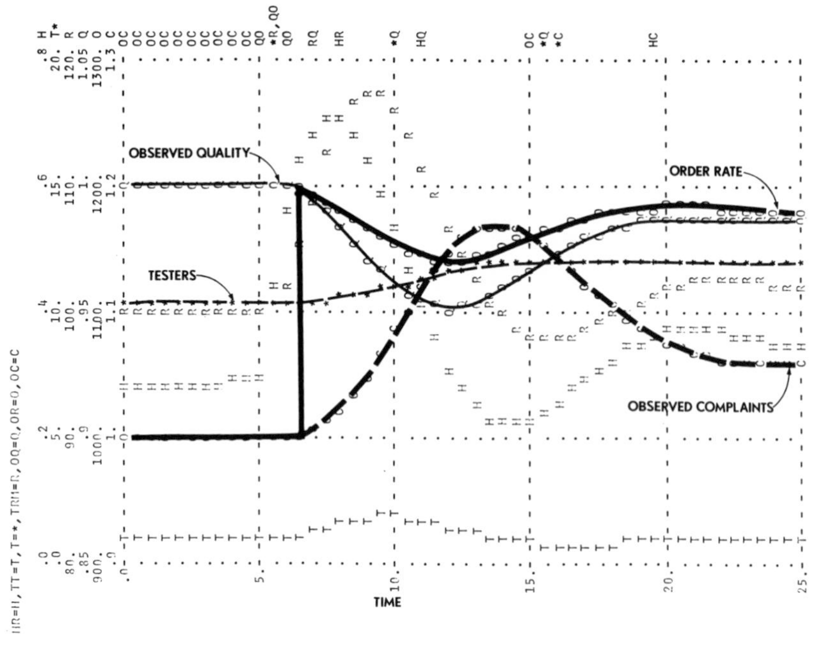

(c)

Detail of new policy response to 20 percent increase in orders

Exercise 14
Yellow-Fever Model

by Kjell Kalgraf

Chapter 4 introduced a one-level model of the growth of an epidemic. This exercise develops a more complex model for an epidemic life cycle. The reader is asked to construct a yellow-fever model from actual data and description. The reader should be familiar with the feedback structure presented in Chapter 3 (section 3.7) and delays (Exercise 9) before undertaking this exercise.

DESCRIPTION--YELLOW-FEVER EPIDEMIC

Yellow fever is a disease caused by a virus transmitted among susceptible hosts by mosquitos. The resulting epidemics have taken many lives, particularly in densely populated areas. In Veracruz, Mexico, a severe epidemic raged from mid-September, 1898, to September, 1899. The death figures for each month appear below:

Month	Sept.	Oct.	Nov.	Dec.	Jan.	Feb.	Mar.
Dead	22	30	40	28	15	8	10
Month	Apr.	May	June	July	Aug.	Sept.	
Dead	35	300	460	220	125	40	

Estimates of the city's population were on the order of 20,000 people. A severe epidemic also struck Philadelphia in 1793. Over 10 percent of the total population died in the course of 4 months.[1]

[1] Public Health Papers and Reports, Volume XXV p. 36, American Public Health Association, Minneapolis, Minnesota, 1899 (Columbus, Ohio: The Berlin Printing Co., 1900).

Two types of yellow fever exist:

1. jungle yellow fever, where man incidentally comes in contact with animal hosts; and
2. urban or classical yellow fever, where transmission from man to man by mosquitos occurs.

We will focus on the second variety.

Yellow fever is self-perpetuating. A human can contract yellow fever only from the bite of an infectious mosquito. A mosquito can become infectious only through biting a contagious human. A bite from an infectious mosquito initiates the disease. The first stage, a period of incubation, lasts 3 to 6 days. After incubation, the victim becomes visibly sick for about 7 days. For the first 3 to 6 days, and only during these days, a human can pass the virus to a mosquito that bites him. Thus, clean mosquitos contract the disease by biting sick humans in their contagious period. After the remainder of the sick period, the victim either dies or recovers. Those who recover stay immune for life.

Yellow fever is spread by the female Aedes Aegypti mosquito. This species has a typical adult life span of 18 days. Once a female has a blood feed, she will lay a batch of eggs. The eggs hatch 3 days later. A week or so after the eggs are laid, the pupae reach adulthood. Two to four days later the new adults take their first blood feeds and the cycle starts again. The following table summarizes the mosquito life cycle.

Mosquito Life Cycle

	Day	
Pre-Adult	0	mother takes blood feed
	1	lay eggs
	2	
	3	
	4	eggs hatch--enter pupae stage
	5	
	6	
	7	pupae enter adult stage

(table continued to next page)

	Day	
Adult	1	
	2	
	3	mosquito takes first blood feed
	4	lays eggs
	5	
	6	
	7	
	8	second blood feed
	9	lays eggs
	10	
	11	
	12	
	13	third blood feed
	14	lays eggs
	15	
	16	
	17	
	18	fourth blood feed; death

Sixty eggs are laid in one batch; half of the survivors will become females. The proportion of eggs that develop into adult mosquitos depends on such factors as temperature, humidity, and sanitation. The adult mosquito population of an area may quickly reach an equilibrium dependent on environmental conditions. Therefore, we reasonably consider the mosquito population remains constant during the epidemic given a particular environment.

A mosquito that bites a human in the contagious period enters a 12 day incubation process. During this time she continues to digest, lay eggs, and perhaps bite again, but does not transmit the disease. After 12 days she infects all vulnerable humans she bites. No difference appears between the mortality and fertility patterns of infected and non-infected mosquitos.

Yellow fever immunizations are available. Vaccinated people have almost certain immunity for 6 to 7 years. In addition, contagious humans may remain isolated from mosquitos and some techniques or changed conditions may eradicate or reduce mosquitos in number. In this exercise, however, we ignore immunizations.

E14.1 Develop a simplified causal-loop diagram from the description.

E14.2 Construct a system dynamics flow diagram from your causal diagram.

E14.3 Develop DYNAMO equations. Supply any necessary constant values not contained in the description.

E14.4 Run your model on the computer. Compare its behavior to the behavior indicated in the description.

Solution 14

S14.1

Causal-loop Diagram. During a yellow-fever epidemic, a sharply rising number of people become ill. The crisis reaches a climax and returns to a low value. Assume that the observed time history corresponds to the period when infectious mosquitos bite vulnerable people who then contract the disease and become contagious. The contagious people are then bitten by clean mosquitos which after a while become infectious and transfer the disease to other vulnerable people. The humans who recover from the disease remain immune for life. These assumptions form the dynamic hypothesis of the model. The figure below offers a simplified causal-loop diagram.

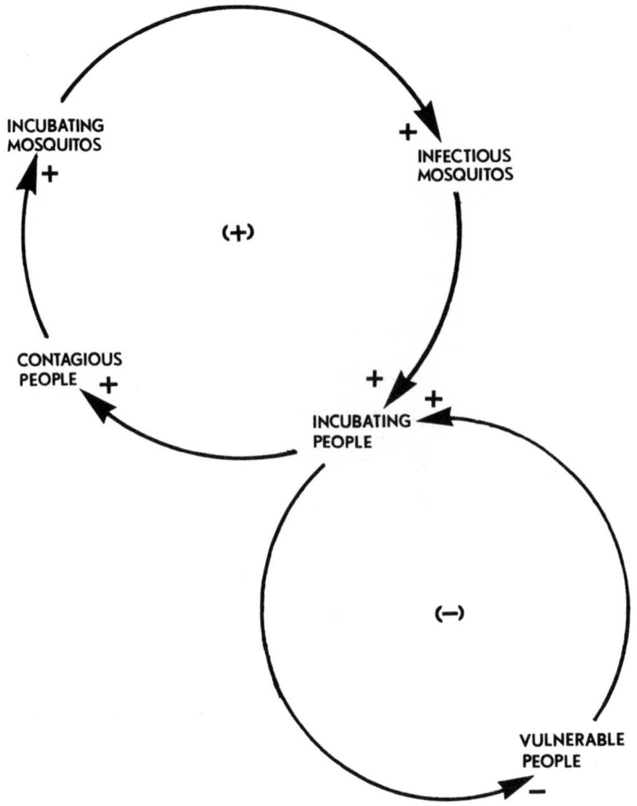

Simplified causal-loop diagram for the
essential yellow-fever structure

S14.2

Flow Diagram.

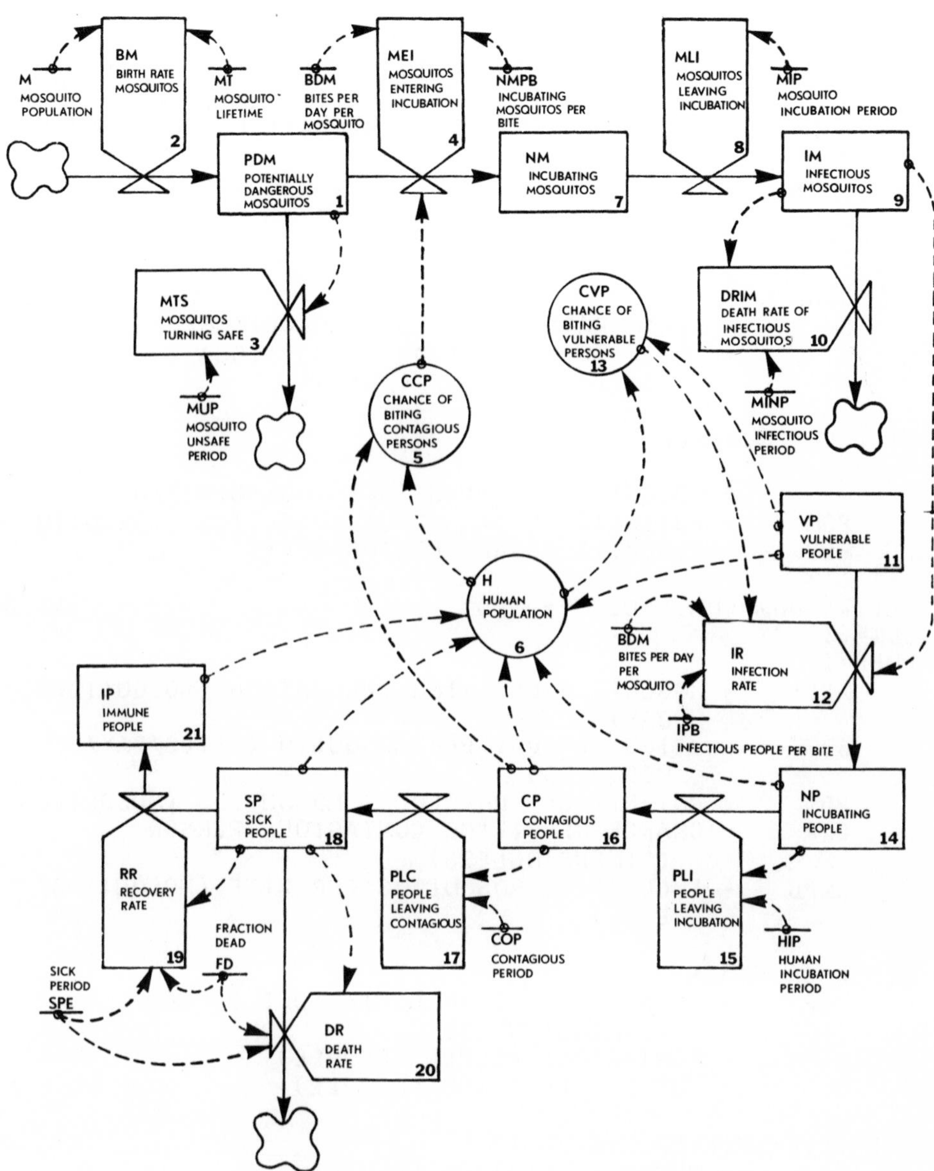

S14.3 DYNAMO Equations.

```
PDM.K=PDM.J+(DT)(BM.JK-MTS.JK-MEI.JK)                      1, L
PDM=SPDM                                                   1.1, N
SPDM=0                                                     1.2, C
    PDM    - POTENTIALLY DANGEROUS MOSQUITOS (MOSQUITOS)
    BM     - BIRTH RATE MOSQUITOS (MOSQUITOS/DAY)
    MTS    - MOSQUITOS TURNING SAFE (MOSQUITOS/DAY)
    MEI    - MOSQUITOS ENTERING INCUBATION (MOSQUITOS/
               DAY)
    SPDM   - INITIAL VALUE FOR PDM (MOSQUITOS)

BM.KL=M/MT                                                 2, R
MT=18                                                      2.1, C
M=500000                                                   2.2, C
    BM     - BIRTH RATE MOSQUITOS (MOSQUITOS/DAY)
    M      - MOSQUITO POPULATION (MOSQUITOS)
    MT     - MOSQUITO LIFETIME (DAYS)

MTS.KL=PDM.K/MUP                                           3, R
MUP=3                                                      3.1, C
    MTS    - MOSQUITOS TURNING SAFE (MOSQUITOS/DAY)
    PDM    - POTENTIALLY DANGEROUS MOSQUITOS (MOSQUITOS)
    MUP    - MOSQUITO UNSAFE PERIOD (DAYS)

MEI.KL=BDM*PDM.K*CCP.K*NMPB                                4, R
BDM=.2                                                     4.1, C
NMPB=1                                                     4.2, C
    MEI    - MOSQUITOS ENTERING INCUBATION (MOSQUITOS/
               DAY)
    BDM    - BITES PER DAY PER MOSQUITO (BITES/DAY/
               MOSQUITO)
    PDM    - POTENTIALLY DANGEROUS MOSQUITOS (MOSQUITOS)
    CCP    - CHANCE OF BITING CONTAGIOUS PERSON
               (DIMENSIONLESS)
    NMPB   - INCUBATING MOSQUITOS PER BITE (MOSQUITOS/
               BITE)

CCP.K=CP.K/H.K                                             5, A
    CCP    - CHANCE OF BITING CONTAGIOUS PERSON
               (DIMENSIONLESS)
    CP     - CONTAGIOUS PEOPLE (PEOPLE)
    H      - HUMAN POPULATION (PEOPLE)

H.K=VP.K+NP.K+CP.K+SP.K+IP.K                               6, A
    H      - HUMAN POPULATION (PEOPLE)
    VP     - VULNERABLE PEOPLE (PEOPLE)
    NP     - INCUBATING PEOPLE (PEOPLE)
    CP     - CONTAGIOUS PEOPLE (PEOPLE)
    SP     - SICK PEOPLE (PEOPLE)
    IP     - IMMUNE PEOPLE (PEOPLE)
```

```
NM.K=NM.J+(DT)(MEI.JK-MLI.JK)                          7, L
NM=SNM                                                 7.1, N
SNM=0                                                  7.2, C
    NM      - INCUBATING MOSQUITOS (MOSQUITOS)
    MEI     - MOSQUITOS ENTERING INCUBATION (MOSQUITOS/
                DAY)
    MLI     - MOSQUITOS LEAVING INCUBATION (MOSQUITOS/
                DAY)
    SNM     - INITIAL VALUE FOR NM (MOSQUITOS)

MLI.KL=NM.K/MIP                                        8, R
MIP=12                                                 8.1, C
    MLI     - MOSQUITOS LEAVING INCUBATION (MOSQUITOS/
                DAY)
    NM      - INCUBATING MOSQUITOS (MOSQUITOS)
    MIP     - MOSQUITO INCUBATION PERIOD (DAYS)

IM.K=IM.J+(DT)(MLI.JK-DRIM.JK)                         9, L
IM=SIM                                                 9.1, N
SIM=0                                                  9.2, C
    IM      - INFECTIOUS MOSQUITOS (MOSQUITOS)
    MLI     - MOSQUITOS LEAVING INCUBATION (MOSQUITOS/
                DAY)
    DRIM    - DEATH RATE INFECTIOUS MOSQUITOS (MOSQUITOS/
                DAY)
    SIM     - INITIAL VALUE FOR IM (MOSQUITOS)

DRIM.KL=IM.K/MINP                                      10, R
MINP=3                                                 10.1, C
    DRIM    - DEATH RATE INFECTIOUS MOSQUITOS (MOSQUITOS/
                DAY)
    IM      - INFECTIOUS MOSQUITOS (MOSQUITOS)
    MINP    - MOSQUITO INFECTIOUS PERIOD (DAYS)

VP.K=VP.J+(DT)(-IR.JK)                                 11, L
VP=SVP                                                 11.1, N
SVP=20000                                              11.2, C
    VP      - VULNERABLE PEOPLE (PEOPLE)
    IR      - INFECTION RATE (PEOPLE/DAY)
    SVP     - INITIAL VALUE FOR VP (PEOPLE)

IR.KL=BDM*IM.K*CVP.K*IPB                               12, R
IPB=1                                                  12.1, C
    IR      - INFECTION RATE (PEOPLE/DAY)
    BDM     - BITES PER DAY PER MOSQUITO (BITES/DAY/
                MOSQUITO)
    IM      - INFECTIOUS MOSQUITOS (MOSQUITOS)
    CVP     - CHANCE OF BITING VULNERABLE PERSON
                (DIMENSIONLESS)
    IPB     - INFECTIOUS PEOPLE PER BITE (PEOPLE/BITE)
```

```
CVP.K=VP.K/H.K                                             13, A
    CVP     - CHANCE OF BITING VULNERABLE PERSON
                (DIMENSIONLESS)
    VP      - VULNERABLE PEOPLE (PEOPLE)
    H       - HUMAN POPULATION (PEOPLE)

NP.K=NP.J+(DT)(IR.JK-PLI.JK)                               14, L
NP=SNP                                                     14.1, N
SNP=100                                                    14.2, C
    NP      - INCUBATING PEOPLE (PEOPLE)
    IR      - INFECTION RATE (PEOPLE/DAY)
    PLI     - PEOPLE LEAVING INCUBATION (PEOPLE/DAY)
    SNP     - INITIAL VALUE FOR NP (PEOPLE)

PLI.KL=NP.K/HIP                                            15, R
HIP=4.5                                                    15.1, C
    PLI     - PEOPLE LEAVING INCUBATION (PEOPLE/DAY)
    NP      - INCUBATING PEOPLE (PEOPLE)
    HIP     - HUMAN INCUBATION PERIOD (DAYS)

CP.K=CP.J+(DT)(PLI.JK-PLC.JK)                              16, L
CP=SCP                                                     16.1, N
SCP=0                                                      16.2, C
    CP      - CONTAGIOUS PEOPLE (PEOPLE)
    PLI     - PEOPLE LEAVING INCUBATION (PEOPLE/DAY)
    PLC     - PEOPLE LEAVING CONTAGIOUS (PEOPLE/DAY)
    SCP     - INITIAL VALUE FOR CP (PEOPLE)
PLC.KL=CP.K/COP                                            17, R
COP=4.5                                                    17.1, C
    PLC     - PEOPLE LEAVING CONTAGIOUS (PEOPLE/DAY)
    CP      - CONTAGIOUS PEOPLE (PEOPLE)
    COP     - CONTAGIOUS PERIOD (DAYS)

SP.K=SP.J+(DT)(PLC.JK-RR.JK-DR.JK)                         18, L
SP=SSP                                                     18.1, N
SSP=0                                                      18.2, C
    SP      - SICK PEOPLE (PEOPLE)
    PLC     - PEOPLE LEAVING CONTAGIOUS (PEOPLE/DAY)
    RR      - RECOVERY RATE (PEOPLE/DAY)
    DR      - DEATH RATE (PEOPLE/DAY)
    SSP     - INITIAL VALUE FOR SP (PEOPLE)

RR.KL=(1-FD)*SP.K/SPE                                      19, R
SPE=2.5                                                    19.1, C
    RR      - RECOVERY RATE (PEOPLE/DAY)
    FD      - FRACTION DEAD (DIMENSIONLESS)
    SP      - SICK PEOPLE (PEOPLE)
    SPE     - SICK PERIOD (DAYS)
```

```
DR.KL=FD*SP.K/SPE                                         20, R
FD=.1                                                     20.1, C
     DR       - DEATH RATE (PEOPLE/DAY)
     FD       - FRACTION DEAD (DIMENSIONLESS)
     SP       - SICK PEOPLE (PEOPLE)
     SPE      - SICK PERIOD (DAYS)

IP.K=IP.J+(DT)(RR.JK)                                     21, L
IP=SIP                                                    21.1, N
SIP=0                                                     21.2, C
     IP       - IMMUNE PEOPLE (PEOPLE)
     RR       - RECOVERY RATE (PEOPLE/DAY)
     SIP      - INITIAL VALUE FOR IP (PEOPLE)

C DT=.5
C LENGTH=300
C PLTPER=6
PLOT IM=I(0,2000)/VP=V,IP=P,H=H(0,20000)/SP=S(0,800)/
X DR=D(0,20)/CCP=*,CVP=+
```

Model Behavior. S14.4

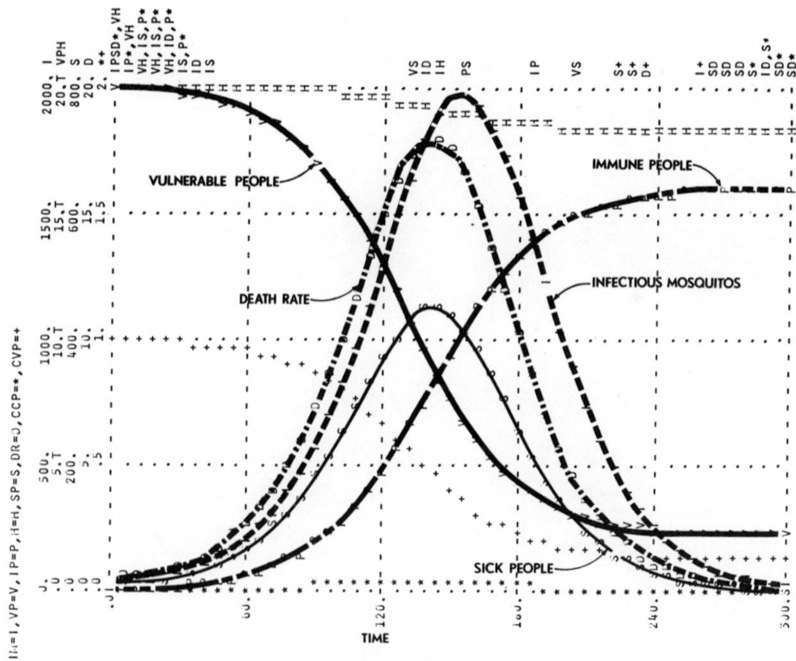

Yellow fever epidemic

Exercise 15
Kaibab Plateau Model

by Michael R. Goodman

In this exercise the reader constructs a simple population-resource model. It recreates the 40-year growth and decline of a deer herd on the Kaibab Plateau. Given a very brief description, the reader must identify underlying feedback loops capable of explaining the observed behavior. Through simulation, he determines if his hypothetical model structure can produce the correct behavior.

DESCRIPTION--KAIBAB PLATEAU

Prior to 1907, the deer herd on the Kaibab Plateau, which consists of some 727,000 acres and is on the north side of the Grand Canyon in Arizona, numbered about 4,000. In 1907, a bounty was placed on cougars, wolves, and coyotes--all natural predators of deer. Within 15 to 20 years, there was a substantial extirpation of these predators (over 8,000) and a consequent and immediate irruption of the deer population. By 1918, the deer population had increased more than tenfold; the evident overbrowsing of the area brought the first of a series of warnings by competent investigators, none of which produced a much needed quick change in either the bounty policy or that dealing with deer removal. In the absence of predation by its natural predators (cougars, wolves, coyotes) or by man as a hunter, the herd reached 100,000 in 1924; in the absence of sufficient food, 60 percent of the herd died off in two successive winters. By then, the girdling of so much of the vegetation through browsing precluded recovery of the food reserve to such an extent that subsequent die-off and reduced natality yielded a population about half that which could theoretically have been previously maintained. Perhaps the most pertinent statement relative to the matter of the interregulatory effect of predator and prey is the following by Aldo Leopold, one of the most significant of recent figures on the conservation scene:

> We have found no record of a deer irruption in North America antedating the removal of deer predators. Those parts of the continent which still retain the native predators have reported no irruptions. This circumstantial evidence supports the surmise that removal of predators predisposes a deer herd to irruptive behavior.

Source: E.J. Kormondy, Concepts of Ecology (Englewood Cliffs, N.J.: Prentice-Hall, 1969), p. 96.

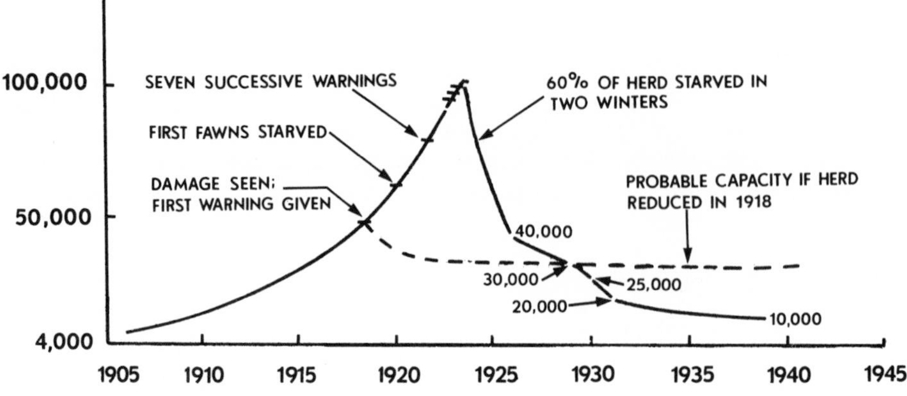

FIGURE E15-1 The effect of removal of natural predators on the deer population on the 727,000 acres of the Kaibab Plateau on the north rim of the Grand Canyon, Arizona.

Source: Wisconsin, Bulletin No.321, Department of Conservation, Madison, Wisconsin, 1943.

E15.1 From the description develop a simple causal-loop diagram.

E15.2 Convert your causal-loop diagram to a system dynamics flow diagram. Restrict yourself to a two or three-level model.

Convert your flow diagram to DYNAMO equations. Sketch out all table functions. **E15.3**

Run the model with parameter values that seem consistent with the growth trend. Your model should qualitatively replicate the observed behavior mode. **E15.4**

Solution 15

Causal-loop Diagram. **S15.1**

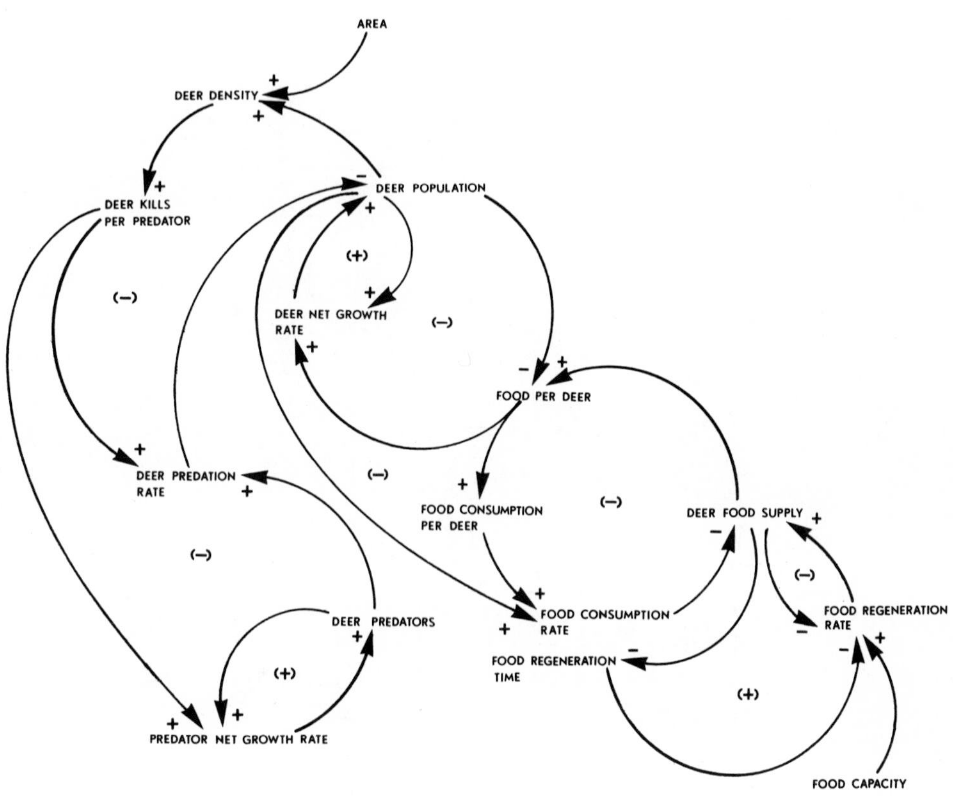

Simplified causal-loop diagram of Kaibab Plateau

S15.2 Flow Diagram.

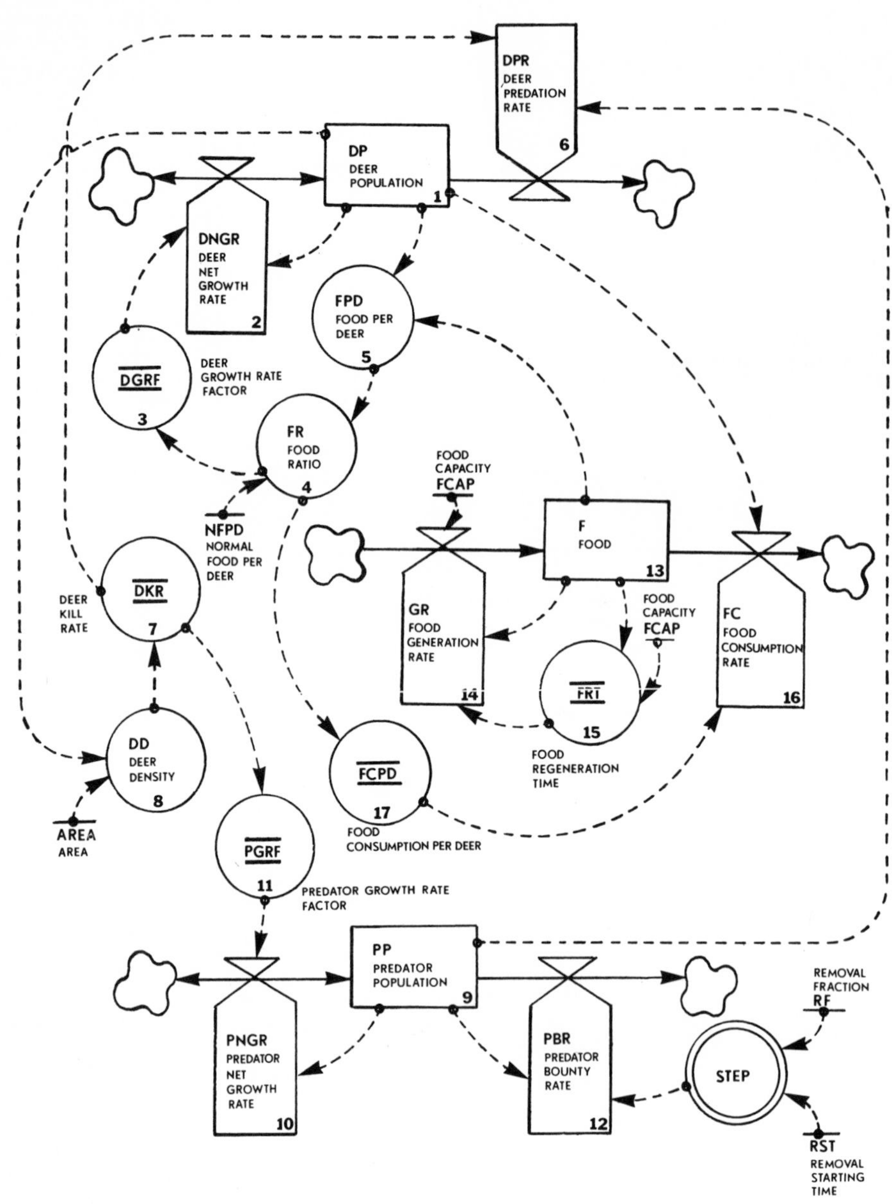

Flow diagram of Kaibab Plateau

DYNAMO Equations.

S15.3

```
         ***DEER POPULATION SECTOR***

DP.K=DP.J+(DT)(DNGR.JK-DPR.JK)                              1, L
DP=DPI                                                      1.1, N
DPI=4000                                                    1.2, C
     DP      - DEER POPULATION (DEER)
     DNGR    - DEER NET GROWTH RATE (DEER/YEAR)
     DPR     - DEER PREDATION RATE (DEER/YEAR)
     DPI     - DEER POPULATION INITIAL (DEER)

DNGR.KL=DP.K*DGRF.K                                         2, R
     DNGR    - DEER NET GROWTH RATE (DEER/YEAR)
     DP      - DEER POPULATION (DEER)
     DGRF    - DEER GROWTH RATE FACTOR (FRACTION/YEAR)
```

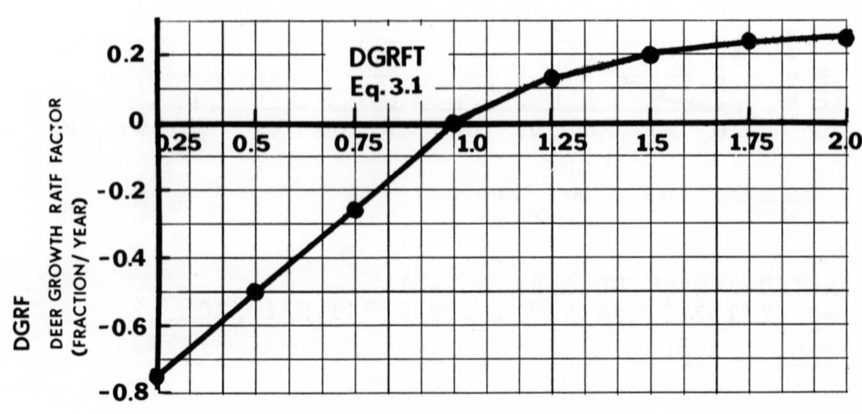


```
DGRF.K=TABHL(DGRFT,FR.K,.25,2.,.25)                         3, A
DGRFT=-.75/-.5/-.25/0/.12/.2/.23/.24                        3.1, T
     DGRF    - DEER GROWTH RATE FACTOR (FRACTION/YEAR)
     DGRFT   - DEER GROWTH RATE FACTOR TABLE
     FR      - FOOD RATIO (DIMENSIONLESS)

FR.K=FPD.K/NFPD                                             4, A
NFPD=1                                                      4.1, C
     FR      - FOOD RATIO (DIMENSIONLESS)
     FPD     - FOOD PER DEER (UNITS/DEER)
     NFPD    - NORMAL FOOD PER DEER (UNITS/DEER)
```

```
FPD.K=F.K/DP.K                                              5, A
    FPD     - FOOD PER DEER (UNITS/DEER)
    F       - FOOD (UNITS)
    DP      - DEER POPULATION (DEER)

DPR.KL=DKR.K*PP.K                                           6, R
    DPR     - DEER PREDATION RATE (DEER/YEAR)
    DKR     - DEER KILL RATE (DEER/PREDATOR/YEAR)
    PP      - PREDATOR POPULATION (PREDATORS)
```

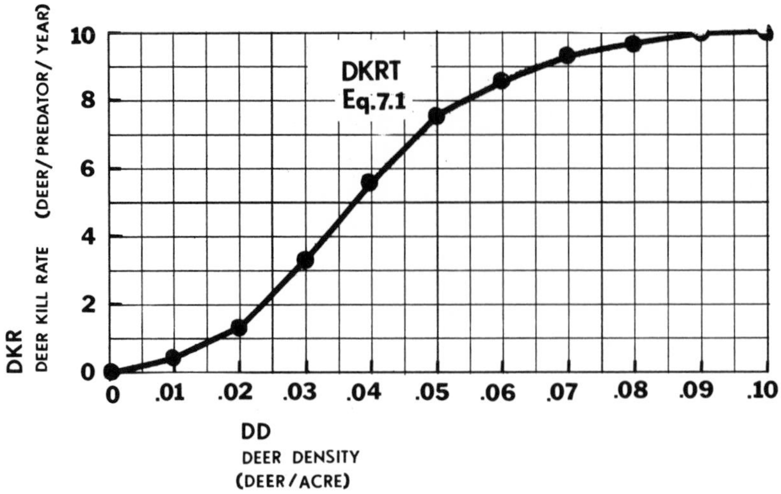

```
DKR.K=TABHL(DKRT,DD.K,0,.1,.01)                             7, A
DKRT=0/.2/1.2/3.2/5.4/7.6/8.6/9.3/9.8/10/10                 7.1, T
    DKR     - DEER KILL RATE (DEER/PREDATOR/YEAR)
    DKRT    - DEER KILL RATE TABLE
    DD      - DEER DENSITY (DEER/ACRE)

DD.K=DP.K/AREA                                              8, A
AREA=800000                                                 8.1, C
    DD      - DEER DENSITY (DEER/ACRE)
    DP      - DEER POPULATION (DEER)
    AREA    - AREA (ACRES)


            ***PREDATOR POPULATION SECTOR***

PP.K=PP.J+(DT)(PNGR.JK-PBR.JK)                              9, L
PP=PPI                                                      9.1, N
PPI=8000                                                    9.2, C
    PP      - PREDATOR POPULATION (PREDATORS)
    PNGR    - PREDATOR NET GROWTH RATE (PREDATORS/YEAR)
    PBR     - PREDATOR BOUNTY RATE (PREDATOR/YEAR)
    PPI     - PREDATOR POPULATION INITIAL (PREDATORS)
```

```
PNGR.KL=PP.K*PGRF.K                                          10, R
    PNGR   - PREDATOR NET GROWTH RATE (PREDATORS/YEAR)
    PP     - PREDATOR POPULATION (PREDATORS)
    PGRF   - PREDATOR GROWTH RATE FACTOR (FRACTION/YEAR)
```

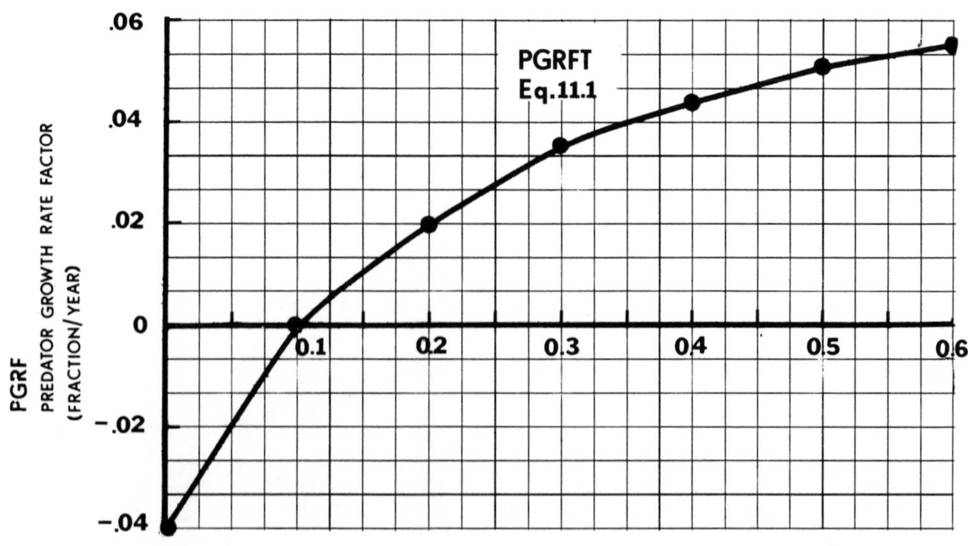


```
PGRF.K=TABHL(PGRFT,DKR.K,0,.6,.1)                            11, A
PGRFT=-.04/0/.02/.035/.045/.05/.055                          11.1, T
    PGRF   - PREDATOR GROWTH RATE FACTOR (FRACTION/YEAR)
    PGRFT  - PREDATOR GROWTH RATE FACTOR TABLE
    DKR    - DEER KILL RATE (DEER/PREDATOR/YEAR)
```

```
PBR.KL=PP.K*STEP(RF,RST)                                     12, R
RF=.2                                                        12.1, C
RST=1905                                                     12.2, C
    PBR    - PREDATOR BOUNTY RATE (PREDATOR/YEAR)
    PP     - PREDATOR POPULATION (PREDATORS)
    STEP   - STEP FUNCTION
    RF     - REMOVAL FRACTION (FRACTION/YEAR)
    RST    - REMOVAL STARTING TIME (YEAR)
```

```
            ***FOOD SECTOR***

F.K=F.J+(DT)(GR.JK-FC.JK)                              13, L
F=FI                                                   13.1, N
FI=350000                                              13.2, C
    F       - FOOD (UNITS)
    GR      - FOOD GENERATION RATE (UNITS/YEAR)
    FC      - FOOD CONSUMPTION RATE (UNITS/YEAR)
    FI      - FOOD INITIAL (UNITS)

GR.KL=(FCAP-F.K)/FRT.K                                 14, R
FCAP=350000                                            14.1, C
    GR      - FOOD GENERATION RATE (UNITS/YEAR)
    FCAP    - FOOD CAPACITY (UNITS)
    F       - FOOD (UNITS)
    FRT     - FOOD REGENERATION TIME (YEARS)
```

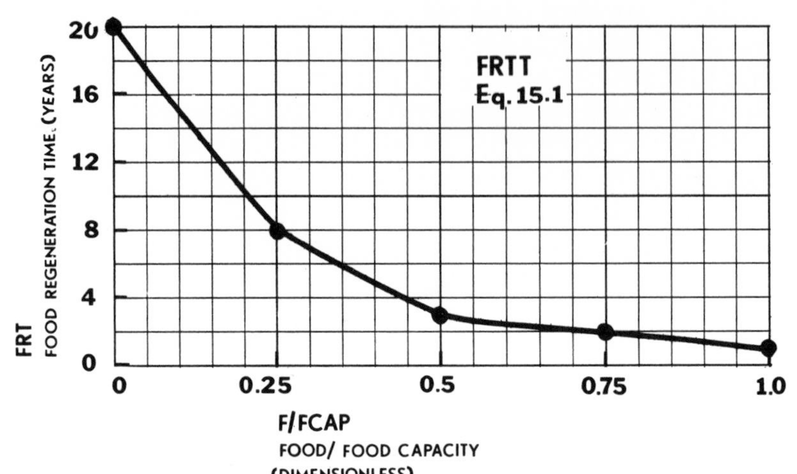


```
FRT.K=TABHL(FRTT,F.K/FCAP,0,1,.25)                     15, A
FRTT=20/8/3/2/1                                        15.1, T
    FRT     - FOOD REGENERATION TIME (YEARS)
    FRTT    - FOOD REGENERATION TIME TABLE
    F       - FOOD (UNITS)
    FCAP    - FOOD CAPACITY (UNITS)

FC.KL=DP.K*FCPD.K                                      16, R
    FC      - FOOD CONSUMPTION RATE (UNITS/YEAR)
    DP      - DEER POPULATION (DEER)
    FCPD    - FOOD CONSUMPTION PER DEER (UNITS/DEER/YEAR)
```

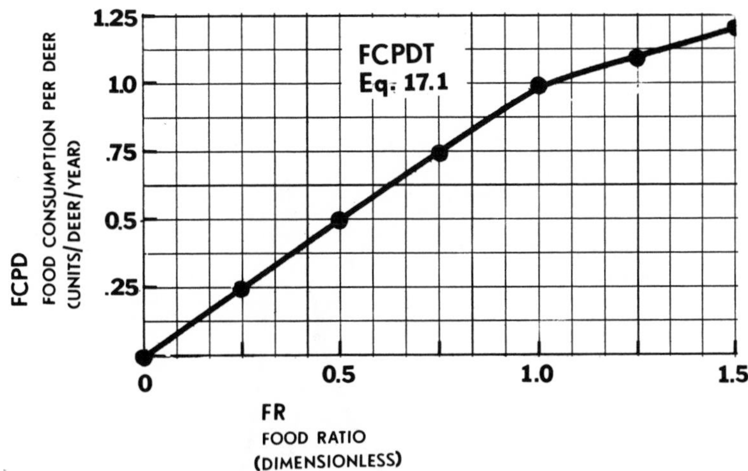


```
FCPD.K=TABHL(FCPDT,FR.K,0,1.5,.25)                      17, A
FCPDT=0/.25/.5/.75/1/1.12/1.2                           17.1, T
    FCPD   - FOOD CONSUMPTION PER DEER (UNITS/DEER/YEAR)
    FCPDT  - FOOD CONSUMPTION PER DEER TABLE
    FR     - FOOD RATIO (DIMENSIONLESS)


NOTE          ***CONTROL STATEMENTS***
NOTE
C DT=.1
C LENGTH=1950
C PLTPER=1
N TIME=TIMEI
C TIMEI=1900
PLOT DP=P(0,120000)/PP=Q/F=F/FR=R(0,8)/DD=D
```

S15.4 Model Behavior.

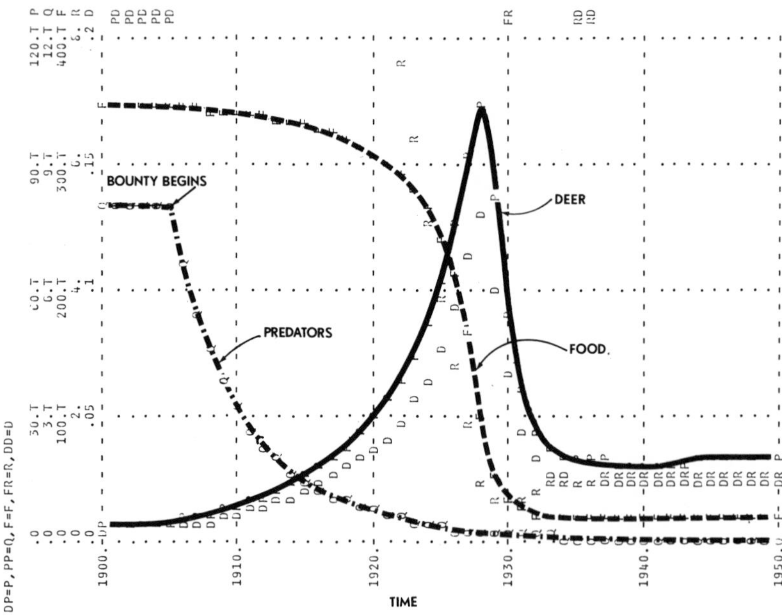

Growth and decline of deer on Kaibab Plateau

Notes

Notes

Notes

Notes

Notes

Notes